T0073766

Flow Networks

Flow Networks

Analysis and Optimization of Repairable Flow Networks, Networks with Disturbed Flows, Static Flow Networks and Reliability Networks

Michael T. Todinov

Oxford Brookes University
Oxford, UK

ELSEVIER

AMSTERDAM • BOSTON • HEIDELBERG • LONDON • NEW YORK • OXFORD
PARIS • SAN DIEGO • SAN FRANCISCO • SINGAPORE • SYDNEY • TOKYO

Elsevier
32 Jamestown Road, London NW1 7BY
225 Wyman Street, Waltham, MA 02451, USA

First edition 2013

British Library Cataloguing-in-Publication Data
A catalogue record for this book is available from the British Library

Library of Congress Cataloging-in-Publication Data
A catalog record for this book is available from the Library of Congress

ISBN: 978-0-12-398396-1

For information on all Elsevier publications
visit our website at store.elsevier.com

This book has been manufactured using Print On Demand technology. Each copy is produced to order and is limited to black ink. The online version of this book will show color figures where appropriate.

To Prolet, Marina and Marin

Contents

Preface

A lot has been written about static flow networks, where repairs and flow disruption caused by failures of components are not considered. Static flow networks have numerous applications and currently they are an inextricable part of the modern education in computer algorithms and operational research. However, despite the years of intensive research on static flow networks, a common fundamental flaw of the classical methods for maximising the throughput flow has been identified. *The classical algorithms for maximising the throughput flow leave undesirable directed loops of flow in the optimised networks.* These parasitic flow loops are associated with wastage of energy and resources and increased levels of congestion in the optimised networks and are associated with big financial losses for the affected sectors of the economy.

Furthermore, an important aspect of real flow networks is that their flows are often disturbed by particular contingency events, for example, overloading, congestion, failures and sudden fluctuations in flow generation and demand. Indeed, in many real flow networks (electrical, production, transportation, manufacturing and computer networks, etc.) components fail or sections of the network suffer congestion or overloading. These events disturb the network flows and in this sense, it can be stated that almost all real networks are, in fact, networks with disturbed flows.

After a contingency event, for example, a congestion or failure, it is important to redirect the flows immediately through alternative paths, so that a new maximum of the throughput flow is quickly reached.

At the same time, on a failure of a component from a production line, computer network, power supply system or water supply system, a repair is initiated and after a particular delay, the failed component is returned to operation. Similarly, after an accident on a road section, a clearing operation is initiated and after a certain delay, the road section is returned to operation at its full capacity. Consequently, it can be stated that real flow networks are almost always repairable networks. The research on networks with disturbed flows and repairable flow networks is an important emerging research area, part of the modern network science.

The potential application of repairable flow networks is huge: oil and gas production networks, computer networks, power networks, telecommunication networks, transportation networks, water supply networks, emergency evacuation networks, supply networks, etc. Repairable flow networks can even be used for conducting reliability analysis of large and complex systems.

An important problem associated with repairable flow networks delivering a particular commodity (gas, oil, electrical power, data, goods, etc.) is the adverse effect of component failures and unavailability of generation sources on the network performance. The extent of this adverse effect is measured by the probability of having a particular level of the throughput flow on demand or by the probability of having a particular expected level of total throughput flow, during a specified time interval. These performance measures are commonly specified in contracts and are a key for evaluating the performance and quality of service of flow networks. They are used for comparing alternative solutions and for making an informed selection among competing network topologies. This is the main reason why oil and gas production companies, for example, have already invested in software for simulating the performance of their repairable oil and gas production systems and determining their production availability. There is also an emerging demand for such software tools from the telecom sector, power distribution sector and distribution logistics sector.

As a result, failures, repairs, fluctuating supply and demand and the flow reoptimisation after these events are an essential part of the network analysis, optimisation and real-time management. This important aspect of the network flow analysis has not yet received the attention it deserves.

Although the analysis and optimisation of networks with disturbed flows and repairable flow networks is extremely important, currently no books are available to provide the much-needed support for researchers and practitioners. By introducing the subject 'networks with disturbed flows and repairable flow networks', this book takes the flow networks a step further. The book has been written with the intent to fill the existing gaps, by developing the theory, algorithms and applications related to networks with disturbed flows and repairable flow networks. The research area *Networks with disturbed flows and repairable flow networks* is new and this is the first book covering the subject.

The theoretical results presented in this book form the foundations of a new generation of ultra-fast algorithms for optimising networks with flows disturbed by failures, congestion or sudden change in demand or flow generation. The high computational speed creates the possibility of optimal control of very large and complex networks in real time. This is of particular importance to large and complex power distribution networks, where the optimisation of the power flows after overloading or failure of a power line needs to be done within the range of milliseconds. Reoptimising the network flows in real time significantly increases the yield from real production networks and reduces to a minimum the lost flow and disruption caused by failures.

The initial chapters introduce basic concepts, conventions and techniques related to static flow networks, where no contingency events disturbing the edge flows exist. The concepts, results and techniques discussed are necessary for introducing the theory of repairable flow networks and networks with disturbed flows.

In order to correct the common flaw of existing classical methods for maximising the throughput flow, an efficient algorithm for identifying and removing directed loops of flow has been developed.

New efficient algorithms for maximising the flow in static, single commodity and multi-commodity networks have also been proposed. In this respect, a new fundamental theorem referred to as '*dual network theorem for static networks*' has been stated and proved. The theorem states that the maximum throughput flow in any static network is equal to the sum of capacities of the edges coming out of the source minus the total excess flow at all excess nodes plus the maximum throughput flow in the dual network. Consequently, a new algorithm for maximising the throughput flow in a network has been proposed. For networks with few imbalanced nodes, the proposed algorithm outperforms all classical algorithms for maximising the throughput flow in a network. An additional stage in the proposed new algorithm guarantees that no parasitic directed loops of flow are left after the throughput flow maximisation.

It is also shown that the computational speed related to determining the maximum throughput flow in a network with merging flows can be improved enormously if the tree topology of the network is exploited directly. Accordingly, for networks with merging flows, an efficient algorithm with linear running time in the size of the network has been proposed for maximising the throughput flow. A theorem which provides the theoretical justification for the proposed algorithm has also been stated and proved.

Chapter 5 introduces the theory of networks with disturbed flows and discusses several fundamental results referred to as 'dual network theorems for networks with disturbed flows'. One of the results states that *in any network with maximised throughput flow, the new maximum throughput flow after choking the flows along several edges is equal to the maximum throughput flow in the original network, minus the total amount of excess flow at the excess nodes, plus the maximum throughput flow in the dual network.*

On the basis of this result, very efficient augmentation algorithms have been proposed for restoring the maximum possible throughput flow in a network with disturbed flows after an edge failure. The proposed algorithms are the fastest available methods for reoptimising the throughput flow after edge failures. In many cases, the average running time of the proposed algorithms is constant, independent of the size of the network or varies linearly with the size of the network. It is shown how the developed algorithms can be applied for optimising the performance of gas production networks.

The high computational speed of the proposed reoptimisation algorithm makes it suitable for optimising the performance of large and complex repairable flow networks in real time. Essentially, the proposed algorithm is a very efficient decentralised algorithm for achieving a global maximum of the throughput flow in a network by independent distributed agents, which possess local knowledge about the network topology but do not necessarily possess knowledge about the entire network topology. Consequently, a very important application of the proposed algorithm has been found in the high-speed control of active power distribution networks after a failure or congestion of power lines or after a sudden change in demand and power generation.

In Chapter 9, an important result has been stated regarding the average production availability of repairable flow networks composed of independently working edges, whose times to failure follow the negative exponential distribution. The average production availability is the ratio of (i) the average of the maximum throughput flow on demand calculated after removing the separate edges with probabilities equal to their unavailabilities, to (ii) the maximum throughput flow in the absence of failures. For the first time, the new algorithm for determining the production availability, created the basis of extremely fast solvers for the production availability of large and complex repairable networks, whose running time is independent of the length of the operational interval, the failure frequencies of the edges or the lengths of their repair times.

A discrete-event solver for determining the production availability of repairable flow networks with complex topology has also been constructed, where the times to failure of the edges could follow any specified distribution. The proposed discrete-event solver maximises the throughput flow rate in the network upon each component failure and return from repair. Maximising the flow rate upon each component failure and return from repair, ensures a larger total throughput flow during a specified time interval.

Methods for assessing the reliability of the throughput flow in a network with complex topology occupy a central space in the book. These methods use Monte Carlo simulation techniques to estimate the probability that the throughput flow will be equal to or greater than a specified threshold value and are superior to alternative methods based on minimal paths and cut sets. They can be applied for assessing the quality of service of computer networks and telecommunication networks. Analytical methods for determining the probability of a source-to-sink flow on demand, have also been introduced and discussed.

By using a specially designed software tool, a study has been presented on the link between performance, topology and size of repairable flow networks. The topology of repairable flow networks has a significant impact on their performance. Two networks built with identical type and number of components can have very different performance levels because of slight differences in their topology.

In Chapter 11, a new topology optimisation algorithm has been proposed for achieving a maximum throughput flow and a maximum production availability within a specified budget for building the network. The algorithm incorporates an efficient search in the space of available alternatives, based on combining the branch and bound method and pruning of the full-complexity network. The algorithm always determines the exact solution and is considerably faster than exact algorithms based on a full exhaustive search. At the heart of the optimisation procedure is a production availability algorithm, whose running time is independent of the length of the operational interval, the failure frequencies of the components, or the lengths of their repair times.

The topology optimisation method has also been applied to reliability networks of safety–critical systems. An important problem has been solved, related to how to build within a fixed budget a safety–critical system, characterised by the smallest risk of failure. The topology optimisation of reliability networks creates the opportunity of increasing the safety margin of common safety–critical systems without increasing the cost of current designs.

Substantial space in the book has been allocated on flow optimisation in non-reconfigurable repairable flow networks. A number of theorems related to non-reconfigurable repairable flow networks have been stated and proved. For a specified source-to-sink path, the difference between the sum of the unavailabilities along its forward edges and the sum of the unavailabilities along its backward edges has been referred to as *path resistance*. This property plays an important role in modelling the properties of non-reconfigurable flow networks. In a non-reconfigurable flow network, the absence of augmentable cyclic paths with a negative resistance is a necessary and sufficient condition for a minimum lost flow due to edge failures.

For a given source-to-sink path, the difference between the sum of the hazard rates of its forward empty edges and the sum of the hazard rates of its backward empty edges is the flow disruption number of the path. In non-reconfigurable networks, the absence of augmentable cyclic paths with a negative flow disruption number is a necessary and sufficient condition for a minimum probability of undisturbed throughput flow by edge failures.

Reliability networks, their design and analysis also received attention. It is shown that a reliability network can be interpreted as a repairable flow network. The probability of system operation is equal to the probability that, on demand, a path with unit flow can be augmented from the source to each of the end nodes. This analogy permits reliability networks to be analysed with the tools developed for repairable flow networks.

Unlike the repairable flow networks, the reliability network does not necessarily match the functional diagram of the modelled system. This is the reason why alongside the analysis of reliability networks an extended discussion has been provided on building reliability networks. It is demonstrated that contrary to a widespread belief, complex reliability networks that cannot be represented with series and parallel arrangements are common. Even simple mechanical systems may have reliability networks that cannot be reduced to a combination of series and parallel arrangements.

It is also shown that the traditional presentation of reliability networks, based on a single start node, single end node, undirected edges and edges with two end points, is insufficient for a correct representation of the logic of operation and failure of some systems. Undirected and directed edges, multiple end nodes and 'edges' with multiple end points are all necessary to represent correctly the logic of operation and failure of these systems.

A powerful algorithm for analysis of reliability networks, which avoids the drawbacks of commonly accepted methods based on cut sets or path sets, has also been introduced.

Finally, a method has been proposed for virtual accelerated testing of complex repairable networks. As a result, the life of a complex repairable flow network under normal operating conditions can be extrapolated from the accelerated life models of its edges and nodes, at elevated levels of the acceleration stresses (temperature, humidity, pressure, vibrations, speed, concentration, corrosion activity, etc.). The proposed method makes building test rigs for complex flow networks unnecessary, which can be an expensive and a very difficult task. It also reduces drastically the amount of time and resources needed for accelerated life testing of complex flow networks.

Building a model of a complex repairable network based on the accelerated life models of its components also reveals the impact of the acceleration stresses on the availability of the network.

The algorithms in the book have been embedded in a software tool with graphics user interface through which the network is drawn on screen by the user and the parameters characterising the components are specified. Functions have been provided for quickly transferring parameters from one edge/node to another. Nodes and edges can be easily deleted and added, which permits easy modifications of an existing network. This is particularly useful for comparing quickly the performance of derivative network topologies and selecting the topology with the best performance.

In conclusion, I gratefully acknowledge the financial support received from The Leverhulme Trust, with the research grant F/00 382/J 'High-speed algorithms for the output flow in repairable flow networks'. I also acknowledge the help of Mr Paul Hansford and Mr Calvin Earp in developing the graphics user interface of the software tool; the helpful discussions with clients from oil and gas companies, the helpful comments from colleagues in the Department of Mechanical Engineering and Mathematical Sciences at Oxford Brookes University and from overseas colleagues.

I acknowledge the editing and production staff at Elsevier for their excellent work and in particular, the help of Ms Tracey Miller, Dr Erin Hill-Parks and Mr Stalin Viswanathan.

Finally, I acknowledge the immense help and support from my wife Prolet, during the preparation of this book.

I hope that the findings, algorithms and examples presented in this book will provide key knowledge, useful techniques and tools to mathematicians, computer scientists, engineers, and operators of power networks, computer networks and production networks. The book has already been used with success as a basis of the module 'Repairable flow networks and networks with disturbed flows' taught by me to final year students in mathematics at Oxford Brookes University, the United Kingdom. The chapter related to building and analysing reliability networks constitutes an essential part of the module 'Engineering Reliability and Risk Management' taught by me to postgraduate students in Oxford Brookes University. I also believe that the book will strongly complement the education in computer algorithms, operational research methods and network science.

Oxford, September 2012

1 Flow Networks – Existing Analysis Approaches and Limitations

1.1 Repairable Flow Networks and Static Flow Networks

There is no exaggeration in saying that most of the real flow networks are, in effect, *repairable flow networks*. Indeed, after a failure of a component or section from a production flow line, computer network, power supply system or water supply system, a repair is initiated and after a particular downtime, the component/section is returned to operation (Figure 1.1).

During a specified time interval with length 'a' (Figure 1.1), the components building a flow network fail and their flow capacity is reduced. After a certain delay for repair, each failed component is repaired to 'as good as new' condition. An essential feature of repairable flow networks is that repair of failed components is taking place and the repair *is part of the analysis and optimisation of the network*. This feature distinguishes repairable flow networks from *static flow networks* and *stochastic flow networks* for which no repair of failed components is ever considered. Another difference is that repairable flow networks are not necessarily stochastic flow networks. Thus, planned maintenance of different sections/components of the networks (e.g. routers in computer networks) may take place at regular (predictable) intervals. During the time of maintenance, the corresponding section of the network will not be working. Usually, planned events, where network components are taken out for routine maintenance, are mixed with unplanned/random events, e.g. random component failures.

By using his network simplex method, Dantzig (1951) first solved the transportation problem which includes the maximum flow problem as a special case. Since then, a large number of algorithms have been proposed for solving the maximum flow problem. Most of this research has been focused mainly on static flow networks, *where no failure of components is considered*. Research related to static flow networks has been comprehensively reviewed in (Ahuja et al., 1993; Asano and Asano, 2000; Ford and Fulkerson, 1962; Goldberg et al., 1990; Hu, 1969; Tarjan, 1983). Description of algorithms related to flow networks can also be found in Cormen et al. (2001), Kleinberg and Tardos (2006), Goodrich and Tamassia (2002), Papadimitriou and Steiglitz (1998), Gibbons (1985), and Lawler (1976).

This research is mainly related to the maximum flow problem (determining the maximum throughput flow transmitted from a source to a sink) and the minimum

Flow Networks. DOI: http://dx.doi.org/10.1016/B978-0-12-398396-1.00001-5

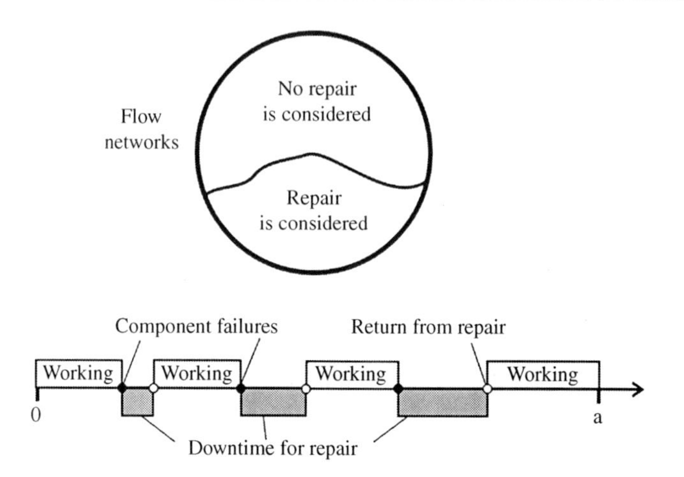

Figure 1.1 Most of the real flow networks are essentially repairable flow networks.

cost problem (minimising the cost of the transmitted flow). The analyses are based on static flow networks where no component failures occur.

The first big category of algorithms for maximising the throughput flow in networks start from an empty network and continue by gradually saturating the edges with flow (Ahuja and Orlin, 1991; Cherkaski, 1977; Dinic, 1970; Edmonds and Karp, 1972; Elias et al., 1956; Ford and Fulkerson, 1956; Goldberg and Tarjan, 1988; Hochbaum, 2008; Karzanov, 1974; Shiloach and Vishkin, 1982; Sleator and Tarjan, 1980).

Within this big category of algorithms two major sub-categories can be distinguished. The first sub-category includes path augmentation algorithms which, at all steps, preserve the feasibility of the network flow until the maximum flow is attained (Dinic, 1970; Edmonds and Karp, 1972; Elias et al., 1956; Ford and Fulkerson, 1956). Thus, the Ford–Fulkerson algorithm (Ford and Fulkerson, 1956) is based on augmenting available source-to-sink paths until no more augmentable paths can be found. An improvement of the Ford–Fulkerson algorithm was the layered network approach proposed by Dinic (1970) based on the new concept of augmenting the shortest paths at once. An algorithm based on the sequential augmentation of the shortest paths was also proposed by Edmonds and Karp (1972).

The second major sub-category of algorithms for maximising the throughput flow by starting from a network with empty edges is based on the *preflow concept* proposed by Karzanov (1974). The preflow differs from a flow in the way that the sum of the edge flows going into a node is allowed to exceed the sum of the flows going out of the node. As a result, the preflow satisfies the capacity constraints on the edges, but the flow conservation law at the nodes may be violated, in the sense that each node other than the source may contain excess flow. This concept was later used as a basis of the preflow-push algorithms by Sleator and Tarjan (1980) and Goldberg and Tarjan (1988). The preflow-push algorithms work to convert the preflow into a feasible flow. This is done by including labels on the nodes indicating their 'height' with respect to the source node. Flow is pushed from an excess node with a higher height to a neighbouring node with a lower height. If this cannot be done for a particular neighbouring node, which is connected with the original node by an augmentable edge, a relabelling operation is initiated. This consists of increasing the height of the excess node. Continuing this process until no internal node (other than the source and the sink) has an excess flow, guarantees that the maximum throughput flow will be reached.

Recently, another distinct category of methods for maximising the flow in a network appeared, based on starting from a fully saturated with flow edges and working backwards, by redistributing and draining flow from the network, until a maximum throughput flow is reached (Dong et al., 2009; Todinov, 2011b, c, 2012a). Depending on whether the sum of the capacities of the edges bringing flow into a node is greater than, equal to or smaller than the sum of the capacities of the edges taking flow out of the node, the node is either an *excess node*, a *balanced node* or a *deficit node* (Dong et al., 2009). The draining algorithm proposed by Dong et al. (2009) starts from a fully saturated with flow network and works backwards, by draining flow from flow paths connecting deficit nodes with excess nodes until all the nodes become balanced. Unfortunately, as it has been demonstrated by Todinov (2012a), the algorithm of Dong et al. (2009) is fundamentally flawed and leads to sub-optimal solutions, because node balancing was done only in a network which includes a backward circulation edge directed from the sink to the source. The correct algorithm has been suggested by Todinov (2011b, c, 2012a). It is based on two sequential stages, involving (i) redistribution of excess and deficit flow in a network which does not include a backward circulation edge followed by (ii) draining flow from the network which includes the backward circulation edge. Fundamental results referred to as dual network theorems for networks with disturbed flows and static networks have been stated and proved in (Todinov, 2011b, c, 2012a), which serve as a basis for the two-stage algorithm.

The minimum cost flow problem has also been treated comprehensively in earlier research. This problem is about finding a feasible flow in the network which minimises the cost for delivering a throughput flow of certain magnitude from the source to the sink (Ahuja et al., 1992; Bennington, 1973; Friesdorf and Hamacher, 1982; Goldberg and Tarjan, 1987, 1989; Klein, 1967; Orlin, 1993; Tardos, 1985). A variety of this problem is about finding a feasible network flow which maximises the gain associated with a flow of certain magnitude (Jewell, 1962; Shigeno, 2004; Truemper, 1977). Issues related to charging, rate control and utility maximisation for flow networks have been addressed by Kelly (1997) and Yi and Chiang (2008).

Finally, multi-commodity flow problems have also been studied. Coverage of multi-commodity networks can be found in Ford and Fulkerson (1958), Hu (1963), Sakarovitch (1973), Assad (1978), Evans (1976b) and Okamura (1983).

Most of the algorithms for maximising the flow in static flow networks have polynomial running time in the size of the network. The running time of the algorithms for maximising the throughput flow, however, can often be improved significantly, if the particular network topology is exploited directly. For example, for

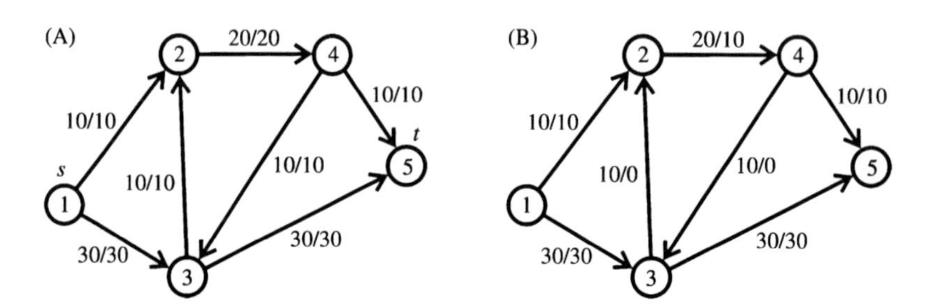

Figure 1.2 (A) A directed parasitic loop of flow (3,2,4,3) left after the termination of preflow-push optimisation algorithm and the draining algorithm and (B) A flow network from which the directed parasitic loop of flow has been removed. The first number on the edge stands for the edge capacity and the second number stands for the actual flow along the edge.

planar networks, very efficient throughput flow maximisation algorithms have been designed (Itai and Shiloach, 1979). These algorithms take advantage of the planar topology and their running time is superior to the running time of algorithms handling networks with arbitrary topology. Another example are the networks with tree topology, which consist of a number of sources giving rise to streams that join into bigger streams, etc., until the streams from all sources join into a single stream. As it will be shown later, the maximum throughput flow in networks with merging flows can be obtained in worst-case running time which depends linearly on the size of the network.

Despite the vast number of methods and algorithms devoted to maximising the flow in static networks, a recent analysis published by Todinov (2013) revealed a striking shortcoming of all the classical methods for maximising the flow in networks. They all yield sub-optimal solutions for a number of networks because they all leave directed loops of flow in the optimised networks.

These parasitic loops of flow (e.g. the directed loop of 10 units flow (3,2,4,3) in Figure 1.2) are *highly undesirable* because they consume unnecessarily residual capacity from the edges of the network. Energy and resources are also unnecessarily wasted for maintaining these parasitic loops. For a number of network topologies, the classical methods lead to 'optimal' network solutions, associated with wastage of resources and congestion.

A network without a directed loop of flow (Figure 1.2B) has the same throughput of 40 flow units from the source s to the sink t and no energy is wasted on maintaining a parasitic flow loop. Later, it will be demonstrated that shortest-path augmentation algorithms also result in directed parasitic loops of flow in the optimised networks.

Consequently, the shortest-path augmentation algorithms, the preflow-push algorithms and the draining algorithms for maximising the throughput flow *should not be used for optimising the edge flows in a network without an additional stage aimed at discovering and removing parasitic loops of flow*.

The problem of parasitic directed loops of flow does not occur in acyclic directed flow networks. The existence of cyclic paths however is unavoidable in undirected

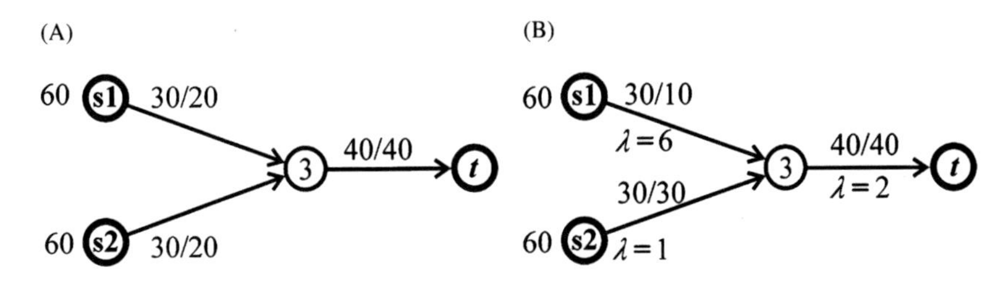

Figure 1.3 Maximising the flow in the static flow network (A) does not necessarily maximise the flow in the repairable flow network (B).

flow networks which do not have a tree topology. This problem has been discussed in detail in Chapter 4, where an algorithm has been proposed, with the sole purpose of identifying and removing parasitic loops of flow.

Furthermore, the algorithms mentioned earlier have been designed for static flow networks, where no failures are considered. Maximising the throughput flow in static flow networks, where no component failures occur, does not necessarily maximise the flow in the corresponding repairable flow network, where components fail and are subsequently repaired to 'as good as new' condition (Todinov, 2009).

Indeed, consider Figure 1.3A, which depicts a static flow network where no component failures occur. Figure 1.3B depicts a repairable flow network where component failures occur and repairs are conducted. On top of each edge of the networks, there is a label '*c/f*', where *c* stands for the flow capacity of the edge and *f* stands for the actual flow through the edge. In Figure 1.3B, below each edge, λ denotes the failure frequency of the edges (expected number of failures per unit time). For the sake of simplicity, the downtime for repair has been assumed to be the same for each edge. Flows are generated at each of the two sources *s*1 and *s*2, and once having been set up, cannot be altered. The maximum possible flow which each source is capable of generating is 60 units per unit time. The generated flows from the sources are collected at the sink *t*.

The flow in the static flow network where no component failures occur ($\lambda = 0$ for each edge) can be maximised by saturating both flow paths (*s*1,3,*t*) and (*s*2,3,*t*) with flows of magnitude 20 units. The maximum throughput flow in the network is therefore 40 units. In the repairable flow network from Figure 1.3B, losses of flow due to edge failures can be reduced significantly by directing more flow through the more reliable flow path (the second flow path *s*2,3,*t*). Edge (*s*2,3) fails less frequently than edge (*s*1,3), and the lost flow caused by the repair downtimes will be smaller for the second flow path (*s*2,3,*t*). The cost of intervention for repair will also be smaller because of the fewer failures and repairs. It is therefore sensible to set up the flows as follows: 10 units of flow through the first flow path (*s*1,3,*t*) and 30 units of flow through the second flow path (*s*2,3,*t*). This flow distribution sets the same maximum throughput flow of 40 units in the network but reduces significantly the lost amount of flow due to component failures and the cost of intervention for repair. The edge flows in Figure 1.3B maximise the throughput flow in the repairable flow network and minimise the lost flow due to failures.

This simple example shows that *maximising the throughput flow in a static flow network (where no component failures occur) does not necessarily guarantee that the throughput flow in the corresponding repairable network will be maximised.*

1.2 Repairable Flow Networks and Stochastic Flow Networks

Stochastic flow networks have already been researched, but they have all been defined in different ways. Thus, in the stochastic flow networks, considered by Zhou et al. (2010), each node in the network was considered to be a splitter. A token can enter a node through an incoming edge and exit on one of the output edges, according to a predefined probability distribution. No repair of failed components has been considered. This type of stochastic flow network has specific applications and is not suitable for modelling the flow in production networks, transportation networks, power distribution networks and computer networks.

Stochastic flow networks, where the flow capacities of components have been treated as random variables, have been considered by Lee (1980), Evans (1976a), Lin et al. (1995), Lin and Yuan (1998), Yeh (2001a, b), and Lin (2001a, b, 2002, 2006). The problem of interest was the probability that, on demand, the throughput flow be equal or greater than a specified level. Maximising the flow in stochastic flow networks was based on methods involving minimal cut sets or minimal paths. A similar approach has been adopted by Jane et al. (1993), where stochastic flow networks with multistate components have been considered. Minimum cut sets have also been used by Fishman (1987) to evaluate the distribution of the maximum flow in a directed network whose edges have random capacities.

An algebraic method for determining the reliability of communication networks, characterised with the same probability of failure of the links, has been proposed by Yarlagadda and Hershey (1991). This method, however, is also based on the existing possible paths in the network.

A survey on reliability evaluation of stochastic flow networks in terms of minimal paths has been provided by Peixin and Xin (2009). Finding all minimal paths or cut sets is an NP-hard problem (Colbourn, 1987).

Although, for small-size networks an approach based on minimal paths or minimal cut sets is acceptable, with increasing the size of the network, the number of minimal paths and cut sets increases exponentially and this approach is no longer feasible. This point can be illustrated with the example shown in Figure 1.4. The flow network in the figure has $N^N + N$ minimal cut sets and N^{N+1} minimal paths. Even for the moderate example $N = 10$ the storage and manipulation of the minimal paths and cut sets is impossible. As a result, an algorithm based on determining all minimal paths or cut sets will be very inefficient because it *will run in exponential time.* These problems caused approaches based on minimal cut sets or minimal paths to be abandoned in the present treatment.

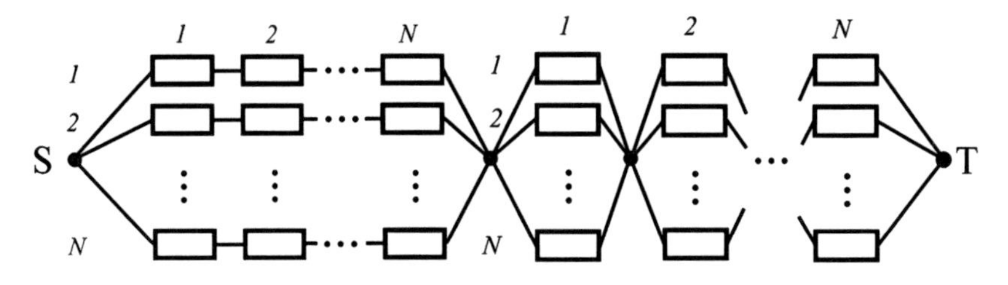

Figure 1.4 An example of a flow network where the number of minimal paths and minimal cut sets increases exponentially with increasing the size of the network.

For a network composed of m edges, the total space of alternatives is huge (2^m) and its enumeration for large networks is impossible. To avoid this predicament, Lee (1980) proposed a method for evaluating the reliability of stochastic flow networks by using the concept of lexicographic ordering. In fact, the lexicographic ordering concept was proposed to avoid a full enumeration. However, distinct, non-intersecting sets of edges fulfilling the condition for a reliable operation still have to be listed, and for large networks, the number of these distinct sets also increases exponentially.

More importantly, in previous research on stochastic flow networks, *renewal of components after a failure, and repair, were not considered.* An essential feature of the repairable flow networks, however, is that a renewal of failed components is taking place after a certain delay for repair. This feature distinguishes repairable flow networks from static flow networks and stochastic flow networks.

On the other hand, existing research dealing with repairable systems (Ascher and Feingold, 1984) does not focus on repairable flow networks. The focus is on binary-state networks where each component may be either in an operational state or in a failed state. This analysis framework is not sufficient to handle repairable flow networks, where one of the main parameters of interest is the variation of the flow output over a specified time interval, not just the existence of a throughput flow in the network. Another feature of flow networks is that different commodities may be transported simultaneously through the network. This is natural for oil and gas production systems, transportation systems and multi-service telecommunication networks.

Furthermore, in previous research on stochastic networks, *no distinction has ever been made between non-reconfigurable flow networks and reconfigurable flow networks.* In the non-reconfigurable flow networks, the flows through the separate branches, once set, cannot be redirected within the network upon component failure. As a result, failure of a component in a non-reconfigurable network leads to a loss of the flow through the component.

Flow networks with merging flows, where the flows from the sources have been set and cannot vary, are a typical example of non-reconfigurable flow networks. In networks with merging flows several sources of flow give rise to separate flow paths (e.g. sub-sea oil production wells connected to production trees, manifolds and pipelines). The flow paths (streams) merge into bigger streams, etc., until the streams from all the sources join into a single stream. Failure of a component along any of the flow streams causes all flows from the sources feeding the steam to be lost during the repair, because there is no possibility for redirecting the flow blocked by the failed component.

Conversely, for reconfigurable networks, failure of a component does not necessarily cause a loss of flow because of the possibility for redirecting the flow through alternative flow paths. A typical reconfigurable section is two components arranged in parallel, where the flow capacity of each component is sufficient to accommodate the entire flow through the section. Upon failure of any component from the section, the entire flow can be redirected through the parallel component and, as a result, no flow is lost.

Analysis and optimisation of repairable flow networks and networks with disturbed flows is an emerging area of research. Central to this research is the theoretical and computational framework for analysis and optimisation of repairable flow networks and networks with disturbed flows, presented in this book. Important characteristics of repairable flow networks are discussed, such as the *throughput flow reliability* and the *expected amount of throughput flow transmitted during a time interval, in the presence of failures*. The throughput flow rate reliability is the probability that upon demand, the maximum throughput flow in the repairable network, in the presence of component failures, will be at least equal to a specified level.

Determining the throughput flow reliability is particularly important for telecommunication networks, power networks, transportation networks and production networks. In computer networks for example, failures of hosts, routers and communication lines, are frequent, and inevitably associated with delays for repair.

Component failures in flow networks lead to disappearance of flow capacity and the expected magnitude of the throughput flow may not be guaranteed. As a result, the quality of service received from the network (which is a key performance characteristic) can be affected seriously. Furthermore, the ever-increasing demand on existing computer networks, power networks and transportation networks requires better management to minimise congestion and ensure better service to customers.

These problems are particularly acute for telecommunication networks, oriented towards media applications, for transportation networks and power distribution networks because they all require a high throughput flow rate. Selecting the shortest path for a data transfer, power transfer or a road traffic, as it is commonly done, is often far from optimal. Consequently, this common-sense strategy often leads to a congestion and overloading of particular sections of the network, and ultimately, to a low throughput flow. In the case of power networks, overloading of power lines degrades severely the reliability of the power grid. The result is a diminished power transmission capacity and possibility for cascading failures which have the potential to bring down the entire network. This point has been illustrated in Figure 1.5, where the labels stand for the flow capacities of the edges.

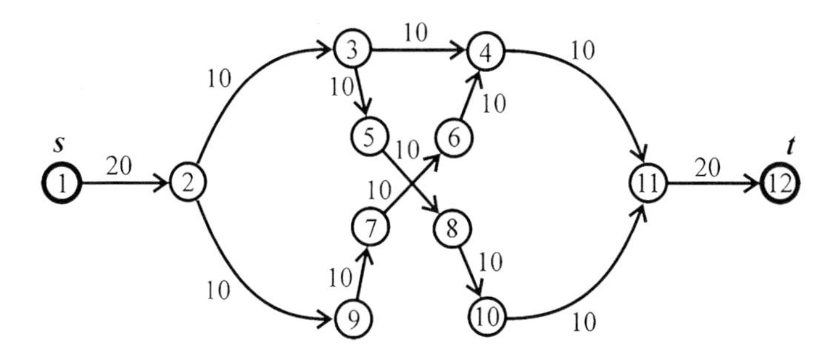

Figure 1.5 An example of a network, where selecting the shortest path for a flow transfer limits the throughput capacity of the network.

Clearly, the shortest path from the source s to the sink t is (1,2,3,4,11,12). Selecting this path for a flow transfer will saturate it with 10 units of flow and, as a result, the entire network will be blocked. The maximum possible flow transfer that can be made by selecting the shortest path is therefore only 10 units. However, if the longest paths (1,2,3,5,8,10,11,12) and (1,2,9,7,6,4,11,12) were selected to transfer 10 flow units each, the maximum throughput flow would increase to 20 units. Clearly, moving away from the shortest-path strategy provides a better management of the network. Note that for the second alternative, edge (3,4) remains empty.

The ratio of the total expected throughput flow in a repairable flow network, in the presence of failures during a specified time interval, and the total throughput flow that could be obtained in the absence of failures, is an important performance measure. For production networks, this performance measure is known as *average availability*. It is specified in contracts and is a key parameter for evaluating the performance of production systems. It is used for comparing alternative solutions, and for making an informed selection among competing design solutions with different network topologies. In production systems, even a relatively small difference (1–2%) in the average availability translates into a big difference in the produced quantities. Such percentage differences may constitute the difference between a profitable and unprofitable production system.

1.3 Networks with Disturbed Flows and Stochastic Flow Networks

Despite the fact that *almost all real networks are networks with disturbed flows*, the focus of existing literature on flow networks is exclusively on static flow networks and stochastic flow networks. The methodology developed for static flow networks does not apply to networks with disturbed flows because the unstated assumption behind the theory of static networks is the lack of edge failures or other events that could disturb the edge flows.

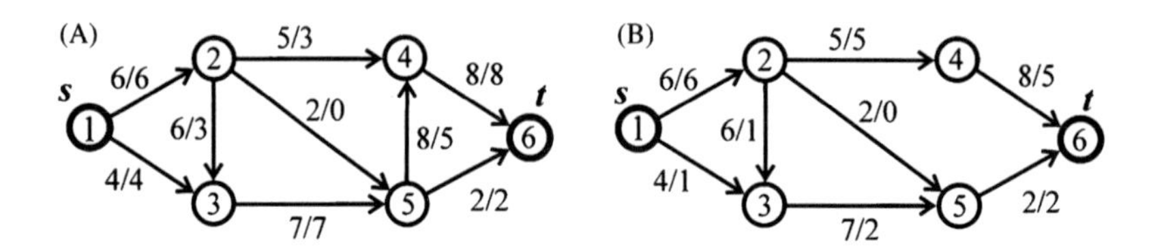

Figure 1.6 The fundamental problem of flow maximisation in networks with disturbed flows.

It is much less obvious why the theory of stochastic flow networks fails to capture the networks with disturbed flows. Although disturbance of edge flows is present in stochastic flow networks, the focus of the theory of stochastic networks is on the reliability of the throughput flow. The central problem is determining the probability that the network will deliver a particular throughput flow on demand.

A network with disturbed flows, however, is not necessarily a stochastic flow network, where the edge capacities are random quantities. Any static flow network can instantly be turned into a network with disturbed flows if a non-empty edge is removed from the network or if its flow capacity is choked to a particular level. There are no stochastic events, just a one-time disturbance. Therefore, *it is inappropriate to treat a network with disturbed flows as a stochastic flow network.* There is no stochastic element present. The central question for networks with disturbed flows is how to re-optimise the network flows after an edge flow disturbance (from edge failure, overloading or a sudden change in flow consumption), so that the new optimal throughput flow is determined quickly.

To the best of the author's knowledge no discussion has ever been presented on this very important problem. By quickly restoring the maximum throughput flow, the lost flow due to component failures is minimised. Consider for example the flow network shown in Figure 1.6. The source s has been connected by directed edges to the sink t.

On each edge label, the first number is the flow capacity of the edge and the second number is the actual flow through the edge. The maximum throughput flow for the network shown in Figure 1.6A is 10 units. Suppose now that edge (5,4), carrying 5 units of flow, has failed. The failure of edge (5,4) does not automatically lead to a loss of 5 units of flow. As can be verified from Figure 1.6B, some edge flows can be reset in such a way that only 3 units of throughput flow are lost due to the failure of edge (5,4). The new maximum throughput flow in the network after the failure of edge (5,4) is 7 units.

Even for such a simple network, it is not immediately clear how the edge flows should be reset, in order to attain the new maximum throughput flow. For large and complex networks, without an appropriate algorithm and the right software tools, the task of resetting correctly the edge flows in order to attain the new maximum throughput flow is almost impossible. The problem of optimal resetting of the edge flows upon component failures is an important problem for all real networks with disturbed flows. It is a problem of *optimal control of the flow network in real time.* This problem is particularly acute for power distribution networks, supply networks and production networks, where *failure to maximise the throughput flow after a component failure entails sub-optimal performance of the entire network.* For power distribution networks in particular, failure to optimise the energy flows after failure often results in overloading of power lines, which could trigger cascading failures. Such events for example occurred in 2003, when large parts of Italy and North East America were plunged into darkness.

Costs associated with component failures in flow networks can be significant, particularly in cases where the cost of intervention for repair is significant. The cost of intervention for repair in many production networks and power distribution networks is a major component of the overall cost associated with a component failure. Losses associated with component failures are associated with financial losses, customer inconvenience and dissatisfaction, and ultimately cause damage to the company's reputation.

For large and complex real networks, closed-form solutions for determining the performance parameters of the networks or for optimal control of the networks, are practically impossible. Instead, high-speed solvers based on efficient network algorithms can be used.

Networks, whose flows have been disturbed by a failure and then re-optimised, are not necessarily repairable flow networks. One of the reasons is that the downtime for repair is not specified during the re-optimisation. It needs to be as short as possible, but it is not part of the analysis and the solution. Furthermore, often, the probability of uninterrupted flows is of interest, not the availability or probability of a specified throughput flow on demand. As a result, there is a difference between repairable flow networks and networks with disturbed flows. The networks with disturbed flows result from a contingency event (e.g. component failure, congestion or fluctuation in flow supply or flow demand).

1.4 Performance of Repairable Flow Networks

The results presented in this book served as a basis to create very fast algorithms revealing the performance of complex repairable networks. The efficiency of the developed high-speed solvers created the possibility of embedding them in simulation loops, performing topology optimisation of large and complex repairable networks. The high-speed solvers are particularly useful for comparing network topologies and selecting the ones characterised by the best performance. The high-speed solvers, revealing the performance of repairable flow networks, have been embedded in a software tool with graphics user interface, which significantly increases the speed of the analysis and reduces the likelihood of input errors. The developed software tool can be used as a decision-support tool in fast-track projects where the performance of many network topologies needs to be assessed quickly in order to identify the network topology with the best performance.

Each flow network topology is also associated with a specific capital cost. An important objective here is to identify the topology which achieves a maximum transmitted flow within a specified budget for building the network. Solving this problem requires a repeated modification of the network topology and calculating the throughput flow in the presence of failures in order to determine the right topology, which combines a minimum cost for building the network, and a maximum throughput flow.

Existing software tools related to flow networks, however, *operate on a fixed network architecture*. They *do not* perform repeated modification of the network topology and *do not* search in a large space of alternatives, in order to determine the network which combines a minimum building cost and losses from failures below a desired level. This is the reason why existing computational tools *are not capable of supporting the optimal design of flow networks*.

In the complex flow networks used today, there are many possibilities for selecting components with different capacities, reliabilities and costs, design configurations, cross-bridges and redundancies. In this huge space of alternatives, identifying the optimum set of alternatives for the components, the optimum network topology and the necessary redundancies is not a trivial task. Without the right models and tools, design alternatives that are far from optimal will be selected. The network designs will either be associated with significant losses from component failures or with a significant cost.

In conclusion, the hallmarks of a well-designed repairable flow network can be summarised as follows:

- A high throughput flow in the presence of component failures and repairs.
- A high probability of delivering a specified throughput flow upon demand.
- A high average availability – a high ratio of the maximum expected throughput flow in the presence of failures to the maximum throughput flow in the absence of failures.
- A high resilience of the throughput flow against individual failures of edges and nodes.
- Minimum lost flow caused by component failures.
- Self-repairing capability upon a contingency event, a possibility of fast resetting of the edge flows, in order to achieve a new maximum throughput flow.
- A maximum throughput flow in the presence of component failures, achieved within a low cost for building the network.

2 Flow Networks and Paths – Basic Concepts, Conventions and Algorithms

2.1 Basic Concepts and Conventions: Data Structures for Representing Flow Networks

At an abstract level, a repairable network with flows from multiple sources can be presented as a number of sources connected to terminals (sinks) through components (edges) characterised by flow capacities.

During a specified time interval of operation (e.g. the life cycle of the network), the components fail independently from one another, their flow capacity is reduced to zero during the downtime for repair, after which they are returned to operation. It is assumed that a sufficient amount of repair resources are available, so that repair starts as soon as a component fails. It is also assumed that the repair fully restores the flow capacity of the failed component.

Formally, a flow network can be presented as an oriented graph $G = (V,E)$, where V denotes the set of nodes and E denotes the set of edges (components). The edge starting at node i and ending at node j will be referred to as edge (i,j) and its flow will be denoted by $f(i,j)$. A source is a node with no ingoing edges and with at least one outgoing edge, whereas a sink is a node with no outgoing edges and with at least one ingoing edge. The letters s and t are reserved for denoting the source and the sink, correspondingly. The source-to-sink flow ($s{-}t$ flow) will be referred to as *throughput flow*. The maximum possible throughput flow from the source s to the sink t will be referred to as *maximum throughput flow*.

Consider a flow network with a single source and a single sink. For all nodes in the flow network different from the source and the sink, *the flow conservation law* is fulfilled: *the sum of the flows that enter a node is always equal to the sum of the flows leaving the node*:

$$\sum_{k \in \delta +} f(k,i) - \sum_{m \in \delta -} f(i,m) = 0 \tag{2.1}$$

where $\delta +$ denotes the set of edges whose flows enter node i and $\delta -$ is the set of edges whose flows come out of node i. For all edges, *the capacity constraint*

Flow Networks. DOI: http://dx.doi.org/10.1016/B978-0-12-398396-1.00002-7

principle is also fulfilled: *the flow through any edge cannot exceed the flow capacity of that edge*:

$$f(i,j) \leq c(i,j) \tag{2.2}$$

where $f(i,j) \geq 0$ is the flow (units of flow per unit time) through the edge (i,j), and $c(i,j) \geq 0$ is the flow capacity of edge (i,j) – the maximum amount of flow per unit time which edge (i,j) can accommodate.

The flow in a network is said to be feasible if (i) the flow conservation law is honoured for all internal nodes (the nodes different from the source s and the sink t) and (ii) the capacity constraint principle is honoured for each edge in the network.

Equations (2.1) and (2.2) are constraints common to both static and repairable flow networks.

2.2 Pseudo-Code Conventions Used in the Algorithms

In describing some basic algorithms in this book, a number of conventions have been used. A common construct is a group of statements placed in braces {*Statement 1*; *Statement 2*; *Statement 3*;...} separated by semicolons (a semicolon marks the end of each statement).

Statements placed in braces are executed as a single block.

Another common construct is the conditional statement. In the conditional statement below, the block of statements {*Statement x1*;...; *Statement xn*;} in the braces following '**then**' is executed only if the specified condition 'Condition' is true. If the condition is false, the block of statements {*Statement y1*;...; *Statement ym*;} following the alternative '**else**' is executed.

If (Condition) **then** {*Statement x1;...;Statement xn;*}
else {*Statement y1;...; Statement ym;*}

Another common construct is the for-loop with control variable *i*, accepting successive values between two fixed numbers (1 and Number_of_trials).

For $i = 1$ **to** Number_of_trials **do**
{
...
}

The for-loop executes the block of statements in the braces a number of times, specified by the variable Number_of_trials. If a statement **break** is encountered in the body of the loop, the execution of statements continues with the next statement immediately after the loop (Holzner, 2001). In the next example, the statement **break** skips all statements until the end of the for-loop, the loop is exited and the program execution continues with *Statement n+1*.

For $i = 1$ **to** Number_of_trials **do**
{

```
Statement 1;
  ...
  break;
  ...
  Statement n−1;
  Statement n;
}
Statement n+1;
```

In the next for-loop, suppose that the condition in the if-statement is true. In this case, the statement **continue** in the body of the loop is executed. The **continue** statement forces transfer of control to the next iteration of the enclosing loop (Holzner, 2001).

The execution of **continue** increments immediately the control variable i of the for-loop and the for-loop immediately continues with the next iteration, skipping the execution of all statements until the end of the loop.

```
for i=1 to Number_of_trials do
{
 if (Condition = TRUE) then continue;
  Statement 1;
  ...
  Statement n − 1;
  Statement n;
}
```

The while-do loop executes a block of statements repeatedly, as long as the specified condition is true. If the variable Condition is false before entering the loop, the block of statements is not executed at all.

```
while (Condition = TRUE) do
 {
 Statement 1;
 Statement 2;
 ...
 Statement n;
 }
```

A similar construct is the repeat-until loop.

```
repeat
 Statement 1;
 Statement 2;
 ...
 Statement n;
until (Condition = TRUE);
```

which repeats the execution of all statements between **repeat** and **until** in its body, as long as the specified condition is true. Unlike the while-do loop, if, before entering the repeat-until loop, the variable Condition is false and is not altered in the loop, the block of statements between **repeat** and **until** is executed only once.

Similar to the for-loop, if a statement **break** is encountered in the body of the while-do loop or the repeat-until loop, the statements after the **break** are skipped and the loop is exited. The programme execution continues with the statement immediately after the loop.

If a statement **continue** is encountered in the body of the while-do loop or the repeat-until loop, the statements after **continue** are skipped, and the loop continues from the beginning with the next iteration.

A *procedure* is a self-contained section which performs a certain task. The procedure may have a number of parameters param_1, param_2,. . ., param_n, listed in its definition, or no parameters.

procedure ***proc***(param_1, param_2,. . .,param_n)
{
Statement 1;
. . .
Statement n;
}

The procedure is called by including its name proc(arg_1,arg_2,. . .,arg_n) in other parts of the algorithm and specifying its arguments arg_1,arg_2,. . ., arg_n (the actual values passed to the procedure). In the case where the procedure has no parameters, the first line of its definition is **procedure** ***proc***().

A *function* is also a self-contained section which returns value. The function may have a number of parameters param_1, param_2,. . ., param_n, listed in its definition, or no parameters.

function ***fn***(param_1, param_2,. . .,param_n)
{
Statement 1;
. . .
Statement n;
return *p*;
}

The function is called by including its name ***fn***(arg_1, arg_2,. . .,arg_n) in other parts of the algorithm and specifying its arguments arg_1, arg_2,. . ., arg_n (the actual values passed to the function). In the case where the function has no parameters, the first line of its definition is **function** ***fn***().

Before returning to the point of the function call, a particular value *p* is assigned to the function name fn, with the statement **return**.

A very large value in the algorithms (practically infinity) will be denoted by 'BigNumber'; text in italic after the symbol '//' denotes comments.

2.3 Efficient Representation of Flow Networks with Complex Topology

The efficiency of an algorithm operating on a flow network depends significantly on the data structure selected for representing the network.

For the sake of simplicity and following the practice in existing literature on flow networks (Ahuja et al., 1993), only directed edges will be assumed. Indeed, an undirected edge connecting two nodes i and j (Figure 2.1A) can always be represented with two directed edges of the same capacity, such as shown in Figure 2.1B (Ahuja et al., 1993).

Note that during the simulations, edges (i,j) and (j,i) in Figure 2.1B are not independent. They fail and get repaired simultaneously, because they represent essentially the same undirected edge (i,j) in Figure 2.1A.

For the sake of simplicity and notational convenience, in the representations to follow, no parallel edges (Figure 2.1C) will be assumed. This is in accordance with the convention adopted in previous textbooks on flow networks (Ahuja et al., 1993).

By convention, if a network contains n nodes, the smallest numbers are reserved for the indices of the sources and the largest numbers are reserved for the indices of the sinks. In particular, if a network with 10 nodes has two sources and three sinks, the sources receive indices '1' and '2' and the sinks indices '8' '9' and '10'. Each of the internal nodes receives a unique index between 3 and 7. For a network with a single source and a single sink, the source receives index '1' and the sink index 'n'.

The main concepts related to the network representation will be explained on the basis of the flow network with complex topology in Figure 2.2. The flow network has been conveniently modelled by a graph, consisting of a set of *nodes* (the open circles in Figure 2.2, numbered from 1 to 8), connected by *edges*. In production networks, the unreliable components are represented by the edges of the graph and the nodes are notional (perfectly reliable). In communication network and energy distribution networks both the edges and the nodes can represent unreliable components. Indeed, in computer networks for example, the nodes represent unreliable routers and the edges represent unreliable data transmission lines.

Each edge is connected to exactly two nodes. If nodes i and j are connected with an edge directed from i to j, the node j is said to be *a successor or a*

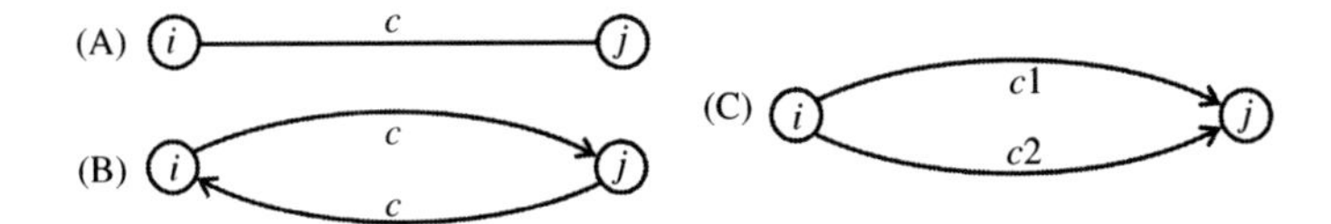

Figure 2.1 (A) An undirected edge with capacity c can always be represented with (B) two directed edges with capacities c. (C) No parallel edges, such as the ones in the figure, are present in the networks.

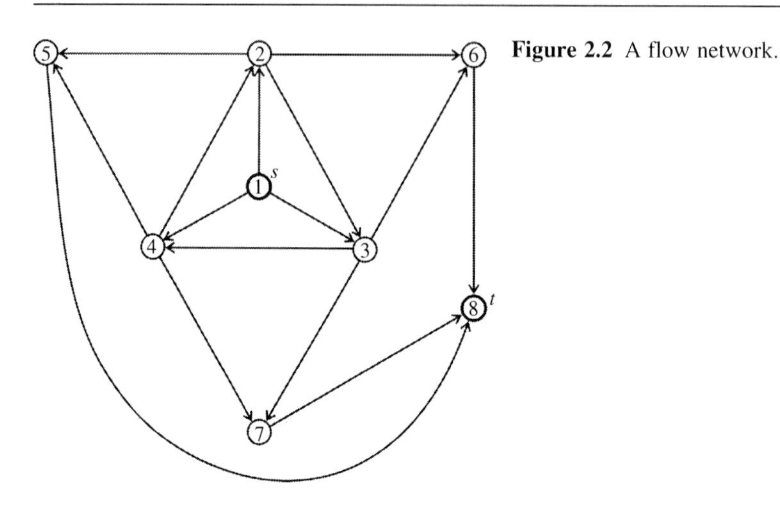

Figure 2.2 A flow network.

descendent of node i. If nodes i and j are connected with an edge, irrespective of its direction, the two nodes are said to be neighbours.

For example, node 4 in the network in Figure 2.2 has three successors/descendents (nodes 2, 5 and 7) and five neighbours (nodes 1, 2, 3, 5 and 7).

2.3.1 Representing the Topology of a Complex Flow Network by an Adjacency Matrix

The conventional adjacency matrix representation of a flow network with directed edges (Ahuja et al.,1993; Gross and Yellen, 2006; Sedgewick, 1992) uses a square matrix A[n,n] with size $n \times n$, where n is the number of nodes in the network. If it is possible to get directly from node i to node j, $A[i,j] = 1$, otherwise, $A[i,j] = 0$. The adjacency matrix representation of the network in Figure 2.2 is

$$A = \begin{pmatrix} 0 & 1 & 1 & 1 & 0 & 0 & 0 & 0 \\ 0 & 0 & 1 & 0 & 1 & 1 & 0 & 0 \\ 0 & 0 & 0 & 1 & 0 & 1 & 1 & 0 \\ 0 & 1 & 0 & 0 & 1 & 0 & 1 & 0 \\ 0 & 0 & 0 & 0 & 0 & 0 & 0 & 1 \\ 0 & 0 & 0 & 0 & 0 & 0 & 0 & 1 \\ 0 & 0 & 0 & 0 & 0 & 0 & 0 & 1 \\ 0 & 0 & 0 & 0 & 0 & 0 & 0 & 0 \end{pmatrix}$$

The space required by an *adjacency matrix* representation is proportional to the square of the number of nodes (n^2). As a result, the adjacency matrix representation requires a significant amount of memory and for large and sparse networks, it may contain a large number of zero elements. This is a rather inefficient representation for sparse networks. For example, a simple network consisting of 999 directed edges, arranged in series, connecting 1000 nodes, will be represented by an adjacency matrix of size 1000×1000 in which only 999 entries will be different from zero!

More importantly, scanning all successors/descendents of a node in a network containing n nodes and represented by an adjacency matrix $A[n \times n]$, requires at least n operations, even if the node has only a single descendent. As a result, the adjacency matrix *is not suitable as a data structure for implementing fast network algorithms.*

2.3.2 Representing the Topology of a Complex Flow Network by Adjacency Arrays

Linked lists can also be used for representing the topology of flow networks (Ahuja et al., 1993; Gibbons, 1985; Sedgewick 1992). The linked lists representation is a significantly more compact representation compared to the adjacency matrix, because for each node it maintains a linked list of its successors only. However, the main advantages of the linked lists such as the possibility to insert and delete a node without the need to move the rest of the nodes will not be used here because there will be no need for such operations. Failure of an edge in the described algorithms is simulated by reducing its flow capacity to zero. Return from repair of an edge is simulated by restoring the flow capacity of the repaired edge to its original level. As a result, the solution described next based on pointers to adjacency arrays is sufficient, simpler and computationally more efficient compared to a representation based on linked lists.

The flow network in Figure 2.2 can be represented by an array of pointers (memory addresses) to adjacency dynamic arrays (Todinov, 2005, 2007). The topology of the network shown in Figure 2.2 is fully represented by the array sb[] of eight pointers which correspond to the eight nodes of the network (Figure 2.3). For each pointer sb[i], the exact amount of memory necessary to accommodate the indices of all successors/descendents of the ith node is reserved. In this way, the topology of the network is described by eight pointers and eight dynamic arrays, where sb[i], $i = 1, 8$ is the address of the ith dynamic array. sb[i][0], which is the first entry in the ith dynamic array, is reserved for the number of descendents of the ith node. Thus sb[3][0] = 3 because node 3 has exactly three descendents. The indices of the actual descendents of the ith node are stored sequentially in sb[i][1], sb[i][2],... (Figure 2.3).

The representation of the network topology by dynamic arrays is a very compact representation. If a pointer or a single node requires 2 bytes computer memory, the space required to define the pointers and the dynamic arrays is $2(2n + m)$, where n is the number of nodes and m is the number of edges. For a sparse network, this

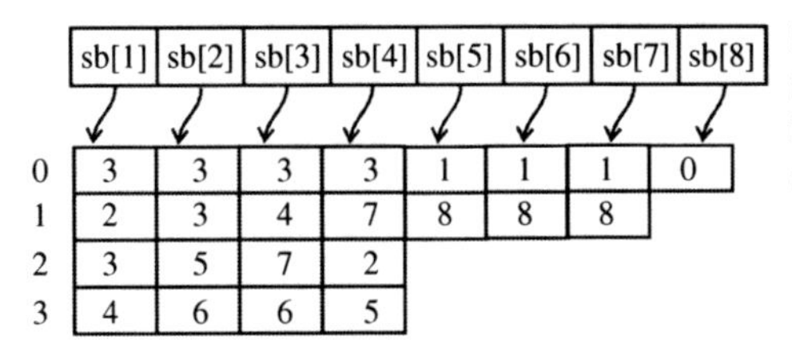

Figure 2.3 Representation of the topology of the network from Figure 2.2 by an array sb[] of eight pointers to dynamic arrays of successors.

storage space is considerably smaller than the space of $2 \times n^2$ bytes required by an *adjacency matrix* representation.

The node adjacency arrays describe fully the topology of any complex flow network, with directed single edges between nodes. If all neighbours are required for a particular node (not only the descendents), pointers to dynamic arrays is again the most efficient representation. In this case, the dynamic arrays list all neighbours of every single node.

2.4 Paths: Algorithms Related to Paths in Flow Networks

A unique sequence of edges between two nodes of the network will be referred to as a *path*. A path between the source s and the sink t will be referred to as *s–t path*. Edges which point into the direction of traversing the path will be referred to as *forward edges*, while edges pointing in the opposite direction will be referred to as *backward edges*. Thus, in Figure 2.4, edge (i,j) is a forward edge, while edge (j,k) is a backward edge.

A path which consists of forward edges only will be referred to as a '*directed path*'. The network in Figure 2.2 for example has many distinct directed paths. Thus, paths (1,2,6,8), (1,2,5,8) and (1,2,3,4,7,8) are directed paths from the source s to the sink t.

The network in Figure 2.2 also contains paths with backward edges. These are for example the paths (1,2,4,7,8) and (1,3,2,6,8). The s–t path with the smallest number of edges will be referred to as '*the shortest s–t path*'.

2.4.1 Determining the Shortest Path from the Source to the Sink

The shortest directed path from the source to the sink is found by using a *breadth-first* search algorithm.

The breadth-first search algorithm for determining the shortest directed path from the source s to the sink t starts from the source s, marks it as 'visited' and considers all its successors which are placed in a queue. The process continues by taking nodes from the head of the queue, and if they have successors that have not been visited, these are placed at the end of the queue. The process continues until there are no more nodes in the queue. If during this process the sink t has not been visited, then there is no directed path from the source s to the sink t. Here is an algorithm in pseudo-code for determining the shortest directed s–t path.

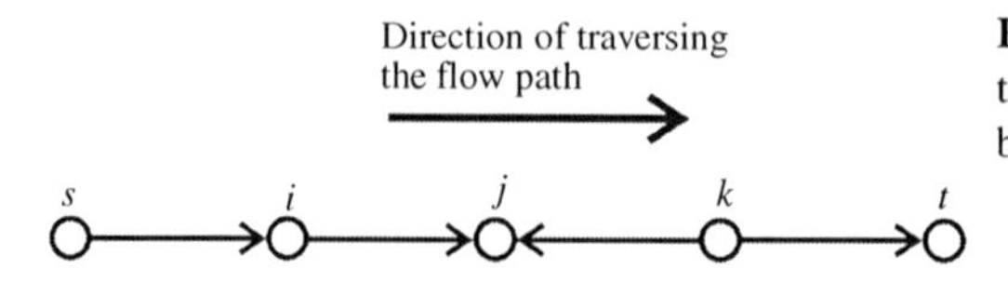

Figure 2.4 A path from the source s to the sink t including forward and backward edges.

Algorithm 2.1

```
function Find_directed_st_path()
{
  visited[1] = −1;
  qhead=0; qend = 1; queue[1]=1;
  for i=2 to n do visited[i]=0;
 while (qhead < qend) do
 {
  qhead=qhead + 1;
  r_node=queue[qhead];
  for i=1 to sb[r_node][0] do
   {
   node=sb[r_node][i];
   if (visited[node]=0) then
        {
        visited[node] = r_node;
        if (node=n) then {
                retrieve_path();
                return 1;
                }
       else {
       qend=qend+1;
       queue[qend]=node;
       }
    }
  }
 }
 return 0;
}
```

The topology of the flow network is coded in the dynamic adjacency arrays sb[]. A single dynamic adjacency array sb[i] is allocated to each node 'i' of the network, where the indices of the successors of node 'i' are listed sequentially as sb[i][1], sb[i][2],... The element sb[i][0] gives the number of successors of node i. The structure of the adjacency array of successors corresponding to node '4' in Figure 2.2 for example, which has three successors (nodes 7, 2 and 5), is the following (Figure 2.3): sb[4][0] = 3; sb[4][1] = 7; sb[4][2] = 2; sb[4][3] = 5.

The algorithm works as follows. A node i is marked as visited, by placing a nonzero value $k > 0$ (visited[i]=k) in the array 'visited[]'. The nonzero value k is the index of the node from which node i has been visited. The indices of all nodes in the network are greater than zero. If node i has not yet been visited, visited[i] will contain zero. In the implementation of the breadth-first search algorithm (Algorithm 2.1) the array 'visited[]' *serves two different goals*. It is used (i) to mark a node as visited and (ii) at the end of the procedure to restore the path from the source to the sink. Otherwise, two different arrays and a set of operations over them need to be maintained – for example, an array marked[] whose elements accept values 0 or 1, for marking the nodes as visited, and array pred[], for recording the predecessor (pred[*i*]) from which node *i* has been visited.

Since the first node (the source) has no predecessors, it is marked as 'visited' by the unique value '−1' assigned by the statement 'visited[1] = −1'. The index of the first node is also placed in the queue by the statement 'queue[1] = 1'.

Next, all nodes except the first node are marked as not visited by the for-loop '**for** i = 2 **to** n **do** visited[*i*] = 0'. The variable 'qhead' marks the head of the queue and the variable 'qend' marks the end of the queue. Initially, the pointer to the head of the queue is set to zero. While the queue is not empty, which is indicated by 'qhead < qend' being true, a node is extracted from the head of the queue with the statement 'r_node = queue[qhead]'. The number of successors of node 'r_node' are stored in sb[r_node][0] and all of the successors are extracted one by one by the statement 'node = sb[r_node][*i*]' in the loop '**for** i = 1 **to** sb[r_node][0] **do**'. For each extracted successor, a check is performed by the statement '**if** (visited[node] = 0) **then**' whether the extracted successor 'node' has been visited. If the node has not yet been visited, the predecessor 'r_node' from which this node has been reached is recorded in the array visited[] by the statement 'visited[node] = r_node'. By the same statement 'visited[node] = r_node', the extracted successor node is marked as visited.

Next, a check is performed by the statement '**if** (node = n) **then**' whether the extracted successor is the sink t (the sink has the largest index). If the sink has been reached, then the function terminates and a value equal to unity is returned, indicating that the sink has been reached. If the considered node is not the sink, it is added at the end of the queue by the statements 'qend = qend + 1; queue [qend] = node'. This process continues until either the sink is reached or the queue is empty. If the queue is empty (qhead = qend) before the sink has been reached, no path exists between the source s and the sink t and the function returns zero.

It is important to mark the considered successors as 'visited' before they are added at the end of the queue. This guarantees that once a node has been considered (and marked as visited) it is never considered again. Furthermore, marking the nodes as 'visited' before they are added to the queue guarantees that the same node is never added more than once in the queue. In other words, the size of the queue will never exceed the number of nodes n in the network.

In effect, the algorithm works by layers. It starts with the source and forms the first layer of nodes which are successors of the source s. If the sink has not been reached, the algorithm forms a second layer of nodes which are successors of the successors of the source s and so on. The nodes are discovered layer by layer, and for each node on the kth layer, the length of the path from the source is exactly 'k' edges.

The number of all considered successors does not exceed the number m of edges in the network. The number of nodes loaded in the queue does not exceed the number of nodes n in the network. Hence, the total number of elementary steps to find the shortest path from the source to the sink does not exceed $m + n$. For each node that has been placed in the queue, at least one edge has been considered – the one through which this node has been visited. Any edge points to exactly one node. Consequently,

	nb[1]	nb[2]	nb[3]	nb[4]	nb[5]	nb[6]	nb[7]	nb[8]
0	3	5	5	5	3	3	3	3
1	2	6	7	1	8	3	3	6
2	3	3	6	3	2	8	4	7
3	4	1	2	7	4	2	8	5
4		4	1	2				
5		5	4	5				

Figure 2.5 The adjacency arrays of neighbours nb[] for the network in Figure 2.2.

the number of considered edges m_c is not smaller than the number of nodes n_q loaded in the queue ($m_c \geq n_q$). Since $m_c \leq m$, the total number of steps does not exceed $2m$ and therefore, the worst-case running time of this algorithm is $O(m)$.

The array 'visited[]' also serves the useful purpose to record and later to retrace the shortest path (in reverse order). This can be done by the following statements:

```
k = n;
while (k > 0) do k = visited[k];
```

The purpose of marking the first node with ' − 1' by the statement visited[1] = −1 is now clear. The negative value assigned to visited[1] serves as a condition for terminating the *while-do* loop.

If the shortest path from source to sink needs to include not only forward edges but also backward edges, the adjacency arrays should contain the indices of all neighbours, not only the indices of the successors. In this case, the structure of the adjacency arrays of neighbours (nb[] in Figure 2.5) is different from the structure of the adjacency arrays of successors (sb[] in Figure 2.3).

Thus, nb[4][0] = 5, because node four has five neighbours. The entries in the nb[4] adjacency arrays of neighbours are nb[4][1] = 1; nb[4][2] = 3; nb[4][3] = 7; nb[4][4] = 2; nb[4][5] = 5.

The algorithm of the function ***Find_undirected_st_path()*** for determining the shortest undirected $s-t$ path is very similar to Algorithm 2.1. The only difference is that instead of the adjacency arrays of successors sb[], the adjacency arrays of neighbours nb[] is used.

2.4.2 Determining All Possible Source-to-Sink Minimal Paths

A minimal source-to-sink ($s-t$) path is a path from the source to the sink with no repeating nodes. All directed minimal $s-t$ paths from a single source to a single sink can be found by using the following backtrack algorithm, given in pseudo-code.

Algorithm 2.2

```
procedure Save_the_path();
procedure Find_all_minimal_paths_from(k)
{
 if (k = n) then {
```

```
            fpath[fp_cnt] = n;
            Save_the_path();
            return;  }
    marked[k] = 1;
    fpath[fp_cnt] = k;
    fp_cnt = fp_cnt + 1;
    for i = 1 to sb[k][0] do
      {
            node = sb[k][i];
            if (marked[node] = 0) then
               Find_all_minimal_paths_from(node);
     }
    marked[k] = 0;
    fp_cnt = fp_cnt - 1;
    }
// Main procedure
  fp_cnt = 1;
  for p = 1 to n do marked[p] = 0;
  Find_all_minimal_paths_from(1).
```

Here is how this algorithm works. The topology of the flow network is coded in the dynamic adjacency arrays sb[]. A single dynamic adjacency array sb[i] is allocated to each node 'i' of the network, where the indices of the successors of the node 'i' are listed sequentially as sb[i][1], sb[i][2],.... Again, element sb[i][0] gives the number of successors of node 'i'. The array fpath[] stores the current minimal path; the variable n contains the number of all nodes in the network. Node k that has been visited is marked by '1' by the statement marked[k]=1. Initially, all nodes are marked as 'not visited' by setting all elements of the marked[] array to zero.

The algorithm is based on a recursive *depth-first search* procedure.

The call of the recursive procedure ***Find_all_minimal_paths_from(1)*** finds all directed paths from the first node with index '1' (the source) to the end node with index 'n' (the sink).

Normally, there will be more than a single minimal path from the source s to the sink t. Because of this, the sink will be reached many times. Consequently, the fragment related to the bottom of the recursive call, ends with a 'return' statement in order to permit other $s-t$ paths to be discovered. If the current node 'k' is not the sink 'n', the path is extended with the node 'k' by the statement 'fpath[fp_cnt] = k', the node 'k' is marked as 'visited' by the statement 'marked[k]=1', and the path length counter 'fp_cnt' is incremented. Next, for all non-visited successors of the node with index 'k', a recursive call is initiated to find a separate path from the non-visited successor, to the sink n. This is achieved by the fragment

```
  for i = 1 to sb[k][0] do
    { node = sb[k][i];
```

```
if (marked[node] = 0)
  then Find_all_minimal_paths_from(node); }
```

where 'node = sb[k][i]' in the kth adjacency array, is the successor 'i' of the node with index 'k'. A zero entry in the marked array (marked[node] = 0) indicates that the successor with index 'node' has not been visited.

With this fragment, the initial problem of finding all minimal paths from the source to the sink is decomposed into problems of the same type as the original problem. Each of these sub-problems is in turn decomposed into sub-problems until the process yields sub-problems with trivial solutions. These are handled by the bottom of the recursive call, given in the next fragment:

```
if (k = n) then {
  fpath[fp_cnt] = n;
  Save_the_path();
  return; }
```

This fragment is executed whenever the sink, with index 'n', is reached.

An essential statement is the assignment 'marked[k] = 0' after the return from a recursive call. This statement effectively changes the state of node 'k' from 'visited' to 'non-visited'. This permits the other recursive calls to explore paths passing through node 'k' which otherwise would be impossible. The statement 'fp_cnt = fp_cnt−1' decrements the path length counter 'fp_cnt', because after the statement 'marked[k] = 0', node 'k' no longer belongs to the current path.

Executing the algorithm for the network in Figure 2.2, which possesses a cycle (2,3,4,2), generates 18 distinct directed minimal paths: (1258, 12368, 123478, 123458, 12378, 1268, 1368, 13478, 134258, 134268, 13458, 1378, 1478, 14258, 142368, 142378, 14268, 1458).

If all minimal paths are to be generated, including non-directed minimal paths, then the adjacency arrays nb[], containing the indices of all neighbours (not only the successors) should be used. The number of all possible minimal non-directed paths from the source node '1' to the sink node '8', for the network in Figure 2.2, then increases to 60.

An algorithm for determining all minimal cut sets in a network can be found in (Yeh et al. 2012).

2.5 Determining the Smallest-Cost Paths from the Source

Section 2.4.1 discusses a breadth-first search algorithm for determining the shortest $s-t$ path in terms of number of edges between the source s and the sink t. There are a number of applications where each edge (i,j) is associated with a particular non-negative weight (cost) $w(i,j)$and it is required to determine the path with the smallest cost from the source to each node of the network.

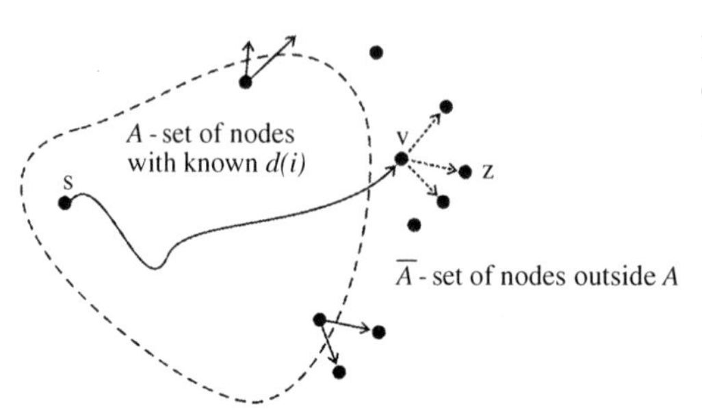

Figure 2.6 Expanding the set A, containing the nodes with known shortest paths to the source s.

The classical Dijkstra's algorithm is appropriate for solving this problem (Cormen et al., 2001; Dasgupta et al., 2008; Dijkstra, 1959). The idea behind the algorithm is to maintain a set A with known smallest-cost paths from the source s. The smallest-cost distance from the source s to any node $i \in A$ will be denoted by $d(i)$.

The set A is subsequently expanded by adding the node with the smallest distance label outside the set A.

The node with the smallest distance from s and outside the set A is determined by testing the distances $d(i)$, $i \in \overline{A}$, and selecting the node v with the smallest distance. The smallest-cost distance $d(v)$ is determined, the node v is added to the set A and the set A is expanded (Figure 2.6). Next, an updating procedure is performed on the neighbours of node v, which has been added *latest* to the set A. For all neighbouring nodes z that can be reached from v and which do not belong to the set A, the distance $d(z)$ is compared with $d(v) + w(v,z)$. If $d(z) > d(v) + w(v,z)$, the distance $d(z)$ is replaced by the smaller distance $d(v) + w(v,z)$, ($d(z) = d(v) + w(v,z)$). Every time an updating is performed, the predecessor of node z is recorded in the 'pred[]' array by the statement pred [z] = v. This record shows that the predecessor of node z, on the smallest-cost path from s, is node v hence the pred[] array can later be used to restore the smallest-cost path from the source s.

The purpose of the updating procedure is to take an old estimate for the smallest-cost distance and check whether it can be improved. The neighbours of the node v, which has been added latest to the set A, are always selected for updating because the only new extensions are those coming from the node, latest added to the set A. All other extensions from other nodes belonging to set A must have been assessed previously and do not need to be assessed again (Dasgupta et al., 2008). The algorithm terminates when all nodes of the network are included in the set A. To find all shortest paths from the source s to each node of the network, the Dijkstra's algorithm can be implemented in $O(m \log n)$ time. The Dijkstra's algorithm can be summarised in pseudo-code.

Algorithm 2.3

```
d(1) = 0; known_nodes = 0; pred[1] = 0;
for (i = 2 to n) do {
        d(i) = BigNumber;
        pred[i] = 0;
    }
```

```
    A = {∅};   Ā = V;

while (known_nodes < n) do
 {
 select the node v∈Ā with the smallest d(v); d(v) = min{d(j) : j∈Ā}
 A: = A ∪ v; //Add node v to the set A;
 Ā: = Ā − {v}; //Exclude node v from the set Ā
 known_nodes = known_nodes + 1;
 for (all neighbours z of node v, which do not belong to set A) do
     {
     td = d(v) + w(v,z);
     if (d(z) > td) then {
                     d(z) = td;
                     pred[z] = v;
                 }
     }
 }
```

A detailed discussion about the correctness of the Dijkstra's algorithm can be found in Cormen et al. (2001).

2.6 Topological Sorting of Networks Without Cycles

Many flow networks do not have cycles and can be represented by directed acyclic graphs (Chartrand et al., 2011). The computational speed of an algorithm for finding a shortest path from the source to the sink can be increased significantly if the topological feature 'flow network with no cycles' is exploited.

Directed networks without cycles can be presented as a directed acyclic graph. Ordering the vertices of directed acyclic networks can be done by using a topological sort, such that for any directed edge (u,v) the index of node v is greater than the index of node u. The idea is to have all the information necessary to compute the distance from the source, by the time the algorithm 'arrives' at a particular node v. If this is the case, *dynamic programming* (Bellman, 1957) can be employed.

The algorithm is based on the return from a recursion, during a depth-first scan of the network.

Algorithm 2.4

```
marked[] // Array where the visited nodes are marked by '1'
sb[n] // Adjacency array representation of the network; n is the total number of
nodes
procedure dfs(i) // Procedure for a depth-first scan of the network
{
  marked[i] = 1;
```

```
   for k = 1 to sb[i][0] do
   {
    node = sb[i][k];
    if (marked[node] = 0) then dfs(k); //where the recursive calls are made
    }
    sp = sp + 1; // Load the node without successors into the stack
    stack[sp] = i;
}
// Main routine
for k = 1 to n do marked[k] = 0; // Mark all nodes as 'not visited'
  sp = 0; // Initialise an empty stack
  dfs(1); // Call the depth-first scan routine starting with the source node '1', which has
         no predecessors
// Print the topologically sorted nodes in the correct order
for i = 1 to sp do
print (stack[i], ←, n − i + 1);
```

The algorithm works as follows. The return from recursion from the depth-first scan occurs when a node with no successors is reached. This will be the sink of the flow network. Immediately after the return from recursion, the sink node '*i*' (with no successors) is pushed into a stack with the statements 'sp = sp + 1'; and 'stack[sp] = i'. Since node '*i*' has been marked as 'visited' after entering the ***dfs***-procedure, there will be no further visits to this node. In this way, node i has essentially been excluded from the network.

The first node to be pushed into the stack is the sink t and it should have a label n, where n is the number of nodes in the network. The next node to be pushed into the stack will be the next node without successors to non-visited nodes. This node will have as a successor the sink t, but because the sink t has already been marked as visited, this route will not be explored because the statement '**if** (marked [node] = 0) **then** ***dfs***(k)' explores only successor nodes marked as 'not visited' (marked[k] = 0). This node will be pushed as a second node into the stack and should have a label 'n − 1'.

Continuing the recursive calls, nodes without successors to non-visited nodes will be pushed into the stack until all nodes of the network are pushed into the stack. The last node to be pushed into the stack should have a label '1'. As a result, in order to produce a topological sort, the nodes loaded in the stack should be relabelled. This is done by the statement '**for** i = 1 **to** sp **do print** (stack[i], ←, $n - i + 1$)', where 'stack[i], ←, $n - i + 1$' stands for 'node stack[i] should have a label $n - i + 1$'. The network in Figure 2.7A is not a topologically sorted network because on the directed edges (4,3) and (8,7), the second index is smaller than the first one.

For the network in Figure 2.7A, which has not been topologically sorted, the execution of the algorithm yields: 9←9, 7←8, 8←7, 6←6, 5←5, 3←4, 4←3, 2←2, 1←1. The topologically sorted flow network is shown in Figure 2.7B.

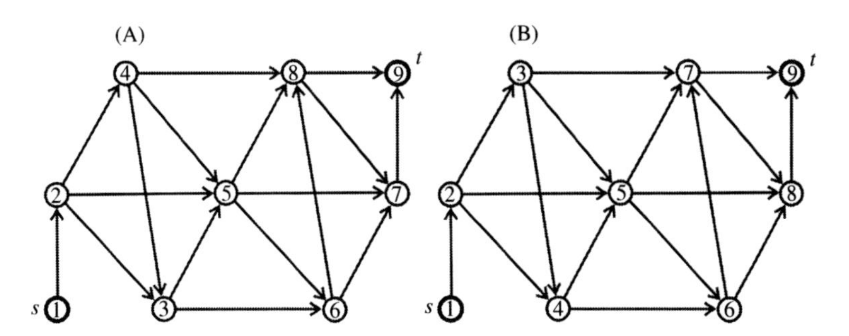

Figure 2.7 (A) A directed network without cycles, that has not been topologically sorted; and (B) topologically sorted network.

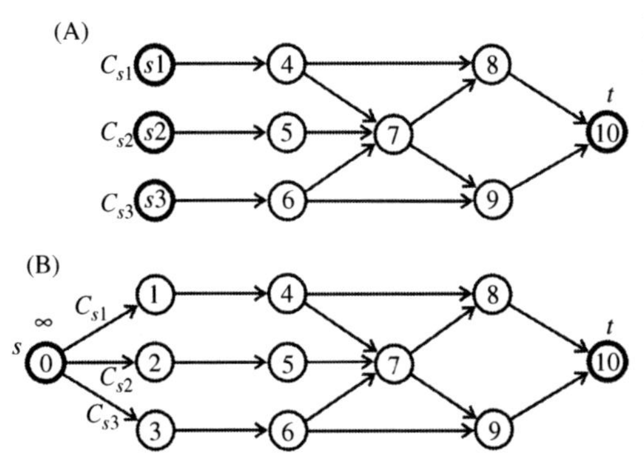

Figure 2.8 (A) An example of a flow network with three sources of flow ($s1$, $s2$, $s3$). (B) The network (A) has been reduced to a flow network with a single source s, characterised by infinite flow generation capacity.

The algorithm for topological sorting of a network is characterised by a linear running time $O(m + \text{n})$, where m is the number of edges and n is the number of nodes in the network.

2.7 Transforming Flow Networks

A flow network with multiple sources (the sources $s1$, $s2$, and $s3$ in Figure 2.8A) with flow generation capacities C_{s1}, C_{s2} and C_{s3} can always be reduced to a flow network with a single source (Figure 2.8B). The multiple sources $s1-s3$ in Figure 2.8A can be replaced by a single source with infinite generation capacity $C_s \to \infty$ feeding each of the multiple sources $s1$, $s2$ and $s3$ through edges with throughput flow capacities C_{s1}, C_{s2} and C_{s3}, correspondingly.

The sources of flow $s1-s3$ become ordinary throughput edges, with flow capacities equal to the maximum possible flow generation capacity of the sources. As a result, the sources 'disappear', and instead, ordinary throughput edges appear. In a similar fashion, multiple sinks can be reduced to a single sink.

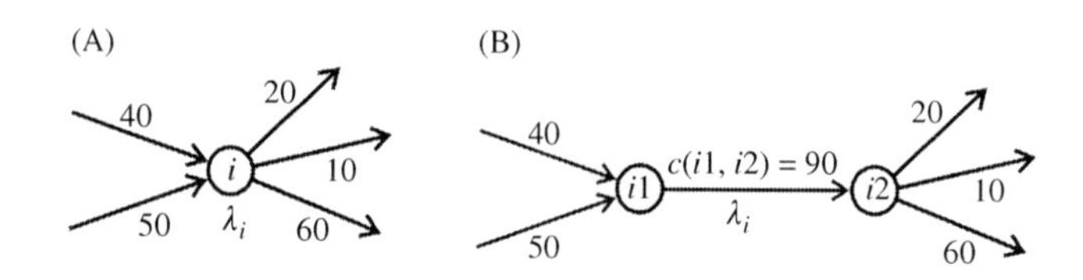

Figure 2.9 Representing an unreliable node '*i*' by splitting it into two perfectly reliable nodes *i*1 and *i*2, connected with unreliable edge (*i*1,*i*2).

Each flow network with unreliable nodes and unreliable edges can always be modelled as a network with perfectly reliable nodes and unreliable edges. Suppose that the nodes are characterised by constant hazard (failure) rates. (A detailed discussion related to the concept *hazard rate* has been provided in Chapter 7.) Each node '*i*' of the repairable network, characterised by a hazard rate $\lambda_i > 0$ (Figure 2.9A), can be represented by a pair of perfectly reliable nodes *i*1, *i*2 ($\lambda_{i1} = 0, \lambda_{i2} = 0$) connected with the unreliable edge (*i*1,*i*2), characterised by a flow capacity equal to the sum of the flow capacities of all edges entering node *i* and hazard rate λ_i, equal to the hazard rate of the unreliable node *i* (Figure 2.9B). For each unreliable node *i*, the first perfect node *i*1 collects the flow from all edges entering node *i*. The second perfect node *i*2 is incident to all edges leaving the unreliable node *i* (Figure 2.7B).

If the unreliable node has a limited capacity C_i, the connecting edge (*i*1,*i*2) has the same capacity C_i.

3 Key Concepts, Results and Algorithms Related to Static Flow Networks

3.1 Path Augmentation in Flow Networks

Comprehensive discussion related to static flow networks algorithms can be found in Ahuja et al. (1993), Asano and Asano (2000), Kleinberg and Tardos (2006) and Cormen et al. (2001).

Here, we selectively review some key results and algorithms related to static flow networks which will later be used for introducing and discussing results related to repairable flow networks, networks with disturbed flows and the new algorithms related to maximising the throughput flow in static networks.

The flow from the source to the sink in a flow network can be increased (augmented) not only along directed paths but also along paths which include backward edges. This is the essence of the Ford–Fulkerson augmentation (Ford and Fulkerson, 1956). A forward edge (i,j) is admissible for augmentation if $f(i,j) < c(i,j)$ holds, where $f(i,j)$ is the edge flow and $c(i,j)$ is the edge capacity. A backward edge (i,j) is admissible for augmentation if $f(i,j) > 0$ is fulfilled. The slacks

$$\Delta = c(i,j) - f(i,j) \geq 0 \tag{3.1}$$

characterising the forward edges (i,j) and the absolute values

$$\Delta = |f(i,j)| \geq 0 \tag{3.2}$$

of the flows through the backward edges (i,j) are compared and the smallest value $\Delta_{\min}$ is selected. The value $\Delta_{\min}$ will be referred to as *bottleneck residual capacity*. If $\Delta_{\min} > 0$, the path is said to be with *nonzero residual capacity*. Paths with nonzero residual capacity are said to be augmentable. In other words, an augmentable path includes only forward edges which are not fully saturated and non-empty backward edges.

Augmenting the path with $\Delta_{\min}$ amount of flow proceeds by increasing the flow along forward edges by $\Delta_{\min}$

Flow Networks. DOI: http://dx.doi.org/10.1016/B978-0-12-398396-1.00003-9

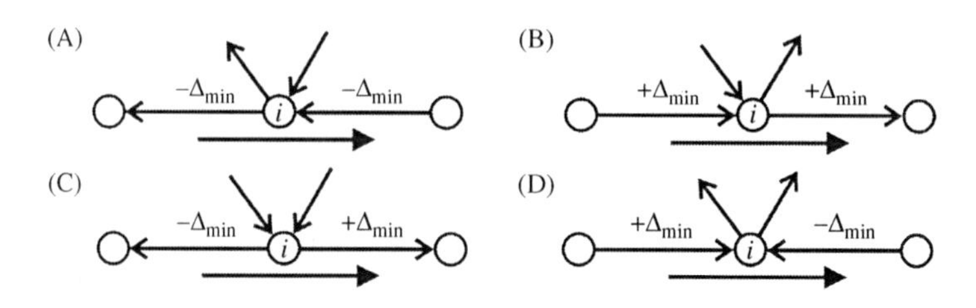

Figure 3.1 The four possibilities for a flow augmentation along a path going through node *i*.

$$f'(i,j) = f(i,j) + \Delta_{\min} \tag{3.3}$$

and decreasing the flow along backward edges by $\Delta_{\min}$,

$$f'(i,j) = f(i,j) - \Delta_{\min} \tag{3.4}$$

This process creates at least one fully saturated forward edge or an empty backward edge. As a result, the path is no longer augmentable. The flow through the path has been increased by $\Delta_{\min}$ without violating the capacity constraints on the edges and the flow conservation law at the nodes along the path.

Indeed, the compliance with the capacity constraints follows directly from the way an augmentation is made. The compliance with the flow conservation law follows from the following reasoning.

The only nodes through which the flow is altered during the path augmentation are the nodes which belong to the augmented path. For any selected internal node *i* on the augmented path (not coinciding with the source or the sink), there are only four possibilities for altering the flow along the incident edges. These possibilities have been listed in Figure 3.1A–D (Gross and Yellen, 2006) and they depend on whether the incident edges are forward or backward edges with respect to the direction of path traversing. The direction of path traversing is shown by an arrow (Figure 3.1). In all four cases (A–D), after the path augmentation, the amount of flow which goes into the internal node *i* is equal to the amount of flow which goes out of the node. In other words, the flow conservation at any node is preserved all the time.

The summary of the algorithm for augmenting a path π is given next.

Algorithm 3.1

$\Delta_{\min}$ = Big_Number;
// *Calculating the bottleneck flow* $\Delta_{\min}$
for (each edge$(i,j) \in \pi$) **do**
{
if (edge (i,j) is a forward edge) **then** $\Delta = c(i,j) - f(i,j)$;
else $\Delta = f(i,j)$;
if $(\Delta_{\min} > \Delta)$ **then** $\Delta_{\min} = \Delta$;
}

// augmenting the π-path with the bottleneck flow $\Delta_{\min}$
for (each edge $(i,j) \in \pi$) **do**
{
if (edge (i,j) is a forward edge) **then** $f(i,j) = f(i,j) + \Delta_{\min}$;
else $f(i,j) = f(i,j) - \Delta_{\min}$;
}

As an illustration of the Ford–Fulkerson augmentation, consider for example the network in Figure 3.2, where the directed $s-t$ path (1,2,3,4,9,10) has been augmented with 10 units of flow. This sets a throughput flow of 10 units in the network. The first number on the edge labels stands for edge capacity and the second number stands for edge flow. Despite that no more directed $s-t$ paths can be augmented, the throughput flow can be increased by augmenting with 10 units of flow the path (1,2,7,5,4,3,6,8,9,10), which includes the backward edge (4,3). The result is the network in Figure 3.2B, whose throughput flow has been increased to 20 units.

3.2 Bounding the Maximum Throughput Flow by the Capacity of $s-t$ Cuts

Let the set V of all nodes in a flow network be partitioned into two sets A and $\overline{A}$ ($A \cap \overline{A} = \varnothing$; $A \cup \overline{A} = V$) in such a way that the source s belongs to the first set and the sink t belongs to the second set ($s \in A$; $t \in \overline{A}$) (Figure 3.3). The as-defined

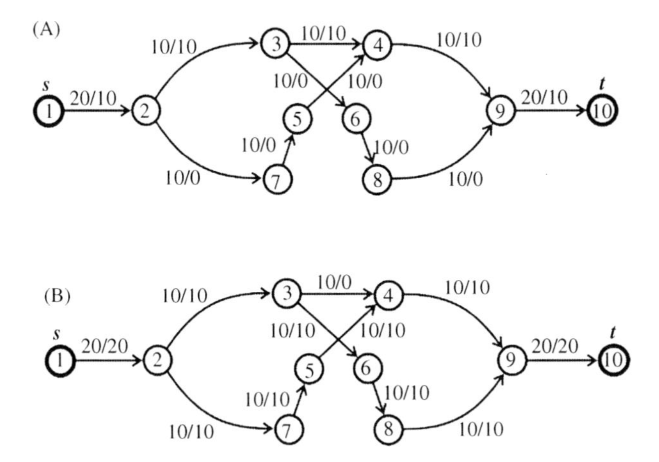

Figure 3.2 (A) The $s-t$ flow through a network can be increased by augmenting the path (1,2,7,5,4,3,6,8,9,10) which includes the backward edge (4,3); (B) a network with a maximum throughput flow.

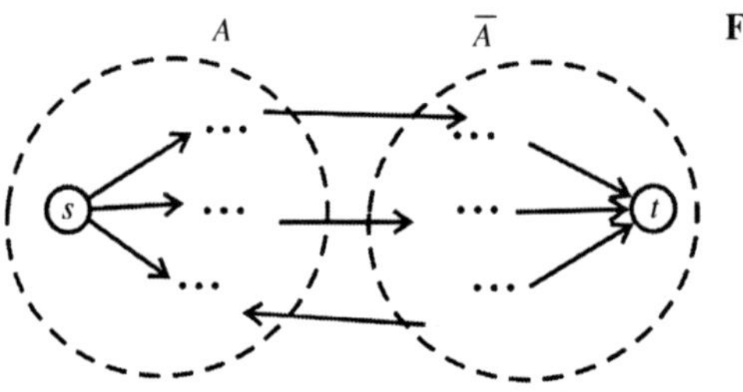

Figure 3.3 An $s{-}t$ cut.

partition $(A,\overline{A})$ is referred to as an $s{-}t$ cut. A capacity of an $s{-}t$ cut $c(A,\overline{A})$ is defined as the sum of the capacities of all edges which start in set A and end in set $\overline{A}$.

$$c(A,\overline{A}) = \sum_{ij} c(i,j); \quad i \in A; \quad j \in \overline{A}; \tag{3.5}$$

Theorem 3.1 *The throughput flow in a network is always equal to the sum of all edge flows which start in subset A and enter subset $\overline{A}$ minus the sum of all edge flows which start in subset $\overline{A}$ and enter subset A.*

Proof (Kleinberg and Tardos, 2006). The throughput flow Q in the network is equal to the sum $f^{out}(s)$ of the flows along the edges going out of the source s: $Q = f^{out}(s)$. Consider the expression $\sum_{v \in A} [f^{out}(v) - f^{in}(v)]$, where $f^{out}(v)$ denotes the sum of the flows along edges which come out of node v and $f^{in}(v)$ is the sum of the flows along edges which enter node v. For a feasible network flow, for all internal nodes in the set A, $f^{out}(v) - f^{in}(v) = 0$ is fulfilled except for the source s, where $f^{out}(s) - 0 = Q$. Consequently, the expression

$$\sum_{v \in A} [f^{out}(v) - f^{in}(v)] = Q \tag{3.6}$$

holds.

Now consider the same edges, incident to the nodes belonging to the set A. In the expression $\sum_{v \in A} [f^{out}(v) - f^{in}(v)]$, the flow along each of these edges is counted twice, once with a positive sign and once with a negative sign. What remains in the expression $\sum_{v \in A} [f^{out}(v) - f^{in}(v)]$ are flows along edges which start in A and end in $\overline{A}$, taken with a positive sign (because these are outgoing edges) and flows along edges which start in $\overline{A}$ and end in A, taken with a negative sign (because these are ingoing edges). As a result,

$$\sum_{v \in A} [f^{out}(v) - f^{in}(v)] = f^{out}(A) - f^{in}(A) \tag{3.7}$$

From Eqs. (3.7) and (3.6), we have

$$Q = f^{\text{out}}(A) - f^{\text{in}}(A) \tag{3.8}$$

This completes the proof.

The next theorem follows from Theorem 3.1.

Theorem 3.2 *The throughput flow in any network does not exceed the capacity of any $s-t$ cut.*

Proof Indeed, from Theorem 3.1, Eq. (3.8) holds. Therefore,

$$Q = f^{\text{out}}(A) - f^{\text{in}}(A) \le f^{\text{out}}(A) \le c(A, \overline{A}) \tag{3.9}$$

This completes the proof.

3.3 A Necessary and Sufficient Condition for a Maximum Throughput Flow in a Static Network: The Max-Flow Min-Cut Theorem

Given is a network with source s and sink t. Consider a process where paths are being augmented sequentially, by the Ford–Fulkerson procedure. The next theorem (Elias et al., 1956; Ford and Fulkerson, 1956) is a necessary and sufficient condition for attaining a maximum throughput flow in a static flow network.

Theorem 3.3 *The necessary and sufficient condition for a maximum throughput flow in a network is the non-existence of augmentable $s-t$ paths.*

Proof The condition 'non-existence of augmentable $s-t$ paths' is a necessary condition for a maximum throughput flow. Indeed, suppose that the converse is true: the throughput flow $Q_{\max}$ is the maximum possible throughput flow and there exists an augmentable $s-t$ path. If an augmentable path exists, then augmenting it with flow will increase the flow from the source to the sink by a positive value $\delta > 0$. As a result, the throughput flow will become $Q_{\max} + \delta$ which contradicts the initial assumption that $Q_{\max}$ is the maximum throughput flow. Consequently, no augmentable $s-t$ path exists, if the throughput flow is the maximum throughput flow.

The condition 'non-existence of augmentable $s-t$ paths' is also a sufficient condition. Let the throughput flow be Q and suppose that no augmentable $s-t$ paths exist. In this case, Q is the maximum possible throughput flow $Q_{\max}$ in the network.

Let us define a set of nodes A, including nodes which can be reached from the source s, through augmentable paths. Along these augmentable paths, none of the forward edges is fully saturated with flow and none of the backward edges is empty.

In the defined node partitioning, the sink t cannot possibly belong to the set A. Indeed, if the sink belonged to the set A, it could be reached from the source s through an augmentable path. Therefore, the flow along this path could be augmented, which contradicts the condition that no augmentable $s-t$ path exists in the network.

Clearly, the sink belongs to the set $\overline{A}$, including the nodes which cannot be reached from the source s through augmentable paths. Hence, the partitioning $(A,\overline{A})$, where $s \in A$ and $t \in \overline{A}$ and $A \cap \overline{A} = \varnothing$, $A \cup \overline{A} = V$, is an $s-t$ cut.

Now the following statement is true. In the as-defined $s-t$ cut, all forward edges starting in set A and ending in set $\overline{A}$ are fully saturated with flow and all backward edges starting in set $\overline{A}$ and ending in set A are empty.

Indeed, suppose that this is not true. Then there will be either: (i) an unsaturated forward edge starting at a node e_i belonging to set A and ending at a node d_i belonging to set $\overline{A}$ or (ii) a non-empty backward edge starting at a node d_i belonging to the set $\overline{A}$ and ending at a node e_i belonging to the set A (Figure 3.4).

The node e_i, however, by the definition of the set A can be reached from the start node s through unsaturated forward edges and non-empty backward edges. According to our assumption, edge (e_i,d_i) is either a non-saturated forward edge or a non-empty backward edge. As a result, node d_i can also be reached from the source s through an augmentable path (s,e_i,d_i) consisting of unsaturated forward edges and non-empty backward edges. Consequently, node d_i should belong to set A and not to set $\overline{A}$. We have arrived at a contradiction. Consequently, neither unsaturated forward edges nor non-empty backward edges exist across the $s-t$ cut. Therefore, all forward edges from the $s-t$ cut are fully saturated and all backward edges are empty.

According to Theorem 3.2, any throughput flow, including the maximum possible throughput flow $Q_{\max}$, cannot exceed the capacity of the as-defined $s-t$ cut $(A,\overline{A})$:

$$Q_{\max} \le c(A,\overline{A}) \tag{3.10}$$

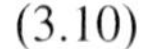

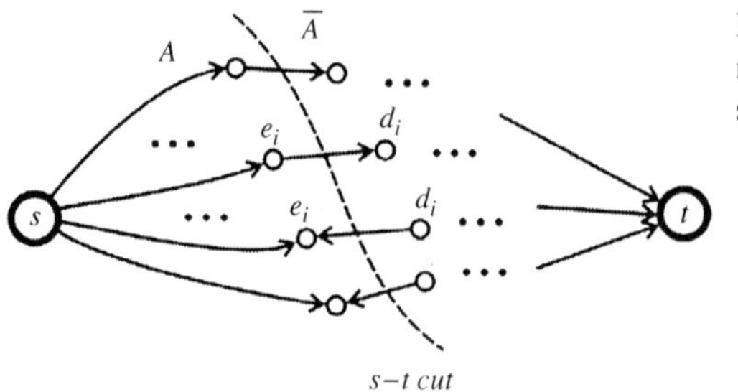

Figure 3.4 An $s-t$ cut, defined by the nodes that can be reached from the source s through augmentable paths.

The throughput flow Q in the as-defined $s-t$ cut $(A,\overline{A})$, however, is exactly equal to the capacity $c(A,\overline{A})$ of the cut because the flows through the backward edges are zero:

$$Q = f^{\text{out}}(A) - f^{\text{in}}(A) = c(A,\overline{A}) \tag{3.11}$$

From Eqs. (3.10) and (3.11), it follows that $Q = Q_{\max}$. Consequently, the flow through the as-defined $s-t$ cut is the maximum possible throughput flow. □

Theorem 3.4 (Max-flow min-cut theorem). *In every flow network, the maximum $s-t$ flow is equal to the smallest $s-t$ cut capacity* (Elias et al., 1956; Ford and Fulkerson, 1956).

Proof Consider the defined $s-t$ cut $(A,\overline{A})$, in the proof of Theorem 3.3. It will be shown that this cut has the smallest capacity among all $s-t$ cuts.

Suppose that there is an $s-t$ cut $(A',\overline{A'})$ with a smaller capacity. In other words, suppose that $c'(A',\overline{A'}) < c(A,\overline{A})$ holds. The maximum possible throughput flow $Q_{\max}$, according to Theorem 3.2 however, does not exceed the capacity of any $s-t$ cut. Therefore, the maximum throughput flow $Q_{\max} \le c'(A',\overline{A'})$ does not exceed the capacity $c'(A',\overline{A'})$ of the $(A',\overline{A'})$ $s-t$ cut. Since, according to our assumption, $c'(A',\overline{A'}) < c(A,\overline{A})$, the inequality

$$Q_{\max} < c(A,\overline{A})$$

also holds. According to Theorem 3.3, however, $Q_{\max} = c(A,\overline{A})$. We have arrived at a contradiction. Consequently, an $s-t$ cut with capacity smaller than $c(A,\overline{A})$ does not exist and the $(A,\overline{A})$ $s-t$ cut has the smallest capacity. This completes the proof. □

3.4 Classical Augmentation Algorithms for Determining the Maximum Throughput Flow in Networks

For the sake of simplicity, augmentation algorithms for maximising the throughput flow in a static network will later be used as a basis for introducing concepts, ideas and results related to repairable flow networks. The main concepts, ideas and results however will not be affected, if other, conceptually different algorithms were used for maximising the throughput flow, for example, if the *push-relabel* algorithms (Goldberg and Tarjan, 1988; Karzanov, 1974) were used.

An augmentation algorithm for maximising the throughput flow can be constructed easily, by a direct implementation of the Ford–Fulkerson theorem. An augmentable $s-t$ path is identified and subsequently augmented with flow and this sequence continues until no more augmentable paths can be found. The algorithm in pseudo-code is given next.

Algorithm 3.2

path_exists = ***Find an augmentable s−t path;***
while (path_exists = TRUE) **do** {***Augment the path with flow;***
Accumulate the augmented flow into the throughput flow;
path_exists = ***Find an augmentable s−t path;***}

The variable 'path_exists' accepts values TRUE or FALSE depending on whether an augmentable $s-t$ path exists or not.

If the capacities are rational numbers, they can always be made integral by using an appropriate scaling transformation.

It can be shown that for integer capacities of the edges, the Ford–Fulkerson algorithm runs in time $O(m|f^*|)$, where $|f^*|$ is the magnitude of the maximum throughput flow.

Indeed, finding an augmentable flow path and augmenting it with flow is an operation of running time $O(m)$. In each iteration, the throughput flow is increased by at least 1 unit; therefore, the maximum possible number of iterations is equal to the magnitude of the maximum throughput flow $|f^*|$. As a result, after at most $O(m|f^*|)$ steps, the maximum throughput flow will be reached.

A significant drawback of the Ford–Fulkerson algorithm, well documented in the literature, is that for some network topologies it could run very slowly. This can be illustrated by the network shown in Figure 3.5.

Suppose that the first identified augmentable path is (1,2,3,4), and it is augmented by 1 unit of flow. Suppose also that the second identified augmentable path is (1,3,2,4), which is also augmented with 1 unit of flow. If this sequence of paths is repeated, at each path augmentation, the output flow to the sink t increases by only 1 unit. There will be 2×10^9 augmentations before the algorithm finds the maximum throughput flow of 2×10^9 units.

This problem is easily avoided by the *Edmonds and Karp algorithm* (Edmonds and Karp, 1972). Unlike the Ford–Fulkerson augmentation algorithm, the Edmonds and Karp algorithm always augments the flow along the shortest $s-t$

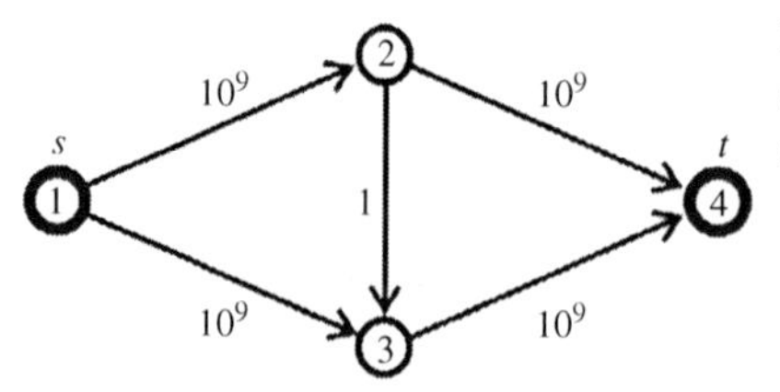

Figure 3.5 A very slow convergence of the solution produced by the Ford–Fulkerson augmentation algorithm.

path, in terms of number of edges, until no more augmentable $s-t$ paths can be found. The algorithm in pseudo-code is as follows:

Algorithm 3.3

> path_exists = ***Find the shortest augmentable s−t path;***
> **while** (path_exists = TRUE) **do** {***Augment the path with flow;***
> ***Accumulate the augmented flow into the throughput flow;***
> path_exist = ***Find the shortest augmentable s−t path;***}

As can be verified, for the network shown in Figure 3.5, the algorithm determines the maximum throughput flow with only two augmentations.

The shortest augmentable $s-t$ path is determined by using the *breadth-first* search algorithm described in the previous chapter. The Edmonds and Karp algorithm will be illustrated by the next example, featuring the network shown in Figure 3.6A, where the edges have been labelled with their flow capacities.

The identified shortest path (1,2,10,11) can be augmented with 5 units of flow. The next identified shortest path (1,3,7,11) can be augmented with 6 units of flow. The result is the network shown in Figure 3.6B, where no directed path can be augmented with flow. The paths (1,4,6,7,3,5,10,11) and (1,4,6,7,3,8,9,11), however, both of which include edge (3,7) as a backward edge, can be augmented with 3 flow units each. The result is the network shown in Figure 3.6C, where no more paths can be augmented. Note that after the augmentations, edge (3,7) remains empty.

While for the network shown in Figure 3.6, the presented optimal solution is unique, other networks admit more than one optimal solution for the throughput flow. In other words, there may be different sets of edge flows which result in the

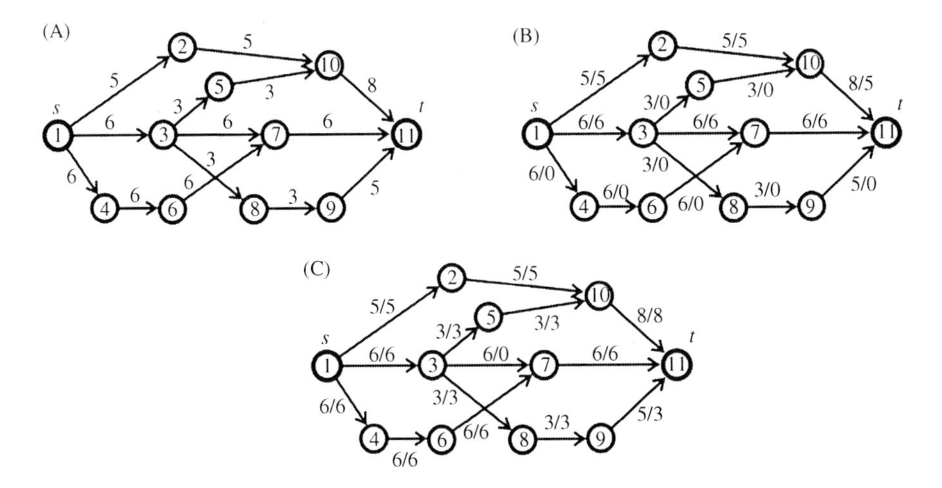

Figure 3.6 A network whose throughput flow can be maximised by using the Edmonds and Karp algorithm.

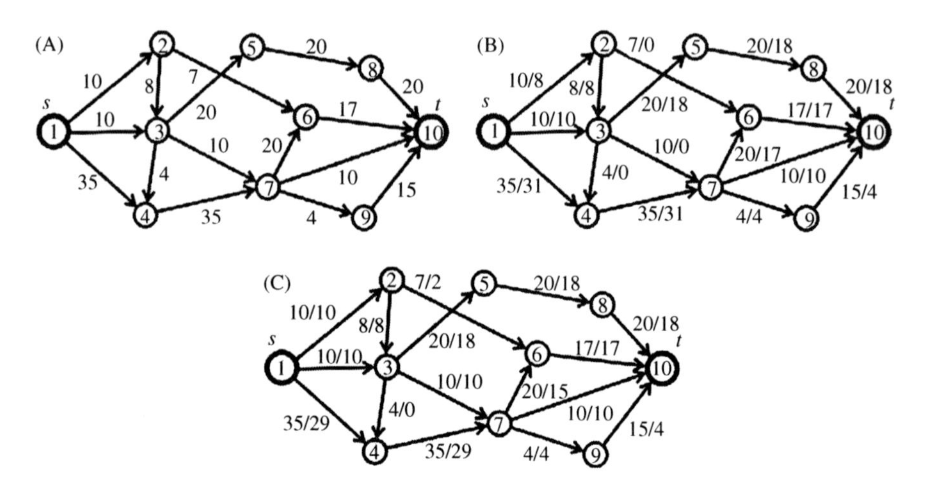

Figure 3.7 Maximising the throughput flow in a network may result in more than a single optimal solution.

same maximum throughput flow. Applying the Edmonds and Karp algorithms for the network shown in Figure 3.7A results in 49 units maximum throughput flow. The edge flows, which correspond to the optimal solution, are shown in Figure 3.7B. Now consider the cyclic path (1,2,6,7,4,1). This cyclic path can be augmented with 2 flow units. The throughput flow remains the same but the augmentation yields a different set of edge flows. The augmentable cyclic path may include the source, the sink or both.

The following result holds:

Theorem 3.5 *The necessary and sufficient condition for more than a single optimal solution of the maximum throughput flow problem is the existence of an augmentable cyclic path.*

Proof Indeed, the existence of an augmentable cyclic path is a sufficient condition because the flow along the cyclic path can be augmented, which results in a different set of edge flows leading to the same throughput flow.

On the other hand, if more than a single set of flows leading to the same throughput flow exist, according to Theorem 4.3 (which has been proved in Chapter 4), one of the sets of edge flows can always be transformed into the other set of flows by augmenting *cyclic paths only*. Consequently, the existence of an augmentable cyclic path is a necessary condition. □

No augmentable cyclic paths exist for the network in Figure 3.6. As a result, the solution presented in Figure 3.6C is unique.

Let $d(i)$ denote the shortest-path distance from the source s to node i, which is part of an augmentable s–t path. The shortest-path distance is measured by *the number of edges* between the source s and node i.

Lemma 3.1 *Let x' be the flow obtained from flow x through augmentation along the shortest path π. Then, the residual distance of any internal node has either increased or stayed the same. In other words, for each internal node v, the inequality*

$$d_{x'}(v) \geq d_x(v) \tag{3.12}$$

holds (Cook et al., 1998).

Proof This lemma can be proved by using the *extreme principle* technique to arrive at a contradiction. Suppose that there exists a node v such that $d_{x'}(v) < d_x(v)$ after the augmentation of a particular path. Without a loss of generality, assume that node v is *the node with the smallest distance* $d_{x'}(v)$ with this property ($d_{x'}(v) < d_x(v)$). Consider now node w, which is the predecessor of node v on the augmentation path π (Figure 3.8).

Because, by assumption, $d_{x'}(v) < d_x(v)$, the following inequality holds:

$$d_x(v) > d_{x'}(v) = d_{x'}(w) + 1 \tag{3.13}$$

For $d_{x'}(w)$, the inequality $d_{x'}(w) > d_x(w)$ holds. If the inequality $d_{x'}(w) < d_x(w)$ was possible then node w and not node v would be the node with the smallest residual distance with the property $d_{x'}(w) < d_x(w)$. Because $d_{x'}(w) > d_x(w)$, from Eq. (3.13), it follows

$$d_x(v) > d_{x'}(v) = d_{x'}(w) + 1 \geq d_x(w) + 1 \tag{3.14}$$

Clearly, edge (w,v) is not augmentable in x. Otherwise, if it was augmentable, for the shortest distance $d_x(v)$ we would have $d_x(v) = d_x(w) + 1$ and not $d_x(v) > d_x(w) + 1$ according to Eq. (3.14). We arrived at a contradiction because all edges belonging to an augmentable path must be augmentable. As a result, with the path augmentations, the residual distance from the source s cannot decrease. □

By using Lemma 3.1, the following theorem can be proved.

Theorem 3.6 *The Edmonds and Karp shortest-path augmentation algorithm runs in time $O(nm^2)$.*

Figure 3.8 Augmenting the shortest $s-t$ paths results in a non-decreasing shortest-path distance $d(v)$.

Proof After each path augmentation, at least one edge from the network, e.g. edge (i,j), with a start node i and end node j, will be fully saturated with flow. During another path augmentation, the flow through edge (i,j) could be reduced and as a result, edge (i,j) could again be fully saturated with flow, during any of the subsequent augmentations. It will be shown that the maximum number of times an edge (i,j) could be fully saturated with flow does not exceed $n/2$, where n is the number of nodes in the network.

Consider the edge (i,j), fully saturated for the first time. No augmenting path can pass through this edge before a flow reversal occurs through it. During a flow reversal, the direction of the path is from node j to node i. Consequently, $d(i) = d(j) + 1$ will hold after the flow reversal and, as a result, the distance $d(i)$ of node i will increase by at least one. The distance $d(j)$ will either remain the same or increase, but, according to Lemma 3.1, it cannot possibly decrease. Because the Edmonds and Karp algorithm always augments the shortest path, during the next augmentation when edge (i,j) is fully saturated, the expression $d(j) = d(i) + 1$ will hold. In other words, the distance of node $d(j)$ will increase by at least 2 because $d(i)$ after the flow reversal has increased by at least one.

Hence, after a single combination of a flow reversal and full saturation, the distance of node $d(j)$ increases by at least 2. The maximum possible distance in any network with n nodes is $n - 1$. Therefore, the maximum number of times a full saturation of an edge can occur does not exceed $(n - 1)/2$. During an augmentation of a path, there is at least a single edge that is fully saturated. The same edge cannot be saturated more than $(n - 1)/2$ times. Because the network has m edges, after at most $m(n - 1)/2$ path augmentations, the algorithm will terminate. Because the augmentation of a path is of complexity $O(m)$, the running time of the Edmonds and Karp algorithm is $O(nm^2)$. □

3.5 General Push-Relabel Algorithm for Maximising the Throughput Flow in a Network

The *general push-relabel algorithm* is based on the *preflow concept* proposed in (Karzanov, 1974). This concept was later used as a basis of the push-relabel algorithms by Sleator and Tarjan (1980) and Goldberg and Tarjan (1988). The algorithm discussed in this section has been proposed by Goldberg and Tarjan (1988). Detailed discussion and analysis of this algorithm can also be found in Kleinberg and Tardos (2006) and Cormen et al. (2001).

The general push-relabel algorithm maintains a preflow, which differs from a flow in the sense that the sum of the edge flows going into a node is allowed to exceed the sum of the flows going out of the node. The push-relabel algorithms work to convert the preflow into a feasible flow. For each node u, the general push-relabel algorithm maintains a preflow defined by

$$\sum_i f(i,u) \geq \sum_j f(u,j) \tag{3.15}$$

A node for which $\sum_i f(i,u) > \sum_j f(u,j)$, is said to be an *overflowing node* or *active node*. In other words, the sum of the flows going into any node is greater than or equal to the sum of the flows leaving the node. The amount of excess flow ef_u, at an active node u, is given by

$$ef_u = \sum_i f(i,u) - \sum_j f(u,j) \tag{3.16}$$

While the augmentation algorithms maintain the flow feasible, at all stages of the throughput flow maximisation, the general push-relabel algorithm maintains a preflow and aims to make it a feasible flow by pushing excess flow from overflowing nodes along augmentable edges towards local neighbouring nodes.

An augmentable edge is either a forward edge which has not been fully saturated or a non-empty backward edge. Excess flow from an overflowing node can only be pushed along augmentable edges.

Labels to all nodes are introduced, indicating the nodes' 'height'. The source s has a height $h(s) = n$, where n is the number of nodes in the network and the sink t has a height zero $h(t) = 0$. These are known as *source–sink conditions*. The source and sink labels are never altered throughout the execution of the algorithm.

For any augmentable edge (u,v), the steepness condition

$$h(u) \leq h(v) + 1 \tag{3.17}$$

must be fulfilled.

Flow is only pushed from overflowing nodes towards nodes with lower heights. The push operation 'push(u,v)' from node u towards node v is executed only if the following conditions are fulfilled:

1. Node u has excess flow (node u is an overflowing node).
2. $h(u) = h(v) + 1$.
3. Edge (u,v) is augmentable.

The maximum possible amount of flow Δ that can be pushed from node u to node v is either equal to the excess flow ef_u at node u or to the residual capacity $c(u,v) - f(u,v)$ of edge (u, v):

$$\Delta = \min\{ef_u,\ c(u,v) - f(u,v)\} \tag{3.18}$$

As a result, the excess flow at the overflowing node u becomes $ef'_u = ef_u - \Delta$ and the excess flow at the recipient node v becomes $ef'_v = ef_v + \Delta$.

The relabel operation 'relabel (u)' is executed if the following conditions are fulfilled:

1. Node u has excess flow ($ef_u > 0$).
2. There are neighbouring nodes v, accessible from node u through augmentable edges.
3. For all neighbouring nodes v, accessible from node u, the condition $h(v) \geq h(u)$ is fulfilled.

The relabelling operation simply consists of increasing the height of node u to exceed by one the smallest among the labels of accessible neighbouring nodes v:

$$h(u) \leftarrow 1 + \min\{h(v)\} \tag{3.19}$$

As can be verified, after each relabel operation, the steepness condition 3.17 is preserved in the network. The relabelling operation gives the overflowing node the greatest height allowed by the steepness constraint 3.17.

Next, a description of the algorithm is given in pseudo-code.

Algorithm 3.4

Set $h(s) = n$ and $h(i) = 0$ for the rest of the nodes in the network.
Fully saturate each edge k coming out of the source $s(f(s,k) = c(s,k))$.

while (there is an overflowing node u) **do**
{
 if (there is a neighbour v towards which flow can be pushed) **then**
  push(u,v);
 else
  relabel(u);
}

Lemma 3.2 *After each push operation of the algorithm, there is no augmentable $s-t$ path in the network.*

Proof (Goldberg and Tarjan, 1988). Suppose that there is an $s-t$ path $(s,v_1,v_2,\ldots, t)$ that can be augmented. According to the source–sink conditions, $h(s) = n$. The path $(s,v_1,v_2,\ldots,v_k = t)$ is augmentable, therefore edge (s,v_1) is also augmentable. Because $h()$ is a valid labelling, the steepness condition $h(s) \leq h(v_1) + 1$ is fulfilled which means also $h(v_1) \geq h(s) - 1 = n - 1$. The same reasoning can now be applied to node v_2. As a result, $h(v_1) \leq h(v_2) + 1$ which means also $h(v_2) \geq n - 2$. Continuing in this fashion, $h(t) \geq n - k$ is obtained for the sink node. Because there are no loops along the path, and the maximum length of a minimal path cannot exceed n − 1, it follows that $k < n$. Therefore $h(t) > 0$, which contradicts the sink condition $h(t) = 0$. As a result, no augmentable $s-t$ path can be found during the operation of the algorithm. The claim in the lemma has been proved. □

After the push operation, when no more excess nodes can be found, the preflow is turned into a feasible flow. According to the Ford–Fulkerson theorem, the absence of augmentable $s-t$ path in the network means that the obtained feasible flow is also the maximum throughput flow.

If no more flow can be pushed towards the sink but overflowing nodes still remain, the push-relabel operations continue and the excess flow from overflowing nodes is sent back to the source s. This is guaranteed because of the next lemma.

Lemma 3.3 *If node u is an overflowing node, then there exists an augmentable path from node u to the source s.*

Proof (Kleinberg and Tardos, 2006). Let A denote the set of nodes from which there is an augmentable path to the source s ($s \in A$). Set $\overline{A}$ includes nodes from which there is no augmentable path to the source s ($A \cup \overline{A} = V$, $A \cap \overline{A} = \emptyset$). It will be shown that all overflowing nodes are in set A and set $\overline{A}$ does not contain any overflowing nodes.

First, note, that no edge (i,j), whose start node i is in set A and whose end node j is in set $\overline{A}$ has a positive flow $f(i,j) > 0$. Otherwise, edge (j,i) would be an augmentable edge and there would be an augmentable path from node j to the source s. However, this is impossible because node j belongs to set $\overline{A}$, which includes nodes from which there is no augmentable path to the source s.

Consider now the sum of the excess flow $\sum_{u \in \overline{A}} ef_u$ at all nodes u belonging to set $\overline{A}$ (irrespective of whether these are overflowing nodes or not). The sum of the excess flow at all nodes from $\overline{A}$ can be presented as

$$\sum_{u \in \overline{A}} ef_u = \sum_{u \in \overline{A}} [f^{\text{in}}(u) - f^{\text{out}}(u)] \geq 0 \tag{3.20}$$

and is non-negative because a node can either have a positive excess flow or a zero excess flow. In Eq. (3.20), $f^{\text{in}}(u)$ is the sum of the flows going into node u and $f^{\text{out}}(u)$ is the sum of the flows coming out of node u. The expression for the excess flow in Eq. (3.20) can be rewritten as a sum of the flows along all edges incidental to nodes from set $\overline{A}$. If both end nodes (u,v) of these edges belong to set $\overline{A}$, $(u \in \overline{A}) \cap (v \in \overline{A})$, then $f(u,v) = -f(v,u)$ and the sum of the edge flows will be zero. However, according to the result established earlier, if the start node belongs to set A and the end node belongs to set $\overline{A}$, the flow along this edge is zero. Consequently, the net sum of the flows along the edges incidental to the nodes belonging to set $\overline{A}$ is given by the expression

$$\sum_{u \in \overline{A}} ef_u = -f^{\text{out}}(\overline{A}) \tag{3.21}$$

Because the sum of the flows $f^{\text{out}}(\overline{A})$ coming out of the set $\overline{A}$ is positive, expression (3.21) can only be compatible with expression (3.20) if $\sum_{u \in \overline{A}} ef_u = 0$. Therefore, there are no overflowing nodes in set $\overline{A}$. The overflowing nodes are all in set A, therefore augmentable paths exist from these nodes to the source s. This proves the lemma. □

The running time of the push-relabel algorithm is $O(n^2m)$. A detailed analysis of the push-relabel algorithm can be found in Goldberg and Tarjan (1988), and Cormen (2001). A variety of this algorithm is the *relabel-to-front algorithm,* discussed in detail by Cormen et al. (2001), whose running time $O(n^3)$ is better for dense networks.

The push-relabel algorithm will be illustrated again by the network shown in Figure 3.6A. Initially, the label of s is set to be equal to the number of nodes in the network. The labels of the rest of the nodes are set to zero and all edges coming out of the source s are fully saturated with flow. This operation creates three overflowing nodes: nodes '2', '3' and '4' (Figure 3.6A). Now, suppose that node 3 is selected as active node. Its label is zero and all neighbouring nodes '5', '7' and '8', reachable through augmentable edges, have also zero labels. Next, a relabelling operation is initiated for node '3' and, as a result, its label becomes equal to '1'. Excess flow of 6 units is pushed towards node '7' and node '7' becomes an overflowing node. A relabelling operation is initiated for node '7' because the only neighbour connected with an augmentable edge is the sink t, whose label is '0'. The label of node '7' is increased to '1' and a flow of 6 units is pushed towards the sink. The resultant state of the network is shown in Figure 3.9A.

Suppose that the next active (overflowing) node is node '4'. The only neigbouring node connected through an augmentable edge is node '6', which has the same label '0' as node '4'. A relabelling operation is initiated for node '4'. Its label is increased to '1' and 6 units of flow are pushed towards node 6, which becomes an

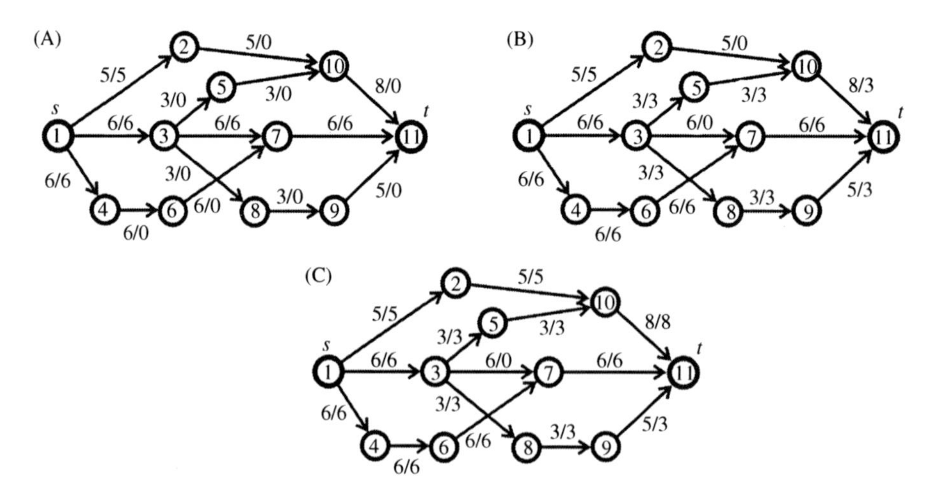

Figure 3.9 Maximising the throughput flow in a network by the general push-relabel algorithm.

overflowing node. Its label is ‘0’, while the label of its only neighbour (node ‘7’) is ‘1’. A relabelling operation is initiated for node ‘6’ and as a result, its label becomes ‘2’. Flow of 6 units is pushed towards node ‘7’, as result of which node ‘7’ becomes an overflowing node again. The only neighbour connected with node ‘7’ through an augmentable edge is node ‘3’. The forward edge (7,11) is fully saturated and therefore not augmentable. The backward edge (7,3) is not empty and is therefore augmentable. Node ‘7’, however, has the same label as node ‘3’ and a relabelling operation is initiated, as a result of which the label of node ‘7’ becomes equal to ‘2’ and 6 units of flow are pushed towards node ‘3’. In a similar fashion, 3 units excess flow are pushed from node ‘3’ to node ‘5’, node ‘10’ and the sink, and 3 units excess flow are also pushed from node ‘3’ towards node ‘8’, node ‘9’ and the sink. The resultant state of the network is shown in Figure 3.9B. The only remaining excess node is node ‘2’. After pushing its excess flow towards node ‘10’ and the sink, the excess flow disappears and the final state of the network is shown in Figure 3.9C.

3.6 Applications

Maximising the flow in a network has various important applications. One of them is the problem of finding the maximum number of directed edge-disjoint s–t paths in a directed graph $G = (V,E)$.

Two paths are said to be *edge-disjoint* if they do not have a common edge (the paths may have common nodes). A set of directed paths are said to be edge-disjoint if no two paths from the set share an edge. This problem is important because it is related to the reliability of communication networks. Increasing the number of directed edge-disjoint communication paths improves the reliability of networks and their resistance to simultaneous edge failures.

To solve this problem, the initial network is transformed into a network where each edge has a unit capacity.

Theorem 3.7 *The maximum number of edge-disjoint directed paths in the original network is equal to the maximum throughput flow in the unit edge capacity network.*

Proof This theorem can be proved by induction. The theorem is obviously true if the throughput flow in the transformed network is equal to unity. In this case, there must necessarily be a directed s–t path, whose augmentation with 1 unit of flow results in a throughput flow also equal to unity.

Suppose now that the theorem is true for a throughput flow of k units. It will be shown that for $k + 1$ units throughput flow, there will be exactly $k + 1$ edge-disjoint directed paths.

An edge coming out of the source is selected, carrying unit flow towards a particular node i. According to the flow conservation law, there must be another edge, carrying a unit flow which goes out of node i. By selecting the edge carrying a unit

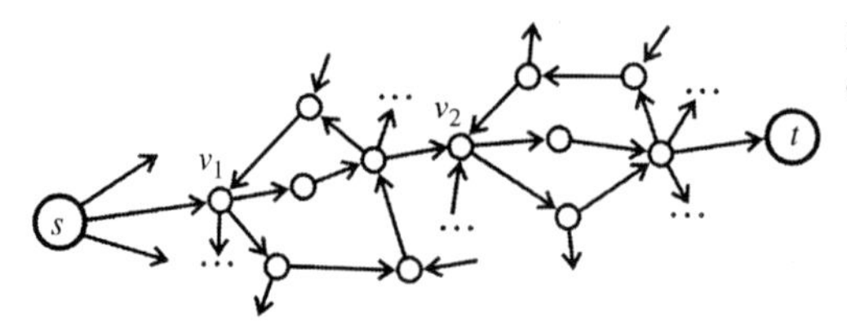

Figure 3.10 Tracing a disjoint directed s–t path.

flow, the node $i+1$ is reached and so on. By continuing this process, we either arrive (i) at a node which has already been visited or (ii) at the sink t (Figure 3.10).

(i) Consider the first alternative. If a node $v1$ has been visited again before the sink t has been reached (Figure 3.10), a directed cycle has been discovered in the network. All edges belonging to the cycle, carrying unit flow, can then be removed from the network. This will result in a new feasible network flow. At the same time, the total throughput flow of $k+1$ units will not be affected because neither an outgoing edge from the source s nor an edge entering the sink t can possibly belong to the cycle. After removing all edges belonging to the cycle, each of which carries unit flow, the process of tracing the selected path continues from the repeated node $v1$. The continuation is guaranteed because exactly one saturated edge entering node $v1$ and exactly one saturated edge leaving node $v1$ had been removed. Therefore, according to the flow conservation law, at node $v1$ there must be a non-empty edge leaving the node. If another cycle is encountered at node $v2$, its edges are removed and the path tracing continues from node $v2$. This way of path tracing guarantees that no edge is scanned twice. By continuing this process, eventually, no more cycles will be encountered and the sink t will be reached. This is guaranteed because the flow conservation law holds at each node and the network has a finite number of edges.

(ii) Consider now the first alternative. If the sink has been reached, the traversed edges, each carrying a unit flow, constitute a directed path that can be removed from the network. Because exactly one outgoing edge from the source s and exactly one edge going into the sink t have been removed, this will result in a network with a new feasible maximum throughput flow of size k units. By the induction hypothesis, there must be exactly k disjoint directed paths for the resultant network which, together with the removed directed s–t path, form $k+1$ disjoint directed paths in the original network. □

All disjoint paths in the original network can be found in $O(nm)$ time.

Indeed, the decomposition procedure can produce a single directed s–t path in $O(m)$ time because no edge is scanned twice before a directed s–t path is produced and removed from the network. The maximum possible number of directed disjoint paths is equal to the maximum possible number of edges emanating from the source s, which is $n-1$, where n is the number of nodes in the network. Consequently, all directed disjoint paths can be found in $O(nm)$ time.

A related application is finding the maximum number of *node-disjoint* s–t paths in a network with directed edges.

Two paths are said to be *node-disjoint* if they do not have a common node. A set of paths are said to be node-disjoint if no two paths from the set share a node. To find the maximum number of node-disjoint paths, the initial network

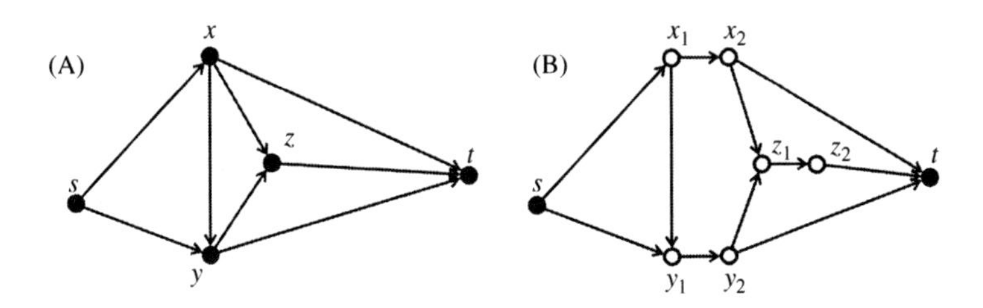

Figure 3.11 (A) Original and (B) transformed network.

(Figure 3.11A) is transformed into a network where each internal node i is split into two nodes i_1 and i_2 connected with an edge (i_1,i_2) (Figure 3.11B). In the transformed network, all edges previously entering node i now enter node i_1, and all edges previously going out of node i now go out of node i_2 (Figure 3.11B). All edges in the network are assumed to be of unit capacity.

Theorem 3.8 *The maximum number of node-disjoint directed paths in the original network is equal to the maximum throughput flow in the transformed, unit edge capacity network.*

Proof If two directed paths are edge-disjoint in the transformed network, they are node-disjoint in the original network. Conversely, if two paths are node-disjoint in the original network, they are edge-disjoint in the transformed network. Hence, there exists a one-to-one correspondence between the number of directed node-disjoint paths in the original network and the number of directed edge-disjoint paths in the transformed network. According to the previous theorem, the maximum number of edge-disjoint paths in the transformed network is equal to the maximum throughput flow. □

The next two results are known as Menger's edge and node connectivity theorems.

Theorem 3.9 *The maximum number of directed edge-disjoint paths in a single source–single sink network with directed edges is equal to the minimum number of edges whose removal separates the source s from the sink t* (Menger's edge-connectivity theorem).

Proof Suppose that the flow in the unit-capacity network has been maximised. Define an $s-t$ cut $(A,\overline{A})$ in the network, where A is the set of all nodes that can be reached from the source s through unsaturated forward edges or fully saturated backward edges. According to the *max-flow min-cut theorem*, this is the $s-t$ cut with the smallest capacity and the maximum flow is equal to the capacity of the cut. Because each forward, fully saturated edge carries exactly 1 unit flow, the capacity of the minimum $s-t$ cut in the unit network is equal to the number of forward edges. Consequently, the as-defined $s-t$ cut is also the $s-t$ cut with the smallest number of forward edges whose removal separates s from t.

According to Theorem 3.7, however, the maximum number of edge-disjoint directed paths in a directed $s-t$ network is equal to the maximum throughput flow in the unit edge capacity network, which is equal to the capacity of the as-defined $s-t$ cut. Hence, the maximum number of edge-disjoint paths in the original network is equal to the minimum number of edges whose removal separates s from t. □

A similar theorem can be stated, related to the number of node-disjoint directed paths in a network. This theorem is known as Menger's node-connectivity theorem.

Theorem 3.10 *The maximum number of node-disjoint directed paths in an $s-t$ network with directed edges is equal to the minimum number of nodes whose removal separates the source s from the sink t* (Menger's node-connectivity theorem).

Proof To prove this theorem, the initial network is transformed into a network where each node i is split into two nodes i_1 and i_2 connected with an edge (i_1,i_2) (Figure 3.11B). In the transformed network, all edges previously entering node i now enter node i_1 and all edges previously going out of node i now go out of node i_2. All edges in the network are assumed to be of unit capacity.

Before presenting the proof of the main theorem, the following lemma will be stated and proved.

Lemma 3.4 *The minimum number of nodes in an $s-t$ separating node set in the original network is equal to the minimum number of edges in an $s-t$ separating edge set in the transformed network.*

Proof First, note that the number of edges in the smallest $s-t$ separating set in the transformed network cannot be larger than the number of nodes $|V^*|$ in the smallest $s-t$ separating node set V^* in the original network. Indeed, let us assume that the number of edges $|E_s'|$ in the smallest set of edges E_s', separating s and t in the transformed network is larger than the number of nodes $|V^*|$ in the smallest $s-t$ separating node set V^* in the original network:

$$|E_s'| > |V^*| \tag{3.22}$$

The number $|V^*|$ of $s-t$ separating nodes in the original network results in the same number of edges in the transformed network, which, if deleted, will constitute an edge separating set E_s^* in the transformed network. In addition, $|E_s^*| = |V^*| < |E_s'|$

will be valid, which contradicts the assumption that the $s-t$ separating set $|E'_s|$ in the transformed network is the one with the smallest number of edges.

It can also be shown that the smallest number of $s-t$ separating edges in the transformed network cannot be smaller than the minimum number $|V^*|$ of $s-t$ separating nodes in the original network. Suppose that the number of edges $|E'_s|$ in the smallest edge separating set E'_s in the transformed network is smaller than the minimum number $|V^*|$ of $s-t$ separating nodes in the original network:

$$|E'_s| < |V^*| \tag{3.23}$$

Consider the set of nodes V'_r towards which the removed edges E'_s point in the transformed network. Clearly, if these nodes are removed from the transformed network there will be no $s-t$ path in the transformed network. Now, consider the set V_r in the original network which corresponds one-to-one to the set of nodes V'_r in the transformed network:

$$|V_r| = |V'_r| = |E'_s|. \tag{3.24}$$

If the set of nodes V_r in the original network is removed, then no $s-t$ path will exist in the original network. From Eqs. (3.24) and (3.23), it follows that $|V_r| < |V^*|$. In words, another $s-t$ separating set of nodes V_r has been found in the original network, with a smaller number of nodes than the $s-t$ separating node set V^*. However, by assumption, V^* was the $s-t$ separating set with the smallest number of nodes – a contradiction. Hence $|E'_s| = |V^*|$ and the lemma is true. Now, Theorem 3.10 can be proved.

Proof According to Theorem 3.8, the maximum number of directed node-disjoint paths in the original network is equal to the maximum number of edge-disjoint paths in the transformed network. According to Theorem 3.9, the maximum number of edge-disjoint paths in the transformed network is equal to the minimum number of edges whose removal separates the source s from the sink t. According to Lemma 3.9, the minimum number of edges in the $s-t$ separating edge set in the transformed network is equal to the minimum number of nodes in an $s-t$ separating node set in the original network. Consequently, the maximum number of node-disjoint paths in the original network is equal to the minimum number of nodes in an $s-t$ separating node set in the original network. □

An important practical problem is related to maximising the supply from a number of sources (suppliers) to a number of consumers. The sources have a limited flow generation capacity and the consumers have a limited flow consumption capacity. The sources are linked to the consumers with transmission paths, each of which is characterised by a maximum throughput capacity. To solve this problem, a super-source s with unlimited flow generation capacity is introduced and connected with the original sources through empty edges with capacities equal

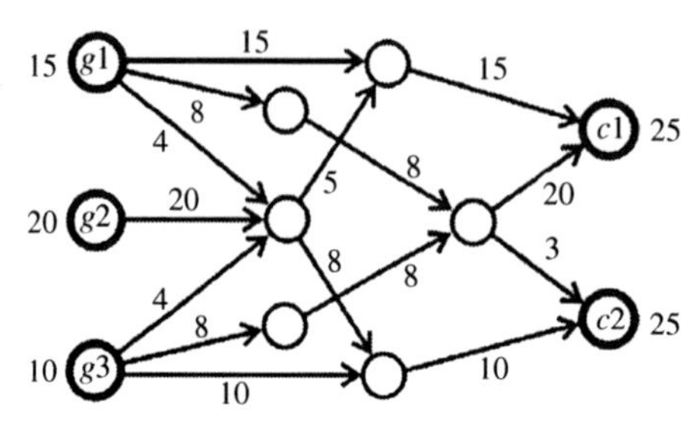

Figure 3.12 The problem of maximising the flow supply from generators g1, g2 and g3 to consumers c1 and c2.

to the flow generation capacities of the separate sources. Similarly, a super-sink t with unlimited flow consumption capacity is also introduced. The super-sink is connected with empty edges to each of the consumer nodes. The capacities of the connecting edges are equal to the flow consumption characterising the consumers. The problem now reduces to maximising the flow in the obtained $s-t$ network.

Let us illustrate this problem with the example shown in Figure 3.12, featuring three generators g1, g2 and g3, and two consumers c1 and c2. The flow generation capacities of the generators (the sources) are 15, 20 and 10 units of flow per unit time, respectively. The flow consumption capacity of each consumer c1 and c2 is 25 units of flow per unit time. The existing transmission paths and their flow capacities are according to Figure 3.12.

The network is first transformed into a network with a single source s and a single sink t, by introducing a super-source s and a super-sink t, as it is shown in Figure 3.13A. The flow capacities of edges (1,2), (1,3) and (1,4), connecting the super-source s with the generators (nodes 2, 3 and 4), are equal to the flow generation capacities of the generators. As a result, all generators disappear and throughput edges appear instead. Similarly, the flow capacities of edges (11,13) and (12,13), connecting the consumers (nodes 11 and 12) with the super-sink t, are equal to the flow consumption capacity of the consumers. As a result, all consumers disappear and throughput edges appear instead (Figure 3.13A).

Next, the throughput flow in the resultant $s-t$ network is maximised by using the Edmonds and Karp algorithm. The maximum throughput flow is 38 units. The edge flows corresponding to the maximum throughput flow are given in Figure 3.13A.

A number of problems, seemingly unrelated to the maximum throughput flow problem, can be reduced to it. Such is for example the personnel assignment problem (Gross and Yellen, 2006). This problem will be illustrated by an example.

Suppose that 8 jobs and 6 people are available. Each person is suited for some jobs but not for others and can perform only one job at a time. The jobs which each person can perform are specified as follows:

p1: {j1, j2};
p2: {j1, j4};
p3: {j2, j4};
p4: {j2, j4, j5, j7};
p5: {j3, j4, j5, j7};
p6: {j4, j5, j6, j7, j8};

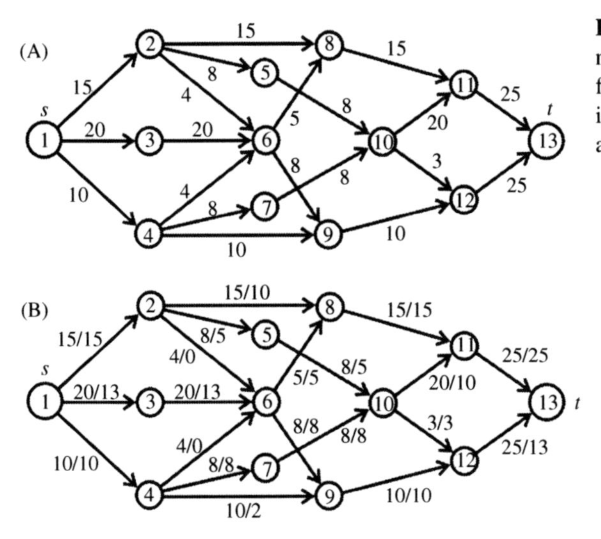

Figure 3.13 The solution for maximising the flow supply from generators g1, g2 and g3 in Figure 3.12 to consumers c1 and c2.

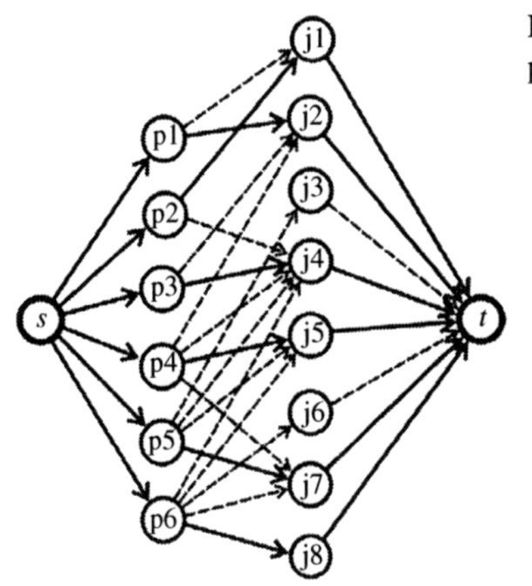

Figure 3.14 Solution of the optimal assignment problem.

The problem is about the optimal assignment of jobs to people, so that a maximum number of jobs can be performed simultaneously.

This problem is a special case of the maximum flow problem and can be solved easily by creating the bipartite graph from Figure 3.14 where the left column of nodes V_P represents people and the right column of nodes V_J represents jobs ($V_P \cap V_J = \varnothing$ and $V_P \cup V_J = V$). A directed edge (i,j), for which $i \in V_P$ and $j \in V_J$, means that the ith person is capable of performing the jth job. Each of the directed edges is characterised by a capacity equal to unity ($c(i,j) = 1, \ \forall (i,j) \in E$).

A source s is introduced, connected by empty edges with unit capacity to each of the nodes representing people. Similarly, a sink t is also introduced, connected by empty edges to each of the nodes representing jobs. The problem now reduces to maximising the flow in the obtained $s-t$ network. The non-empty edges (i,j), $f(i,j) = 1$, $i \in V_P$, $j \in V_J$ after the flow maximisation, define the optimum assignment of people to jobs.

3.7 Successive Shortest-Path Algorithm for Determining the Maximum Throughput Flow at a Minimum Cost

Often, the edges of a flow network are associated with non-negative costs $w_{ij} \geq 0$ for carrying a unit flow. The total cost for transporting the flow through an edge (i,j) is then $w_{ij} \times f(i,j)$, where $f(i,j)$ is the flow through the edge. A problem which often arises in important practical applications is the minimum cost flow problem: *how to transmit the maximum possible throughput flow at a minimum cost.* A variation of this problem is *how to transmit a throughput flow of specified magnitude, at a minimum cost.*

A common algorithm for determining the minimum cost flow is the successive shortest-path algorithm, a detailed discussion about which can be found in Ahuja et al. (1993) and Goodrich and Tamassa (2002).

All forward edges in the network have non-negative costs. Suppose that a particular $s-t$ path is augmented by a unit flow. If the augmentable path contains a forward edge (i,j), the term w_{ij} is taken with a positive sign because the flow through edge (i,j) is increased, and therefore the cost of flow transportation is also increased. If the augmentable path passes through a backward edge (j,i), the term w_{ji} is with a negative sign, because a unit flow is prevented from being transported through the backward edge (j,i) $(w_{ji} = -w_{ij})$. Let F denote the set of forward edges along an augmentable path from the source s to node i and B denote the set of backward edges along the same augmentable path. The cost $D(i)$, associated with augmenting the path $(s, \ldots, i)$ is therefore

$$D(i) = \sum_{ij \in F} w_{ij} - \sum_{ij \in B} w_{ij} \tag{3.25}$$

The smallest cost among all costs $D(i)$, associated with augmenting possible paths from the source s to node i, will be referred to as *distance* $d(i)$. The distances $d(i)$ are an important part of the successive shortest-path algorithm. They are kept in the array $d[]$ and initially, they are all set to zero. The number of nodes and the number of edges of the network are kept in the variables n and m, respectively.

Backward edges also appear on an augmentable $s-t$ path. Depending on the direction of scanning an edge, the cost term can appear with a positive or negative sign. Edges with negative weights *are not fixed in the network.* The same edge of the network can appear as both an edge with a positive or a negative cost term, depending on the direction of traversing the edge.

The Dijkstra's algorithm considered in Chapter 2, for determining the shortest distances from one node to every other node in a network, is capable of handling only edges with positive weights and cannot be applied directly. This predicament, however, can be overcome by introducing modified weights:

$$w_{ij}^{m} = w_{ij} + d(i) - d(j) \tag{3.26}$$

where w_{ij} is the cost per unit flow along edge (i,j); $d(i)$ and $d(j)$ are the smallest possible costs for sending a unit flow from the source s to node i and node j, respectively. The following lemma can then be formulated (Goodrich and Tamassia, 2002).

Lemma 3.5 *For each edge (i,j), the inequality* $w_{ij}^{m} \geq 0$ *holds for the modified weights.*

Proof There are two mutually exclusive cases here: (i) edge (i,j) is augmentable and (ii) edge (i,j) is not augmentable.

Consider the first case. The smallest cost associated with augmenting a path from the source s to node j does not exceed the distance $d(i)$ plus the cost w_{ij} of sending a unit flow along edge (i,j): $d(j) \leq d(i) + w_{ij}$. This inequality can be rearranged as $w_{ij} + d(i) - d(j) \geq 0$ which proves that $w_{ij}^{m} \geq 0$.

Now consider the second case. If edge (i,j) is not augmentable, then it must be fully saturated. Because edge (i,j) has been fully saturated, it must belong to a shortest path from the source s, which passes through both node i and node j. Any part of a shortest path is a shortest path itself; therefore, the path $(s,\ldots, i, j)$ is the shortest path from s to node j. Consequently,

$$d(j) = d(i) + w_{ij} \tag{3.27}$$

holds. For the modified weight we have $w_{ij}^{m} = w_{ij} + d(i) - d(j)$, which, after the substitution of Eq. (3.27) becomes $w_{ij}^{m} = w_{ij} + d(i) - d(i) - w_{ij} = 0$. □

Lemma 3.6 *The shortest s−t path in modified weights* w_{ij}^{m} *is also the shortest s−t path in the original weights* w_{ij}.

Proof Because the modified weights are $w_{ij}^{m} = w_{ij} + d(i) - d(j)$, along any $s-t$ path, the smallest-cost path is $\sum_{ij}[w_{ij} + d(i) - d(j)]$. With the exception of the source s and the sink t, the smallest distance $d(\cdot)$ of each internal node from the source s enters this sum once with a positive sign and once with a negative sign. As a result, the shortest path, in terms of modified weights, becomes $\sum_{ij} w_{ij} - d(t)$, where $d(t)$ is the shortest $s-t$ path to the sink t, irrespective of the selected path. Because $\sum_{ij} w_{ij}$ is the cost of the $s-t$ path in original weights, the two costs differ by the constant $d(t)$. If $\sum_{ij} w_{ij} - d(t)$ is the smallest possible value then $\sum_{ij} w_{ij}$ is also the smallest possible value. □

Because of *Lemmas 3.5* and *3.6*, all modified weights during the path augmentation will be non-negative irrespective of the direction of traversing the edges. As a result, the Dijkstra's shortest-path algorithm (Algorithm 2.3) can be applied to find the $s-t$ path with the smallest cost and to label the nodes with the smallest distances from the source s.

The algorithm in pseudo-code is presented next.

Algorithm 3.5

for i = 1 **to** n **do** d[i] = 0;// *sets all minimal distances from the source to zero.*
exit_condition = FALSE;

repeat
  for (each augmentable edge (i,j)) **do**
  $w_{ij}^m = w_{ij} + d(i) - d(j)$
    Run the Dijkstra's algorithm by using the modified weights w_{ij}^m
    //After running the Dijkstra's algorithm, each node will be labelled with the shortest distance from s.
if (an augmentable s-t path exists) **then** // *an augmentable $s-t$ path has been found*
    Augment the $s-t$ path with the bottleneck flow;
    else exit_condition = TRUE**;**
until (exit_condition = TRUE)

This algorithm can be implemented to run in $O(|f^*|m \log n)$ time (Ahuja et al., 1993).

A number of algorithms have already been proposed for the minimum cost flow problem (Ahuja et al., 1992; Bennington, 1973; Friesdorf and Hamacher, 1982; Goldberg and Tarjan, 1987, 1989; Klein, 1967; Orlin, 1993; Tardos, 1985), some of which are characterised by a strictly polynomial running time (Orlin, 1993; Tardos, 1985).

3.7.1 Solved Example

The algorithm will be illustrated with the network shown in Figure 3.15A. The capacities of all edges are 100 flow units; the labels on the edges stand for the cost of transporting a unit flow. It is required to transfer 160 flow units from the source s to the sink t, with minimal cost.

The problem of sending the specified throughput flow of 160 units from the source s to the sink t, at a minimum cost, is first reduced to the problem of maximising the throughput flow at a minimum cost. This can be done by detaching the source s and connecting it with the rest of the network through the extra edge (1,2) with flow capacity 160 units and cost of transporting a unit flow along the edge equal to zero (Figure 3.15B).

The shortest-path distances from the source s to the sink t are calculated first. They are shown as numbers close to the nodes. Thus, the shortest distance of node 6 is 4 because the lowest-cost path of sending a unit flow from the source s to node 6,

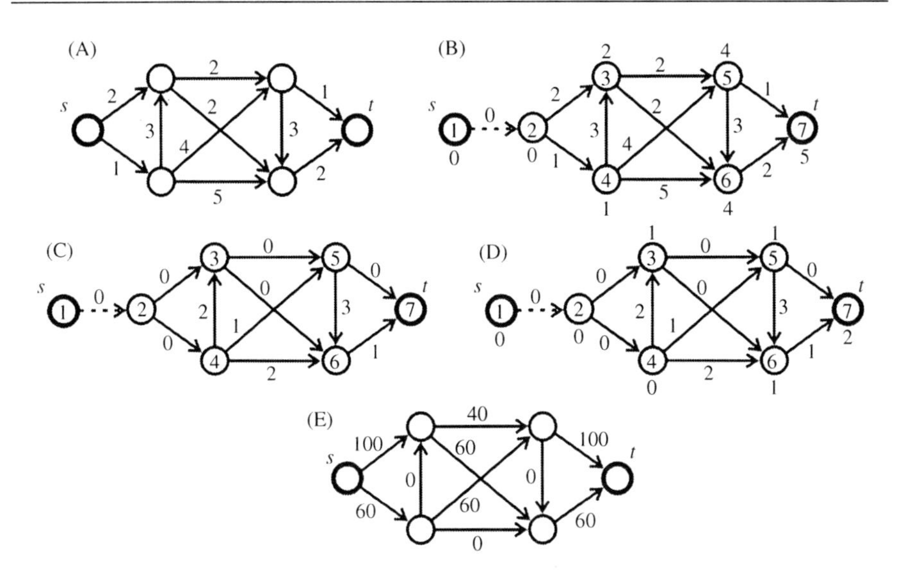

Figure 3.15 Determining the minimum cost flow in a network.

through the empty network, is path (1,2,3,6). The modified weights are calculated according to Eq. (3.26). They are shown as labels on the edges of the network shown in Figure 3.15C. The smallest-cost s–t path in the network with modified weights is the path (1,2,3,5,7), whose modified cost is zero. Following the algorithm, this path is augmented with the maximum possible value of 100 flow units. Next, the new distances of all nodes from the source s are calculated by running the Dijkstra's algorithm. They are shown in Figure 3.15D as numbers close to the nodes of the network. Because some of the edges are fully saturated with flow, forward edges become inadmissible while backward edges can now be augmented. For example, because of the fully saturated forward edge (2,3), the shortest distance of node 6 from the source s is not zero, because the path (1,2,3,6) is not augmentable. The smallest-cost path is the path (1,2,4,5,3,6), because the backward edge (5,3) has been saturated with flow and therefore the edge is augmentable. The distance of node 6 from the source s is therefore equal to one.

The smallest-cost s–t path is (1,2,4,5,3,6,t) with distance $d(t)$ equal to 2. Path (1,2,4,5,t) is not augmentable because the forward edge (5,7) is fully saturated. Paths (1,2,3,6,t) or (1,2,3,5,t) are not augmentable either, because of the fully saturated edge (2,3).

The smallest-cost s–t path (1,2,4,5,3,6,t) is augmented with 60 flow units. The resultant edge flows, minimising the total cost for sending 160 units throughput flow, are shown in Figure 3.15E.

4 Maximising the Throughput Flow in Single- and Multi-Commodity Networks: Removing Parasitic Directed Loops of Flow in Networks Optimised by Classical Algorithms

4.1 Eliminating Parasitic Directed Loops of Flow in Networks Optimised by Classical Algorithms

We must raise awareness of the fact that the network flow optimisation achieved by using classical algorithms results in unwanted parasitic loops of flow in some of the optimised networks. This fundamental flaw, which is present in the classical algorithms for maximising the throughput flow, has been discussed in Todinov (2013).

Consider the counterexample network in Figure 4.1, where the classical Edmonds and Karp shortest-path algorithm proceeds with saturating the shortest path (1,2,3,13) with 10 units of flow, followed by saturating the next shortest path (1,5,6,3,4,13) with 10 units of flow and finally, with saturating the path (1,8,9,10,11,4,2,7,12,13) with 10 units of flow. As a result, a parasitic directed loop (2,3,4,2) appears, carrying 10 units of flow.

Next, consider the network in Figure 4.2 whose throughput flow has been maximised by using a classical general push-relabel algorithm (Cormen et al., 2001; Goldberg and Tarjan, 1988; Kleinberg and Tardos, 2006).

The first step begins with assigning the highest label to the source s and zero to the rest of the nodes. All edges emanating from the source are also fully saturated with flow (Figure 4.2B). The excess node 3 is then selected. Edges (3,2) and (3,t) are augmentable. Because nodes 2 and t have the same label 0, a relabelling operation is initiated for node 3 which increases its label from 0 to 1. Flow of 10 units is then pushed towards node 2 and flow of 20 units towards the sink t (Figure 4.2C). The next selected node with excess flow is node 2. It has the same label 0 as its eligible neighbouring node 4, therefore a relabelling operation is initiated, which increases its label to 1. A flow of 20 units is then pushed towards node 4 (Figure 4.2D). Similarly, a relabelling operation is initiated for node 4 and its label becomes 1. Flow of 10 units is then pushed towards the sink. Node 4 still has excess flow of 10 units, but its neighbour, node 3, has the same label 1. Another relabelling operation is initiated for node 4, whose label is increased to 2 and 10 units of flow is

Flow Networks. DOI: http://dx.doi.org/10.1016/B978-0-12-398396-1.00004-0

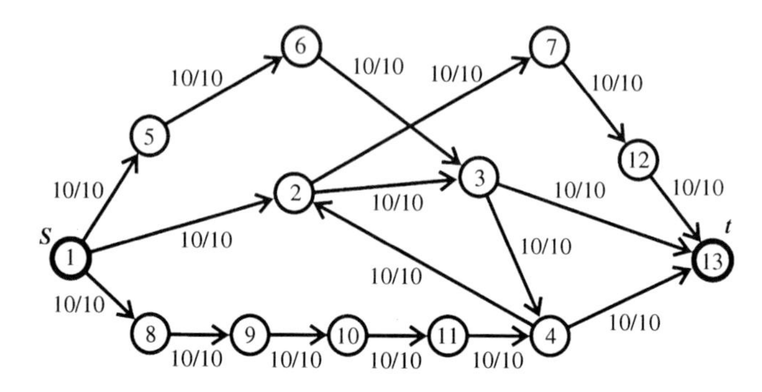

Figure 4.1 A counterexample network, demonstrating that the classical Edmonds and Karp (1972) shortest-path algorithm leaves a directed loop of flow (4,2,3,4) in the optimised network. All edges have a flow capacity of 10 units.

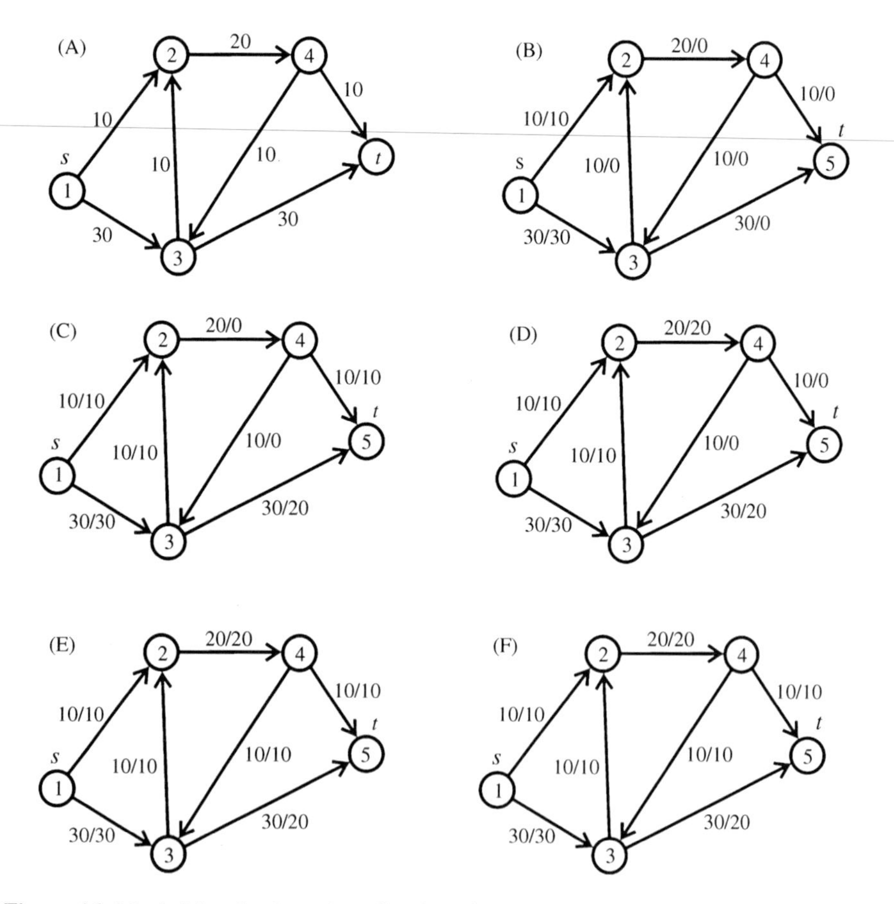

Figure 4.2 Maximising the throughput flow by using the preflow-push algorithm. Note the directed loop of flow of 10 units (3,2,4,3) after the termination of the preflow-push algorithm.

pushed towards node 3 (Figure 4.2E). Now node 3 has 10 units excess flow which is pushed towards the sink. The final state of the edge flows is shown in Figure 4.2F. The maximum throughput flow of 40 units has been obtained correctly, but a loop (3,2,4,3) carrying 10 units parasitic flow now appears in the network.

These directed loops of flow are *highly undesirable* because: (i) they increase unnecessarily the cost of transportation of the flow in the network, (ii) they consume residual capacity from the edges of the network and (iii) energy is unnecessarily wasted for maintaining these parasitic loops of flow. The presence of directed loops of flow in the optimised flow networks entails big financial losses in the affected sectors of the economy. In electric networks, the parasitic loops of flow lead to significant electrical losses and unnecessary consumption of power line capacity which leads to congestion and overloading of power lines and diminishing the reliability of the grid. In computer networks, parasitic loops of flow consume bandwidth from the communication lines, unnecessarily increase data traffic and ultimately lead to congestion and delayed data transmission which negatively affects the quality of service of the network. For a supply network of a particular commodity, the existence of parasitic loops of flow means that the transportation costs are unnecessarily increased and energy is wasted on circulating the commodity unnecessarily.

For undirected large networks, the presence of directed loops of flow is very likely. In large and complex networks with directed edges, the presence of directed loops of flow is also very likely. Consequently, classical shortest-path augmentation algorithms for maximising the throughput flow should not be used for optimising the edge flows in a network, without an additional stage aimed at removing the directed loops of flow.

The procedure described in this section provides such an algorithm, *in a network with feasible flow*, where the capacity constraints on the edges and the flow conservation law at the nodes are honoured.

The procedure identifies directed loops of flow in the network by using a *modified depth-first-search* (*mdfs*) procedure), whose algorithm in pseudo-code is given as Algorithm 4.1. Nodes encountered by the mdfs-procedure are marked as visited by the statement 'marked[r_node] = 1'. Initially, all entries in the array 'marked[]' are set to zero. Apart from the array 'marked[]', another array 'done[]' is also maintained. Initially, all entries of the 'done[]' array are set to zero. The mdfs-procedure is called with the index of the source node as a parameter. We need to point out that the mdfs-procedure *does not scan all successors* of the current node. It scans all *eligible* successors. A successor node *i* of the current node r_node is eligible if: (i) there is an edge directed from node r_node to node *i* and the edge (r_node,*i*) carries *non-zero flow* (there is some flow $f(r\_node,i) > 0$ along the edge). If edge (r_node,i) is empty, the successor *i is not considered* by the mdfs-procedure.

Algorithm 4.1

Algorithm of the mdfs-procedure for discovering a directed loop of flow in a network with directed edges

procedure *retrieve_directed_flow_loop*(*cur_node*)
{*// retrieves and eliminates the identified directed loop of flow*}

```
procedure mdfs(r_node)
{
marked[r_node] = 1;
for i = 1 to all eligible successors of r_node do
{
cur_node = current eligible successor;
if(marked[cur_node] = 0) then {
    pred[cur_node] = r_node;
    mdfs(cur_node);
      }
  else {
    if (done[cur_node] = 0) {  pred[cur_node] = r_node;
            retrieve_directed_flow_loop(cur_node);
            break;
        }
     }
   }
  done[r_node] = 1;
}
```

Statements before the call of the mdfs-procedure:

for i = 1 **to** n **do** {marked[i] = 0; done[i] = 0; pred[i] = 0;}

mdfs(1).

A directed flow loop can only be discovered if both conditions are fulfilled: (i) a node cur_node already marked as visited has been encountered during the search and (ii) the call ***mdfs***(cur_node) is still active, in other words, its activation record is still in the stack. After the end of the ***mdfs*** (cur_node) call, node r_node is marked as 'done' by the statement 'done[r_node] = 1'. This is why, when both marked[cur_node] = 1 and done[cur_node] = 0 are encountered during the search, a directed loop of nonzero flow has been discovered. The directed loop of flow is subsequently retrieved and eliminated by the procedure ***retrieve_directed_flow_loop***(cur_node). The array pred[] records the predecessors of the visited nodes and helps retrieve the identified directed flow loop.

The procedure ***retrieve_directed_flow_loop***(cur_node) retrieves the discovered loop of flow by starting with the statement 'k = cur_node' followed by a loop, where the statement k = pred[k] is repeatedly executed and followed by a check whether k is equal to a descendent of the 'cur_node'. This check is used for identifying the node which closes the identified directed loop carrying nonzero flow.

After discovering the directed flow loop carrying nonzero flow, the procedure determines the edge in the loop carrying the smallest amount of flow and subtracts this flow from the flows of all edges belonging to the loop. As a result, at least one edge in the loop becomes empty. Once an edge becomes empty, it remains empty until the end of the procedure for removing directed loops of flow. This loop will not be discovered again during subsequent searches with the mdfs-procedure, because one of the edges of the loop is empty and therefore will not be traversed by the mdfs-procedure. The empty edge essentially breaks the directed loop.

Subtracting the bottleneck flow from the edges of a directed loop eliminates the directed loop of flow without affecting the throughput flow in the network.

Identifying a directed loop of flow with the mdfs-procedure and subtracting the bottleneck flow from all of its edges, has a worst-case complexity $O(m)$, where m is the number of edges in the network. Because after each flow subtraction, at least a single edge remains empty, after at most m steps (equal to the number of edges in the network), all directed loops of flow will be removed. Consequently, the procedure for removing all parasitic loops of flow in a network has a worst-case running time $O(m^2)$.

Applying the procedure to the network in Figure 4.1 discovers the directed loop (4,2,3,4) carrying 10 units flow. Subtracting this bottleneck flow from the flows of all edges belonging to the loop, transforms the network in Figure 4.1 into the network in Figure 4.3A. The maximum throughput flow is still 30 units, but no parasitic loops of flow are present. Similarly, the procedure discovers the parasitic loop of flow (3,2,4,3) in the network from Figure 4.2, which has been optimised by using the general push-relabel algorithm. After removing the parasitic loop of flow of 10 units, the optimised network is shown in Figure 4.3B.

The directed loops of flow can also be eliminated if the following procedure is used for maximising the throughput flow. Each edge is assigned a cost of transportation per unit flow equal to unity. Then, the problem of maximising the throughput flow at a minimal cost is solved, e.g. by using the successive shortest-path algorithm in Chapter 3. It can be shown that the next theorem holds.

Theorem 4.1 *Maximising the throughput flow at a minimum cost, leaves no directed loops of flow in the network.*

Proof Suppose that the throughput flow has been maximised and the sum of the costs of transportation along all edges is the smallest possible. Suppose that there exists a directed loop of flow $(k, k+1, \ldots, k+c, k)$ in the network. The smallest edge flow $\Delta_{\min}$ along this directed path is then determined. This is the bottleneck flow characterising the cyclic path. The directed cyclic path $(k, k+1, \ldots, k+c, k)$ can be augmented with the bottleneck flow $\Delta_{\min}$ in the opposite direction $(k, k+c, \ldots, k+1, k)$. During this augmentation, all edges belonging to the cyclic path will appear as

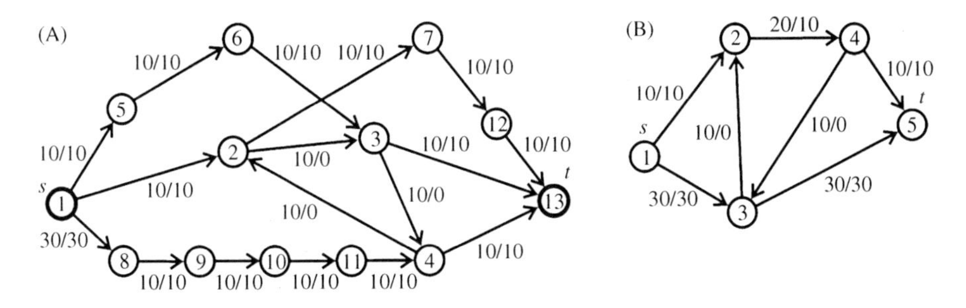

Figure 4.3 Optimised networks from Figures 4.1 and 4.2 by the proposed algorithm. The directed loops of flow have been removed.

backward edges. Because the augmentation is along a cyclic path, it does not affect the maximum throughput flow in the network. The maximum throughput flow will remain unchanged after the augmentation.

Because, during the augmentation with the bottleneck flow, the actual flows along the backward edges are reduced (the bottleneck flow is prevented from being circulated), the overall costs of transportation along the edges will be reduced. However, this contradicts the assumption that the cost associated with the initial edge flows was the smallest possible. Consequently, there can be no directed loops of flow in a network where the maximum throughput flow has been maximised at a minimum cost. □

4.2 A Two-Stage Augmentation Algorithm for Determining the Maximum Throughput Flow in a Network

The flow maximisation presented in this section implements an efficient two-stage algorithm based on a sequential path augmentation (Todinov, 2011a). The algorithm builds on the classical Edmonds and Karp shortest-path algorithm introduced in Chapter 3. The key idea behind the proposed two-stage algorithm is to augment the shortest paths in two stages. The first stage consists of augmenting only the shortest directed $s-t$ paths, consisting of forward edges only. The second stage uses the resultant edge flows from the first stage as initial edge flows. It consists of further augmentation of the shortest $s-t$ paths, which now may also include backwards edges. Here is a concise description of the two-stage algorithm.

During the first stage, the algorithm augments the shortest directed $s-t$ paths only, until no more augmentable directed paths can be found. The execution of function ***find_shortest_directed_st_path()*** returns '1', if an augmentable directed $s-t$ path has been found, with forward edges only. If no augmentable directed $s-t$ path exists, the function ***find_shortest_directed_st_path()*** returns zero and the execution of the statements in the first while-do loop is terminated. A similar structure is implemented in the execution of the function ***find_shortest_undirected_st_path()***, from the second stage of the algorithm. This function returns '1' if an augmentable undirected $s-t$ path has been found. The shortest undirected $s-t$ path may include both forward and backward edges.

Algorithm 4.2

Stage I:

while (*find_shortest_directed_st_path()* = 1) do // *while there exists an augmentable $s-t$ path, with forward edges only*

{

Augment the identified directed shortest s−t path;

Accumulate the augmented flow into the throughput flow;

}

Stage II:

while (*find_shortest_undirected_st_path()* = 1) do // *while there exists an augmentable s−t path*

{

Augment the identified shortest s−t path;

Accumulate the augmented flow into the throughput flow;

}

Identify and remove all directed parasitic loops of flow by running until no more directed loops of flow are present Algorithm 4.1.

During the execution of function ***find_shortest_directed_st_path()***, an edge is admissible for augmentation only if it is a forward edge and has a greater than zero residual capacity. In other words, if the difference $\Delta = c(i, j) - f(i, j)$ between the edge capacity $c(i, j)$ and the flow $f(i, j)$ through it, is greater than zero. During the execution of the function ***find_shortest_undirected_st_path()***, a forward or a backward edge is admissible for augmentation if it has residual capacity. In other words, if the edge is a forward edge, the difference $\Delta = c(i, j) - f(i, j) > 0$ must be greater than zero. If the edge is a backward edge, the flow $f(i, j)$ through it must be greater than zero ($f(i, j) > 0$).

The second stage of the algorithm (Stage II) is essentially an implementation of the Edmonds and Karp shortest-path augmentation algorithm. Running solely the second stage of Algorithm 4.2 will also yield the maximum throughput flow. Running the first stage before the second stage however, and using the network flows obtained at the end of the first stage as initial edge flows for the second stage, significantly increases the speed of the throughput flow maximisation. At the end, Algorithm 4.1 is run to identify and remove all directed loops of flow in the optimised network until no more directed loops of flow are present in the network.

The principal advantage of the described two-stage algorithm for maximising the throughput flow is that *often, the flow blocking all directed paths determined at the end of the first stage is also the maximum throughput flow in the network.*

Often, there is either no need to augment the blocking flow through paths including backward edges, or there is a need for only a few augmentations. As a result, the second stage may never be executed, or may only be executed few times, comparable with the number of times the first stage has been executed. Indeed, the worst-case running time of the algorithm for finding an eligible path to augment, irrespective of whether the path is directed or undirected, is $O(m)$. Finding the bottleneck flow and augmenting the flow along the path has also a worst-case running time $O(m)$, so that the overall worst-case running time is $O(m)$. After each augmentation of a directed path, at least a single bottleneck is created (an edge fully saturated with flow), and at least one edge is essentially excluded from the network. As a result, after at most m augmentations, all directed paths will be saturated. Therefore, the worst-case running time of reaching the flow blocking all directed paths in the network is $O(m^2)$, where m is the number of edges in the network.

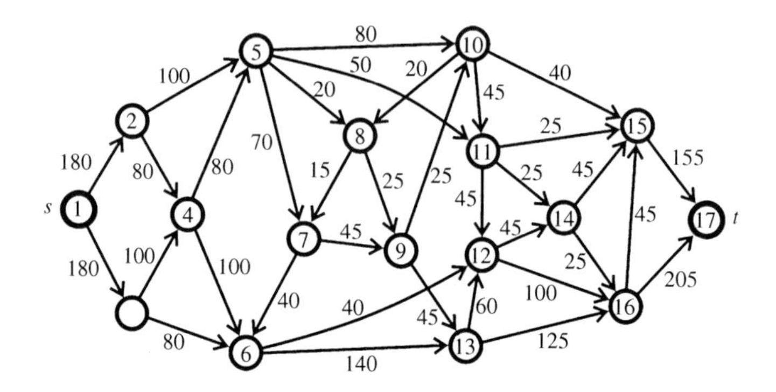

Figure 4.4 A benchmark flow network used to demonstrate the performance of the proposed two-stage algorithm.

Table 4.1 Time in Seconds for Executing the Two-Stage Algorithm one million times on a Computer with Processor *Intel(R) Core(TM) 2 Duo CPU %9900 @ 3.06 GHz*

Network Size m	Execution Time $f(m)$ (s)	Network Size m	Execution Time $f(m)$ (s)
19	4.15	66	23.9
22	7.9	70	19.5
27	4.98	79	10.76
29	9.17	81	20.4
40	11.06	89	55.8
46	10.9	91	22.0
53	12.1	96	15.5
54	16.7	115	16.65
60	9.93	120	106.2
64	13.14	131	30.1

Note: The networks each have a different topology and size.

To illustrate the clear advantage of the two-stage algorithm over the classical Edmonds and Karp algorithm, a comparative study has been conducted on the basis of the network in Figure 4.4.

For the network in Figure 4.4, the maximum throughput flow of 360 was obtained by running the Edmonds and Karp algorithm one million times, which is essentially running only stage 2 of Algorithm 4.2. Subsequently, the two-stage algorithm was also run one million times. On a computer with processor *Intel(R) Core(TM) 2 Duo CPU T9900 @ 3.06 GHz*, the execution of the Edmonds and Karp algorithm one million times took 27.9 s, 57% more time than the two-stage algorithm whose one million runs took 17.78 s. For larger networks this difference increases significantly.

To test the performance of this algorithm, the time for running the algorithm one million times has been recorded on networks with different topologies and

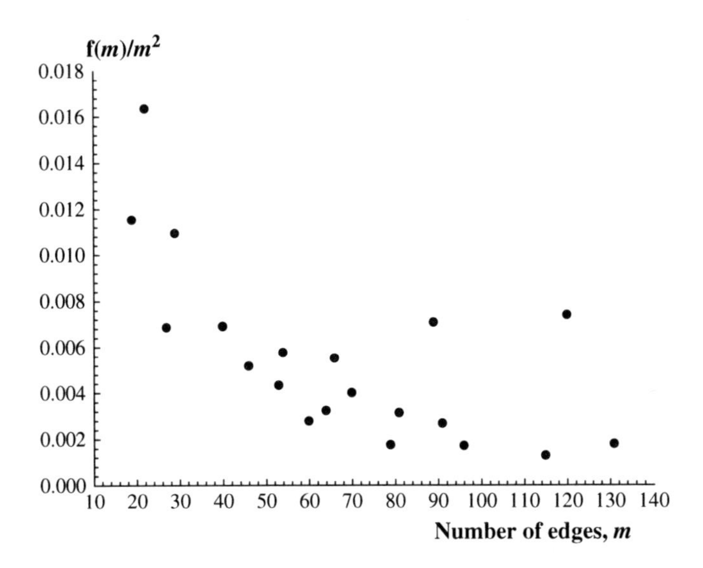

Figure 4.5 A ratio test for different network topologies with different number of edges.

different number of edges m. The time for execution of the two-stage algorithm one million times has been recorded in Table 4.1.

In Figure 4.5, the ratio $f(m)/m^2$ of the running time $f(m)$ versus the square of the size of the network m^2 has been plotted. As can be verified, for most of the tested networks this ratio is in the range (0,0.008). This shows that the average running time does not increase faster than the square m^2 of the network size m.

4.3 A New, Efficient Algorithm for Maximising the Throughput Flow of the Useful Commodity in a Multi-Commodity Flow Network

Flow maximisation algorithms have important applications to multi-commodity networks, where the important task is to guarantee a maximum throughput flow for one of the commodities. Oil production networks are a typical example of multi-commodity networks. Water is produced from the wells and water from injection wells is also pumped into the oil reservoirs to boost production. Consequently, water forms an inextricable part of the production fluid and each well contains a different water cut (percentage of water in the mixture). To maintain oil pressure, water is injected in the oil reservoirs and as the wells mature, the fraction of water in the produced oil/water mixture increases.

Suppose that an oil field contains a number of wells, each of which is characterised by a specific water content. Maximising oil production from the field then requires solving the important problem of setting the production levels from the separate wells in such a way that the throughput flow of oil is maximised.

Let the oil field consists of M wells, each characterised by unique oil fractions $0 \leq \alpha_i \leq 1$, $i = 1, M$. The fraction of water in each well is $1 - \alpha_i$, $i = 1, M$, correspondingly. Without loss of generality, let us assume that the wells have been arranged in descending order, starting with the one with the largest oil fraction: $\alpha_1 \geq \alpha_2 \geq \cdot \geq \alpha_M$.

It can be shown that by preferentially augmenting paths from the wells with the largest oil content, the throughput flow of oil in the oil production network can be maximised.

Augmentation of paths from a new well is not initiated before all possible paths starting from the current well have been augmented. The augmentation of paths starting from the current well is the two-stage augmentation described in the previous section. For the well with the current largest fraction of oil, the shortest directed paths (including forward edges only) starting from the well are augmented first. Next, the shortest undirected paths (which may include forward and backward edges) starting from the same well are augmented.

Note that the oil production network can be transformed from a multi-source network (Figure 4.6A) to a single-source network (Figure 4.6B). The production wells $s1$, $s2$ and $s3$ are treated as throughput edges $(s, s1)$, $(s, s2)$ and $(s, s3)$, with capacity equal to the production capacity of the corresponding wells (300 mbd, where 1 mbd = 1000 barrels per day). The source s is assumed to be with unlimited flow generation capacity.

The maximum throughput flow of oil to the sink t can be guaranteed by guaranteeing simultaneously: (i) a maximum throughput flow of oil/water mixture to the sink and (ii) a maximum fraction of oil in the mixture.

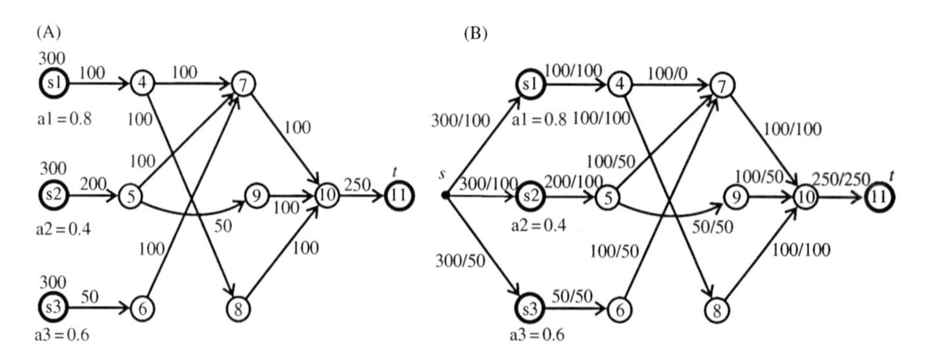

Figure 4.6 Maximising the throughput flow of oil in a production system based on three wells, with production capacities of 300 mbd (thousand barrels per day).

The shortest paths passing through the different wells are saturated with oil/water mixture and the flow in the network is maximised with respect to the mixture, *as a single commodity, without calculating along the edges the actual oil content contributed from the different wells.*

Theorem 4.2 *The throughput flow of oil in an oil production network can be maximised by preferentially augmenting paths from the wells with the largest oil content.*

Proof To prove the theorem, it is sufficient to prove that (i) the flow of oil/water mixture to the sink is maximum and (ii) the oil fraction set up in the mixture by following the steps specified in the theorem is the maximum possible oil fraction and cannot be increased.

Let the flows from the wells, set up in the saturated network be $\Delta_{a1}, \Delta_{a2}, \ldots, \Delta_{ak}$ units of flow per unit time. Then,

$$\sum_{i=1}^{k} \Delta_{ai} = Q_0 \tag{4.1}$$

holds, where Q_0 is the maximum throughput flow in the static flow network (with no edge failures). According to the Ford–Fulkerson theorem, augmenting paths sequentially, until no more paths can be augmented, guarantees a maximum throughput flow of mixture to the sink. Proving the theorem now reduces to proving that the total oil quantity in the mixture, by following the steps specified in the theorem, is the maximum possible oil quantity and cannot be increased.

First, it can be shown that the quantity of oil in the mixture cannot be increased if the flow from a single well is decreased and the flows from several other wells are increased. Indeed, consider a well i whose flow has been decreased by δ. Also consider several other wells $j_1, j_2, \ldots, j_m$ whose flows have been increased by $\delta_1, \delta_2, \ldots, \delta_m$, until the same maximum possible throughput flow of Q_0 has been set in the static flow network. None of the indices of wells $j_1, j_2, \ldots, j_m$ can be smaller than i, because, by following the conditions of the theorem, all possible paths starting from wells with indices smaller than i have already been augmented, and therefore, the flows from wells with indices smaller than i cannot be increased. Consequently, the inequalities

$$\alpha_i \geq \alpha_{j1}, \quad \alpha_i \geq \alpha_{j2}, \ldots, \alpha_i \geq \alpha_{jm} \tag{4.2}$$

hold. After augmenting the flow along paths starting from wells $j_1, j_2, \ldots, j_m$, the maximum throughput flow in the static network Q_0 has been reached. Before the decrease δ of the flow from well i, the throughput flow in the network was also Q_0. Therefore, the flow increments $\delta_1, \delta_2, \ldots, \delta_m$ from wells $j_1, j_2, \ldots, j_m$ go to fill residual capacity released by the flow decrease δ from the ith well. In other words, the equality

$$\delta = \delta_1 + \delta_2 + \ldots + \delta_m \tag{4.3}$$

holds. Taking into consideration inequalities (4.2), the inequality

$$\alpha_i\delta \geq \alpha_{j1}\delta_1 + \alpha_{j2}\delta_2 + \ldots + \alpha_{jm}\delta_m \tag{4.4}$$

is obtained, which means that the oil fraction in the mixture has not been increased by decreasing the flow from a single well and increasing the flows from several other wells.

Now consider n wells whose flows have been simultaneously decreased and several other wells whose flows have been simultaneously increased. Let the oil fractions characterising the wells whose flows have been decreased be $\alpha_{i1} \geq, \ldots, \geq \alpha_{in}$. Note that, for every well 'j' whose flow is increased, residual capacity for increasing the flow from this well can only be freed from reducing the flow from wells characterised by an oil fraction larger than the oil fraction α_j of the well with index j. This is because of the way the network has been saturated: by preferentially saturating paths starting from the wells with the largest fraction of oil. Residual capacity (space) for increasing the flow from well 'j' cannot possibly be released from decreasing the flows from wells with oil fraction smaller than α_j because the paths originating from these wells are augmented *only after* all possible paths originating from well 'j' have been augmented.

As a result, a flow increase δ_j from well 'j' always goes to compensate residual capacity released from decreasing flows from wells, with a larger oil content than the well with index 'j'. In other words, the inequalities

$$\alpha_j \leq \alpha_{i1}, \quad \alpha_j \leq \alpha_{i2}, \ldots, \alpha_j \leq \alpha_{in} \tag{4.5}$$

hold. The relationship

$$\delta_j = \delta_{j,i1} + \delta_{j,i2} + \cdots + \delta_{j,in} \tag{4.6}$$

also holds, where δ_j goes to compensate $\delta_{j,ik}$ − residual capacities from the flow decrease of well 'ik', $ik = 1,2, \ldots ,n$. (Some of the $\delta_{j,ik}$ may be zero.) From the inequalities (4.5), it follows that

$$\alpha_j\delta_j \leq \alpha_{i1}\delta_{j,i1} + \alpha_{i2}\delta_{j,i2} + \ldots + \alpha_{in}\delta_{j,in} \tag{4.7}$$

Relationships similar to Eqs. (4.6) and (4.7) hold for all wells whose flows are increased. Consequently, the oil content in the mixture cannot be increased by decreasing the flow from several wells and increasing the flow from several other wells. This completes the proof of the theorem. □

Theorem 4.2, in fact, defines a method for maximising the flow of one particular commodity in a multi-commodity network. This method will be illustrated with an example, involving a simple production network based on three wells only ($s1$, $s2$ and $s3$ in Figure 4.6A), each characterised by a maximum production capacity of

300 mbd (thousands of barrels a day). The oil fractions of the wells are $\alpha_1 = 0.8$, $\alpha_2 = 0.4$ and $\alpha_3 = 0.6$, respectively.

The paths passing through the richest in oil well $s1$ are saturated first. There is a single directed path (s,$s1$,4,7,10,11) and it is saturated first, with the maximum possible amount of 100 mbd. There are no more augmentable paths (directed or undirected) passing through well $s1$. The next richest well is $s3$. There is no directed augmentable path starting from well $s3$, so that the first augmentation stage is skipped. There exists however an undirected augmentable path passing through this well – the path (s,$s3$,6,7,4,8 10,11). It can be augmented with 50 mbd – the maximum possible amount of flow which is allowed by the bottleneck edge ($s3$,6).

As a result, the amount of flow along edge (4,7) has been reduced by 50 mbd. There are no more augmentable paths passing through well $s3$. The augmentation continues with well $s2$. There is a single augmentable directed path (s,$s2$,5,9,10,11) that is augmented by 50 mbd to saturate the bottleneck edge (5,9). There are no more augmentable directed paths passing through well $s2$. There exists however an augmentable path containing backward edges. Another 50 mbd can be directed from well $s2$ to augment path (s,$s2$,5,7,4,8,10,11). As a result, edge (4,7) becomes empty. The final state of the edge flows is shown in Figure 4.6B.

The state of the wells at the end of this procedure is 100 mbd from well $s1$, 50 mbd from well $s3$ and 100 mbd from well $s2$. The amount of oil in the mixture is $\alpha_1 \times 100 + \alpha_2 \times 100 + \alpha_3 \times 50 = 150$ mbd and this is the maximum possible throughput flow that can be delivered to the sink t.

Note that all flows along the edges are treated as 'mixture' and no attempt was made to calculate the oil and water content along the edges. This is an advantage of the proposed augmentation method. Such a calculation is not necessary. The real oil content at the sink (node 11) is determined easily from the actual flows from the individual wells.

4.4 Network Flow Transformation Along Cyclic Paths

The next theorem concerns the transformation of one set of feasible edge flows into another, by sequentially augmenting cyclic paths only.

Theorem 4.3 *If $f_{ij}^{(1)}$ and $f_{ij}^{(2)}$ are two different feasible edge flows, resulting in the same throughput flow $|f^*|$, then there exists a set of cyclic paths, whose augmentation transforms the flow $f_{ij}^{(1)}$ into the flow $f_{ij}^{(2)}$.*

The cyclic paths are not necessarily directed flow paths. They can include forward as well as backward edges. The cyclic paths *can also include separately or simultaneously* the source s and the sink t.

Proof (Todinov, 2012c). An *extended flow* $f'_{ij} = f_{ij}^{(1)} - f_{ij}^{(2)}$ is formed, by subtracting the flow $f_{ij}^{(1)}$ from the flow $f_{ij}^{(2)}$. Some of the edges will be left with negative extended flow. After the flow subtraction, all extended edge flows f'_{ij} will vary within the limits $-c_{ij} \le f'_{ij} \le c_{ij}$, where c_{ij} is the capacity of edge (i,j). Because the real flows $f_{ij}^{(1)}$ and $f_{ij}^{(2)}$ are feasible, the flow conservation law will hold for each internal node i. Therefore,

$$\sum_{k \in \delta +} f^{(1)}(k,i) - \sum_{k \in \delta -} f^{(1)}(i,k) = 0 \tag{4.8}$$

and

$$\sum_{k \in \delta +} f^{(2)}(k,i) - \sum_{k \in \delta -} f^{(2)}(i,k) = 0 \tag{4.9}$$

hold. Subtracting the two equations results in

$$\sum_{k \in \delta +} f'(k,i) - \sum_{k \in \delta -} f'(i,k) = 0 \tag{4.10}$$

where $f'(k,i) = f^{(1)}(k,i) - f^{(2)}(k,i)$ and $f'(i,k) = f^{(1)}(i,k) - f^{(2)}(i,k)$. As a result, under the operation 'subtraction of flows', at each node, the flow conservation law for the extended flows holds.

If 'extended edge capacity' c'_{ij} is defined, varying between the original flow capacities of the edges and their original flow capacities taken with a negative sign ($-c_{ij} \le c'_{ij} \le c_{ij}$), the extended flow in the resultant network will be feasible everywhere. In other words, the 'extended flow' along the edges f'_{ij}, will be within the extended edge capacity limits: $-c_{ij} \le f'_{ij} \le c_{ij}$ and the extended conservation law at the nodes will hold. The throughput extended flow in the resultant network will be zero.

Indeed, the extended flow coming out of the source s of the resultant network is

$$\sum_k f'(s,k) = \sum_k f^{(1)}(s,k) - \sum_k f^{(2)}(s,k) = |f^*| - |f^*| = 0 \tag{4.11}$$

Note that if it could be proved that there exists a set of cyclic paths whose augmentation yields a resultant network with zero extended edge flows, the statement of the theorem will have been proved. Indeed, let the extended flow $f'_{ij} = f_{ij}^{(1)} - f_{ij}^{(2)}$ be turned into zero after an appropriate augmentation along cyclic flow paths. In other words, the equality

$$f'_{ij} + \Delta_{ij} = f_{ij}^{(1)} - f_{ij}^{(2)} + \Delta_{ij} = 0 \tag{4.12}$$

can be achieved through augmentations along cyclic paths only, where Δ_{ij} are the changes in the edge flows resulting from the augmentations along cyclic paths. However, this essentially means that

$$f_{ij}^{(1)} + \Delta_{ij} = f_{ij}^{(2)} \tag{4.13}$$

In other words, by augmenting the first real feasible flow $f_{ij}^{(1)}$ along cyclic paths only, the second real feasible flow $f_{ij}^{(2)}$ is obtained.

There are two main cases, which will be considered sequentially.

1. *There are edges with positive extended flow in the network.*

   Select an edge with a positive extended flow. Suppose that the edge starts at node i. Because the extended flow conservation law holds at each node, at node i, there must either be an ingoing edge carrying positive extended flow (Figure 4.7A, case 1.1) or an outgoing edge with negative extended flow (Figure 4.7B, case 1.2).

   Note that if no ingoing edge carrying positive extended flow exists, there *must be* an outgoing edge with negative extended flow. Otherwise, all ingoing edges will carry negative extended flows and all outgoing edges will carry positive extended flows. However, the negative sum of the ingoing extended flows cannot possibly be equal to the positive sum of the outgoing extended flows and the extended flow conservation law will be violated (Figure 4.7C). By following the edge carrying the negative extended flow (Figure 4.7B), we arrive at node $i+1$ (Figure 4.7B). For node $i+1$, there must either be an ingoing edge carrying positive extended flow or an outgoing edge carrying negative extended flow (Figure 4.7). Indeed, if this is not the case, for node $i+1$, there will only be either outgoing edges carrying positive extended flow or ingoing edges carrying negative extended flow or both. As can be verified, in all cases, the extended flow conservation law at node $i+1$ will be violated (Figure 4.7D). By following an ingoing edge carrying positive extended flow into node $i+1$, we arrive at the already considered case 1.1. By following an outgoing edge carrying negative extended flow from node $i+1$, we arrive at the already considered case 1.2. This process continues until a visited node is reached. Note that if the sink t has been reached (which is only possible through a forward edge carrying negative extended flow), the process continues. Because the sum of the extended flows into the sink is equal to zero, and the sink can only be visited through an edge carrying negative extended flow, there must be an ingoing edge carrying positive extended flow, after the sink has been visited. This edge is then followed to get out of the sink t.

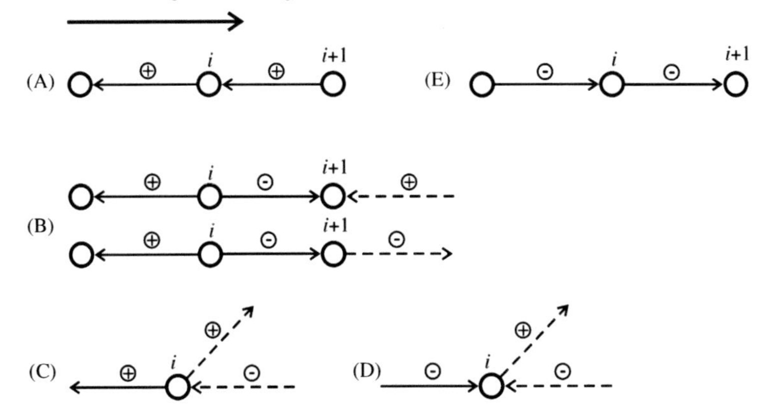

Figure 4.7 The main cases after selecting an edge carrying a positive extended flow.

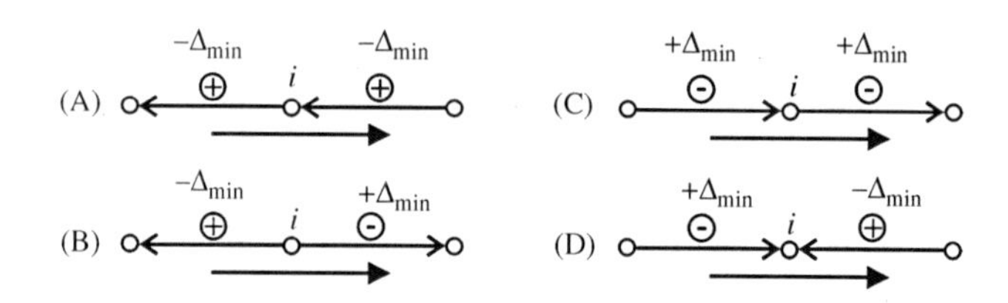

Figure 4.8 Under the defined path augmentation, the extended flow conservation law at the nodes is preserved.

If the source s has been reached (which is only possible through a backward edge carrying positive extended flow), the process also continues. Because the sum of the extended flows going out of the source s is equal to zero, there must be an outgoing edge carrying negative extended flow after the source s has been visited. This edge is then followed, to get out of the source s.

By continuing this process, it is guaranteed that a node will be visited again. This essentially defines a cyclic path, composed of only forward edges carrying negative extended flow and backward edges carrying positive extended flow. The absolute values of the flows in all edges of the cyclic path, are determined and the edge with the smallest absolute value of the flow is selected. Let the smallest absolute value be $|\Delta_{\min}|$. The identified cyclic path is then augmented by adding the positive flow $|\Delta_{\min}|$ to all forward edges carrying negative extended flow and by subtracting the flow $|\Delta_{\min}|$ from all backward edges carrying positive extended flow. As can be easily verified by examining all possible cases (Figure 4.8), under this operation, the extended flow conservation law at the nodes is preserved.

After this operation, the extended flow in at least one edge will become zero. After turning the flow in one or more edges into zero, the process continues with selecting another edge carrying positive extended flow.

2. *There are no edges with positive extended flow in the network.*

In this case, an edge carrying a negative extended flow is selected. Suppose that the edge points at node i (Figure 4.7E). At node i, there must be an outgoing edge with negative extended flow. Indeed, by definition, there can be no edge (ingoing or outgoing) carrying positive flow. If the outgoing edges from node i carry only zero flow, the flow conservation law will be violated. Consequently, the existence of an outgoing edge carrying negative extended flow is guaranteed. Continuing this reasoning for the next node also guarantees an outgoing edge carrying negative flow and so on, until a visited node is visited again. This defines a directed cyclic path, consisting of forward edges, carrying only negative extended flow. The absolute values of the flows along all edges from the cyclic path are then determined, and the edge with the smallest absolute value $|\Delta_{\min}|$ of the extended flow selected. Positive flow of magnitude $|\Delta_{\min}|$ is then added to all edges of the cyclic path. After this operation, the extended flow in at least one of the edges will become zero, while the extended flow in the rest of the edges still remains negative. After turning the flow in at least one of the edges into zero, the procedure continues with selecting another edge carrying negative extended flow and so on.

After conducting step 1 or 2, it is always guaranteed that the extended flow of at least one edge will become zero. Once the flow of an edge is turned into zero, it is never changed again. As a result, after at most m repetitions of step 1 or 2 (where m is the number of edges in the network), all edges in the network will be empty. Because identifying and augmenting a cyclic path with flow is an operation of worst-case time complexity $O(m)$, the worst-case time complexity of the algorithm turning the extended flow in all edges into zero is $O(m^2)$. This completes the proof. □

5 Networks with Disturbed Flows Dual Network Theorems for Networks with Disturbed Flows

Reoptimising the Power Flows in Active Power Networks in Real Time

5.1 Reoptimising the Flow in Networks with Disturbed Flows After Edge Failures and After Choking the Edge Flows

A large number of algorithms have been proposed for solving the maximum flow problem by using a graph representation of the flow network. These have been reviewed comprehensively in Ahuja et al. (1993) and Asano and Asano (2000). The best of these methods have polynomial complexity. None of these methods however is suitable for maximising the throughput flow in a large and complex network *in real time*, after an edge failure. The main reason is that the classical methods start the throughput flow maximisation from a network with empty edges.

After a component failure, there often exists a possibility for redirecting the flow through alternative paths with residual capacity, so that a new maximum of the throughput flow is reached quickly. Consider for example the flow network in Figure 5.1A, which includes a single source of flow s with unlimited capacity.

The parameters of each edge are given in the format 'c/f', where c is the capacity of the edge and f is the actual flow through the edge. The maximum throughput flow in the network shown in Figure 5.1A is 10 units. Suppose that edge (5,4), carrying 5 units of flow, fails. The edge failure, however, does not necessarily lead to a loss of 5 units of flow. As can be verified from Figure 5.1B, the network flows can be reset in such a way, that only 3 units of throughput flow are lost and the new maximum throughput flow after the edge failure is 7 units.

Even for this simple network, it is not immediately obvious how the edge flows should be reset in order to attain the new maximum throughput flow. Without an appropriate algorithm, the task of resetting correctly the edge flows in order to attain the new maximum throughput flow is almost impossible for large and complex flow networks. The optimal resetting of the edge flows after an edge failure is particularly important for production networks, active power networks and supply

Flow Networks. DOI: http://dx.doi.org/10.1016/B978-0-12-398396-1.00005-2

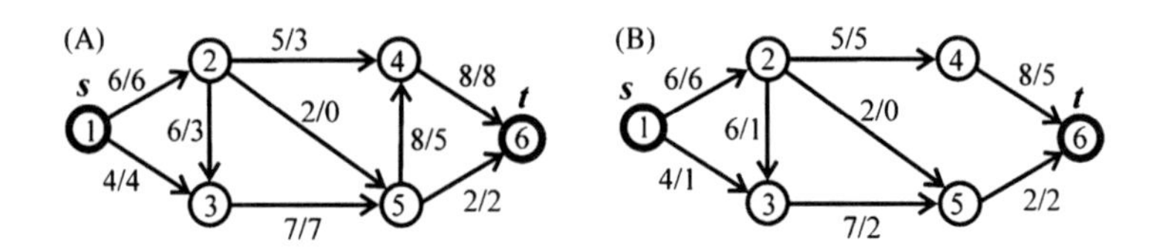

Figure 5.1 The fundamental problem of throughput flow maximisation after an edge failure.

networks. Lack of optimisation of the network flow after an edge/component failure entails congestion and overloading, cascading failures or suboptimal performance of the entire network.

Failure of a component leads to a severe disruption of traffic, suboptimal performance and loss of throughput flow. A fast recovery mechanism for reoptimising the network flows after a component failure is critically important for large flow networks including thousands of edges and nodes (e.g. power distribution networks and computer networks). Restoring quickly the maximum throughput flow after an edge failure *minimises the flow disruption and optimises the network performance in real time.* A pilot study conducted by the author on a computer with a processor *Intel (R) Core (TM) 2 Duo CPU T9900 @ 3.06 GHz*, has indicated that for a network with $m = 10{,}000$ edges, an augmentation algorithm with average running time proportional to m^2 needs many seconds to maximise the throughput flow after an edge failure, which is unacceptable for a real-time control of large power networks and telecommunication networks. For an augmentation algorithm with average running time proportional to m, the time needed for maximising the throughput flow after an edge failure is only few milliseconds.

5.2 A Fast Augmentation Algorithm for Reoptimising the Flow in a Repairable Network After an Edge Failure

The flow reoptimisation speed can be increased significantly, by exploiting the key idea that, after an edge failure, the starting state is not an empty network but a network with feasible flows. After the edge failure (e.g. edge (e,d) in Figure 5.2A) the network flow has been disturbed only locally, at nodes e and d to which the failed edge (e,d) is connected (Figure 5.2A).

If edge (e,d) is not empty, after the edge failure, a momentary excess flow exists at one of the nodes (node e in Figure 5.2B), equal to the flow $f(e,d)$ through the edge before its failure. In other words, the sum of the edge flows going into node e is greater than the sum of the edge flows leaving the node. This difference will be referred to as *excess flow ef*:

$$ef = \sum_{i \in \delta+} f(i,e) - \sum_{j \in \delta-} f(e,j) > 0 \tag{5.1}$$

and node e will be referred to as an *excess node*.

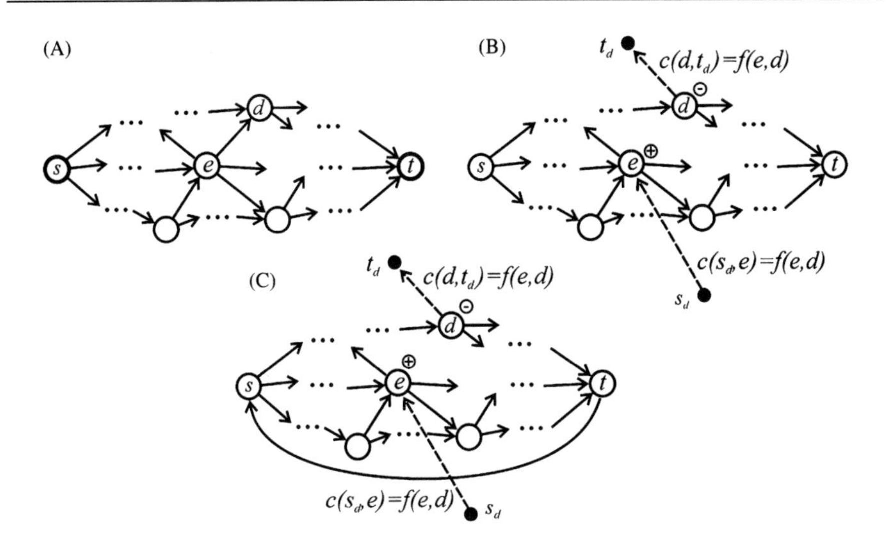

Figure 5.2 A fast reoptimisation method for maximising the throughput flow in a network after the failure of edge (e,d).

Alternatively, deficit flow will be created at node d of the failed edge (e,d), equal to the flow $f(e,d)$ through the edge. The sum of the edge flows going into node d is smaller than the sum of the edge flows leaving node d. The difference between the sum of the ingoing flows and the sum of the outgoing flows is negative and will be referred to as *deficit flow df*:

$$df = \sum_{i\in\delta+} f(i,d) - \sum_{j\in\delta-} f(d,j) < 0 \tag{5.2}$$

Accordingly, node d will be referred to as *deficit node*.

If the sum of the flows going into a node is equal to the sum of the flows leaving the node, the node will be referred to as *balanced node*. For a balanced node,

$$\sum_{i\in\delta+} f(i,d) - \sum_{j\in\delta-} f(d,j) = 0$$

It is important to note, that the definition of excess nodes and deficit nodes given here is *different* from the definition of excess and deficit nodes given in Dong et al. (2009). The definition in Dong et al. (2009) is based on the capacities of the ingoing and outgoing edges, while the definition given here is based on the actual flows through the nodes. According to the method proposed here, excess and deficit flows do not exist anywhere in the network before an edge failure occurs. Excess and deficit nodes are created only *after* edge failures.

The fast reoptimisation algorithm presented here starts with a network where the throughput flow has already been maximised by a classical method, e.g., by using the Edmonds and Karp algorithm (1972).

The essence of the new reoptimisation algorithm consists of the following steps, valid if edge (e,d) is not empty ($f(e,d) > 0$). If edge (e,d) is empty ($f(e,d) = 0$), its failure will not have any effect on the maximum throughput flow in the network.

A new source s_d, with unlimited capacity, is introduced, connected with the excess node e (Figure 5.2B). The connecting edge (s_d,e) has a flow capacity equal to the flow $f(e,d)$ through edge (e,d), before its failure. A new sink t_d is also introduced, connected *with the deficit node d*. The connecting edge (d,t_d) has a capacity equal to the flow $f(e,d)$ through the failed edge. The network in Figure 5.2B, will be referred to as a *dual network*. It consists of the original network where the throughput flow had been maximised before the edge failure, and edges (s_d,e) and (d,t_d).

Suppose that a maximum throughput flow Q_{max} has been established in the original network before the failure of edge (e,d) (Figure 5.2A). Let us introduce a fully saturated with flow auxiliary circulation edge (t,s) that connects the original sink t with the original source s. Its flow capacity is equal to the maximum throughput flow Q_{max} (Figure 5.2C). In effect, the capacity of the circulation edge (t,s) equals the sum of the flows $\sum_i f(s,i)$ coming out of the source s or the sum of the flows $\sum_j f(j,t)$ entering the sink t, $\sum_i f(s,i) = \sum_j f(j,t) = Q_{max}$. By introducing the circulation edge, a maximum circulation flow Q_{max} is set in the network.

The network obtained from the dual network, where the original sink t has been connected with the original source s through a circulation edge (t,s), will be referred to as *dual circulation network* (Figure 5.2C).

A Ford–Fulkerson type augmentation of the shortest paths is initiated, starting at s_d, ending at t_d and passing through the dual network. Let Q_{max} be the maximum throughput flow characterising the original network before any edge failure. The following theorem, whose proof is given in Appendix A, specifies the maximum flow in the original network after the failure of edge (e,d).

Theorem 5.1 (Dual network theorem for a single-edge failure) *The maximum throughput flow in any network after an edge failure is equal to the maximum throughput flow before the edge failure, minus the flow through the edge before its failure, plus the maximum throughput flow in the dual network* (Todinov 2011b,c, 2012a).

If q_d^{max} is the maximum throughput flow in the dual network, the theorem states that the maximum possible throughput flow in the original network, after the failure of edge (e,d), is $Q_{max} - f(e,d) + q_d^{max}$.

Theorem 5.1 establishes a very important link between the maximum throughput flow in a flow network after an edge failure and the maximum throughput flow in its dual network. *The maximum flow after an edge failure theorem replaces the task of determining the maximum throughput flow in the original network with the task of determining the maximum throughput flow in the dual network.* Because there are only two unbalanced nodes after the edge failure, determining the maximum throughput flow in the dual network is a task which is significantly easier than the task of determining the maximum throughput flow in a network with empty edges. The main reason for this important tradeoff is that *the dual network is already saturated with flow*. As a result, in case of very few imbalanced nodes, few path augmentations are normally needed for maximising the throughput flow in the dual network.

The proof of Theorem 5.1 (Todinov 2012a) also provides an algorithm for determining the edge flows, corresponding to the new maximum throughput flow after an edge failure.

In the case where only the maximum throughput flow in the network after the edge failure is required, and not the actual edge flows, Theorem 5.1 provides a very efficient algorithm for determining the new maximum throughput flow. The shortest paths starting from the new source s_d and ending at the new sink t_d are augmented until no more augmentable paths can be found. If the maximum throughput flow in the dual network is $q_d^{\max}$, the new maximum throughput flow $Q'_{\max}$in the original network, after the edge failure, is $Q'_{\max} = Q_{\max} - f(e,d) + q_d^{\max}$. Because of its very high computational speed, the reoptimisation algorithm based on Theorem 5.1 can be used for building ultra-fast, discrete-event solvers for determining the performance parameters of repairable flow networks. If alongside the maximum throughput flow, the edge flows corresponding to the maximum throughput flow are also required; the next theorem which is a corollary of Theorem 5.1 shows how these can be determined.

Theorem 5.2 *After an edge failure, the edge flows corresponding to the maximum throughput flow can be obtained by a two-stage procedure which consists of (i) augmenting with flow the dual network, until no more augmentable paths can be found, followed by (ii) augmenting with flow the dual circulation network, until no more augmentable paths can be found (Todinov 2011b, 2012a).*

The initial edge flows for the augmentation of the dual circulation network are the edge flows obtained after the augmentation of the dual network. The first stage is essentially a *flow redistribution stage* and the second stage is essentially a *flow draining stage*.

This theorem can be proved with a reasoning similar to the one used for Theorem 5.1 (see Appendix A or (Todinov 2012a)). Each augmented $s_d - t_d$ path includes the circulation edge (*t*,*s*) and, in effect, subtracts a certain amount of flow from one of the edges coming out of the source *s* and, simultaneously, subtracts the same amount of flow from one of the edges going into the sink *t*. By continuing the $s_d - t_d$ path augmentation through the circulation edge, there will eventually be an augmentation where the excess flow ef' at the excess node *e* in the original network disappears ($ef' = 0$), after removing the connecting edge (s_d,*e*). Simultaneously, the deficit flow df' at the deficit node *d* will also disappear ($df' = 0$) in the original network, after removing the connecting edge (*d*,t_d). In short, the connecting edges (s_d,*e*) and (*d*,t_d) serve as reservoirs where the initially existing excess flow and deficit flow in the original network now remains.

After removing the connecting edges at the end of the second stage, when they are fully saturated with flow, no excess and deficit nodes will exist in the optimised network. Because, at this point, all flows along the edges will be feasible flows, this marks a state where the maximum throughput flow in the optimised network (after the failure of edge (e,d)) will be attained.

Note that if the flow augmentation starts with the dual circulation network instead of the dual network, there is no guarantee that the maximum throughput flow will be determined. Instead, a suboptimal solution may be found. This is why the draining algorithm proposed in Dong et al. (2009) fails to find the maximum throughput flow in some networks (Todinov, 2012a). The reason why the draining algorithm proposed in Dong et al. (2009) yields suboptimal solutions is that this algorithm *always does node balancing on a network with a circulation edge* (t,s).

For networks nearly saturated with flow, the fast reoptimisation algorithm proposed here is more efficient than algorithms starting from a network with empty edges. It is difficult to give a precise estimate of the average running time of the proposed reoptimisation algorithm. Often, the excess flow at the excess node and the deficit flow at the deficit node are eliminated through a single $s_d - t_d$ path augmentation, and the length of $s_d - t_d$ path *does not depend on the size of the network.* The length of the augmentable $s_d - t_d$ path depends *on the local topology of the network in the vicinity of the failed edge* and is often a constant value. In this case, the algorithm runs in constant time.

This important feature of the algorithm has been illustrated by the application example in Figure 5.3A, featuring part of a gas production network based on three sources (initial injection stations) $s1$, $s2$ and $s3$. Each injection station (source) has a production capacity of 70×10^3 m^3/day. From the initial injection stations, through a system of pipelines, compressors and valves, the gas is delivered to the final delivery station (sink) t. The maximum possible throughput gas flow to the sink is 150×10^3 m^3/day. Suppose that each edge models a pipeline section, with unreliable compressor transporting the gas through it. Failure of a compressor will therefore cause the flow through the pipeline section to stop.

Redundant edges $(s1,5)$, $(s2,4)$, $(s2,6)$, $(s3,5)$, (5,7), (5,9), (8,10), (7,12) and (9,10) have been provided between the three main flow paths $(s1,4,7,10,13,14)$, $(s2,5,8,11,13,14)$ and $(s3,6,9,12,13,14)$. Their purpose is to provide alternative paths for redirecting the gas flow if a compressor (an edge) fails. In addition, in order to accommodate the extra gas flow, only part of the maximum capacity of the edges/pipelines is being used.

The first step of the network analysis is to transform the multi-source network in Figure 5.3A into the single-source network in Figure 5.3B, by replacing the sources (injection stations) $s1$, $s2$ and $s3$ with edges leading to a source s with unlimited capacity (Figure 5.3B). As a result, each of the connecting edges (0,1), (0,2) and (0,3) has a maximum flow capacity equal to the maximum production capacity of the injection stations $s1$, $s2$ and $s3$, correspondingly.

Now, suppose that edge (12,13), carrying flow of 50×10^3 m^3/day, fails. For comparison, the throughput flow in the network after the edge failure (Figure 5.3B) was

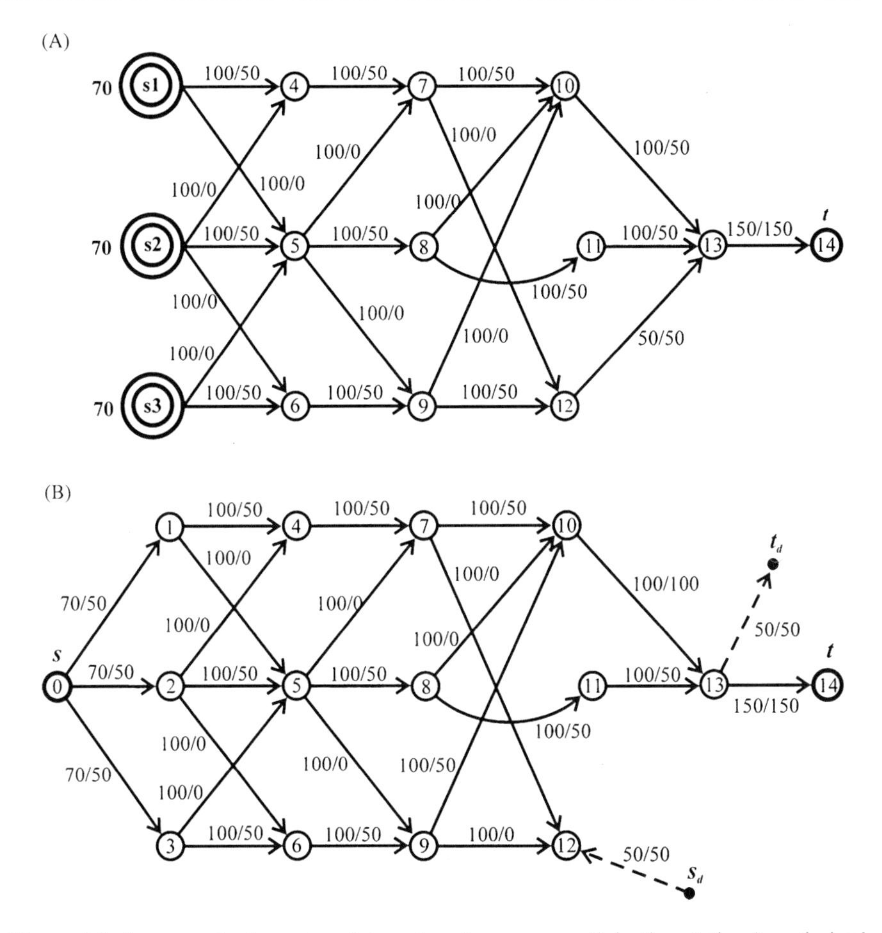

Figure 5.3 A gas production network based on three sources (injection stations), optimised by using the fast reoptimisation algorithm (Todinov, 2012a).

maximised by using the classical Edmonds and Karp algorithm, starting from an empty network. One million runs of the Edmonds and Karp algorithm, on a computer with processor *Intel(R) Core(TM) 2 Duo CPU T9900 @ 3.06 GHz*, took 6.7 s.

The throughput flow in the network from Figure 5.3B was also maximised by running the proposed reoptimisation algorithm, after introducing an extra source s_d and an extra sink t_d. Because the flow through edge (12,13) was 50×10^3 m^3/day, after the edge failure, 50×10^3 m^3/day excess flow will exist at node 12 and the same amount of deficit flow will exist at node 13. Following the reoptimisation algorithm, the extra source s_d is connected to the excess node 12 through the edge (s_d,12) with capacity 50×10^3 m^3/day. The extra sink t_d is connected to the deficit node 13, through the edge (13,t_d), with capacity 50×10^3 m^3/day. The throughput flow in the dual network is maximised by augmenting the shortest paths starting at s_d and ending at t_d. One million runs of the reoptimisation algorithm on the network in Figure 5.3B were executed in only 1.1 s, six times faster than the classical Edmonds and Karp algorithm, starting from an empty network. The reoptimisation algorithm essentially augments the single shortest path (s_d,12,9,10,13,t_d) with 50×10^3 m^3/day flow. After removing the edges incident to the new source s_d and the new sink t_d, the deficit and excess flow at nodes 12 and 13 disappears. The maximum throughput flow in the network (150×10^3 m^3/day) remains the same. Increasing the size of the network, by adding more edges between the source s and nodes 12, 9, 10 and 13, makes the reoptimisation algorithm even more efficient. The new algorithm runs in constant time, independent of the size of the network!

Consider another application example, featuring the supply network in Figure 5.4, where a large amount of goods is transmitted from two sources, $s1$ and $s2$, with supply capacities 180 and 150 units per day, to two consumers $t1$ and $t2$ with demand 150 and 180 units per day, respectively. The capacities of the supply channels are shown as edge labels.

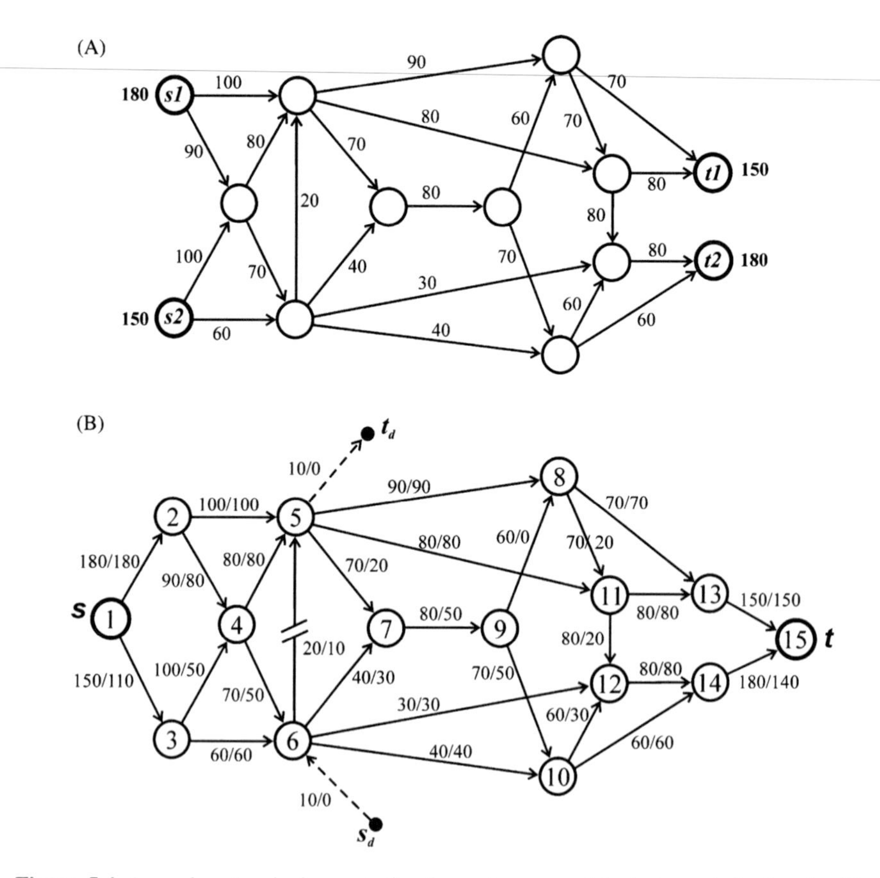

Figure 5.4 A supply network, demonstrating the performance of the fast reoptimisation algorithm.

In Figure 5.4B, the network has been transformed into a single-source, single-sink network. Its maximum throughput flow of 290 units has been obtained by using a classical augmentation algorithm. The edge flows leading to this maximum throughput flow are shown by the number after the '/' symbol.

Suppose that the supply channel (6,5), carrying 10 units of flow, has been blocked by an accident. Again, the throughput flow in the network after the blockage of the edge flow (Figure 5.4B) was reoptimised by using the classical Edmonds and Karp algorithm, starting from an empty network. For the network in Figure 5.4B, one million runs of the Edmonds and Karp algorithm on a computer with processor *Intel(R) Core(TM) 2 Duo CPU T9900 @ 3.06 GHz*, took 11.3 s.

The fast reoptimisation algorithm was also run on the network in Figure 5.4B, after introducing an extra source s_d and extra sink t_d. Because the flow through edge (6,5) before its blockage was 10 units, after its blockage, 10 units excess flow exist at node 6 and 20 units deficit flow exist at node 5. Following the proposed algorithm, the throughput flow in the dual network was maximised by augmenting the shortest paths starting at s_d and ending at t_d. One million runs of the reoptimisation algorithm were executed in only 0.98 s, more than an order of magnitude faster than the running time of the classical Edmonds and Karp algorithm optimising on a network with empty edges! The reoptimisation algorithm augments essentially the shortest path (s_d,6,7,5,t_d) with 10 units of flow. After the augmentation and the removal of the connecting edges, the deficit and excess flow at nodes 6 and 5 disappear and feasible edge flows are set everywhere in the network. The maximum throughput flow in the network after the reoptimisation is still 290 units. Even if the network is increased significantly in size, by adding many nodes between node 7 and the sink t, for example, the running time of the reoptimisation algorithm will remain the same. While the running time of all classical reoptimisation methods based on maximising the throughput flow always increases with increasing the size of the network, *the running time of the fast reoptimisation algorithm does not necessarily increase with increasing the size of the network.* Often, the algorithm reoptimises the network in constant time, without actually 'considering' the rest of the network. The reoptimisation is therefore done locally, but this is also a global optimisation, as long as the edge flows are feasible before the edge failure. This property of the proposed method is the key factor for its high computational speed.

Another significant advantage of the proposed method is that upon failures of edges, the edge flows can be simultaneously re-optimised locally, by several independent agents which possess knowledge about the local network topology but do not necessarily possess knowledge about the entire network topology. In this case, achieving a global maximum throughput flow is guaranteed, as long as no imbalanced nodes remain in the network after the re-optimisation from the independent agents. *The proposed method is essentially a very efficient decentralised algorithm that allows a network locally optimised by independent distributed agents, to achieve a global maximum of the throughput flow, after contingency events.* This property makes the proposed method very attractive for reoptimising the power flows in power networks with active control, by multiple, independent agents, upon contingency events (failures, congestion or fluctuations in energy supply and demand).

For some flow networks, where no residual capacity or redundant edges are present, failure of an edge leads to a loss of the entire flow through the edge. In this case, no excess flow can be redistributed after the edge failure. This state however will be discovered by the reoptimisation algorithm in $O(m)$ running time. Indeed, discovering whether an augmentable path exists is a procedure of running time $O(m)$. The augmented flow q_d^{max} in the dual network is therefore zero ($q_d^{max} = 0$) and according to Theorem 5.1, the new maximum throughput flow in the network is $Q_{max} - f(e,d)$. In other words, as a consequence of the edge failure, the maximum throughput flow Q_{max} has been reduced by the magnitude of the flow $f(e,d)$ through the failed edge (e,d).

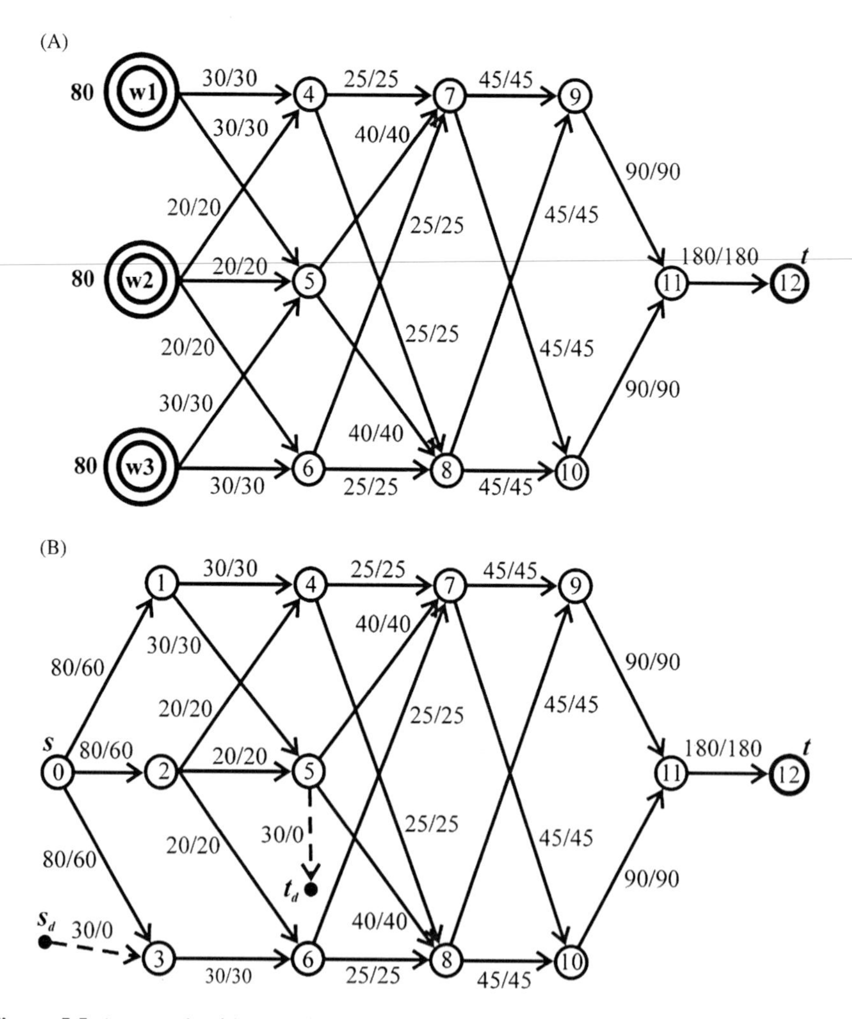

Figure 5.5 A network with no residual capacity or redundant edges. The maximum flow of 150 after the failure of component (3,5) is obtained instantly, after discovering in $O(m)$ time that no augmentable paths exist in the dual network.

As an illustrating example, consider the gas production network in Figure 5.5A, where each well has a maximum production capacity of 80×10^3 m^3/day. In Figure 5.5B, the network has been transformed into a single-source network. After the failure of edge (3,5), the new maximum throughput flow in the network can be found quickly. Because there are no augmentable $s_d - t_d$ paths in the dual network, the new maximum throughput flow after the failure of edge (3,5) is $Q_{\max} - f(3,5) = 180 \times 10^3 - 30 \times 10^3 = 150 \times 10^3$ m^3/day. One million runs of the Edmonds and Karp shortest-path algorithm for determining the maximum throughput flow by starting from an empty network took 9.3 s. In contrast, one million runs of the proposed reoptimisation algorithm took only 0.53 s.

Because augmenting a single path or few paths is a procedure of worst-case complexity $O(m)$, in the cases where the excess flow is eliminated after augmenting a single path or few paths, the running time of the fast reoptimisation algorithm is proportional to the number of edges in the network. Because the excess and deficit node are adjacent nodes, in many cases, particularly where redundant edges are present, the excess and deficit flow are eliminated after a single augmentation along a single path or after augmenting few paths. Numerous experiments with different network topologies showed that apparently, only in extreme, deliberately designed cases, the running time of the proposed re-optimisation algorithm approaches the running time of the classical Edmonds and Karp algorithm. Experiments with networks of different size and topology indicated that the average running time of the reoptimisation algorithm appears to be increasing approximately linearly with increasing the number of edges in the network.

The high computational speed of the reoptimisation algorithm makes it suitable for optimising the performance of networks with disturbed flows, in real time. This is particularly important for large power networks with active control, where the power flows need to be controlled upon failure or a sudden change in generation and demand, within the range of milliseconds.

For n failed edges (e_i,d_i), $i = 1,2, \ldots ,n$, where e_i and d_i are the first and the second node of the ith failed edge, respectively, the fast reoptimisation algorithm maximises the throughput flow equally well.

Suppose that the flows $f(e_i,d_i)$, $i = 1,2, \ldots ,n$, along n edges have either been constrained (choked) to zero in the case of edge failure, or to a smaller value due to edge capacity degradation. Choking the flows along particular edges could also be done deliberately, to relieve particular edges from congestion and overloading (e.g. for power lines and transportation channels). In this case, the capacity $c\,(i,j)$ of the edge (i,j) whose flow has been choked is reduced to a value $c'(i,j)$, which is smaller than the actual flow $f(i,j)$ through the edge $(c'(i,j) < f(i,j))$. In this case, the flow $f'(i,j)$ through the edge after the choking becomes equal to the reduced edge capacity $(f'(i,j) = c'(i,j) < f(i,j))$.

The edge flow constraint (edge flow choking) creates *M1* excess nodes and *M2* deficit nodes.

Note that the number of excess and deficit nodes *is not necessarily equal* to the number n of choked edges, although the total amount of created excess flow is always equal to the total amount of created deficit flow. The reason is that two or more choked edges may share the same end node and the state of the shared node (excess, deficit or balanced node) depends on the actual amount of choked flows through the incident edges and their direction.

Each excess node can be connected with the new source s_d, by an edge with capacity equal to the excess flow at the node. Similarly, each deficit node can be connected with the new sink t_d, by an edge with capacity equal to the deficit flow at the node. The flows through the edges, before their choking, are $f(e_i,d_i)$, $i = 1,2, \ldots ,n$. The flows through the edges after their choking are $f'(e_i,d_i)$, $i = 1,2, \ldots ,n$; $f'(e_i,d_i) < f(e_i,d_i)$. If the flow along an edge (e_i,d_i) has been fully choked, $f'(e_i,d_i) = 0$.

The next theorem (whose proof is given in Appendix A to this chapter) provides the basis of a reoptimisation algorithm which determines the maximum throughput flow in the network after choking the flow along several edges. The maximum throughput flow after choking the flow along several edges can be obtained by a simple procedure that consists of augmenting the dual network with flow until no more augmentable paths can be found.

Theorem 5.3 (Dual network theorem related to choking the flow along multiple edges) *The maximum throughput flow in any flow network after choking the flow along several edges is equal to the maximum throughput flow in the network before choking the edge flows, minus the total amount of excess flow, plus the maximum throughput flow in the dual network* (Todinov 2011b,c, 2012a).

Let $q_d^{\max}$ be the maximum throughput flow in the dual network and $Q_{\max}$ be the maximum throughput flow in the network before choking the edge flows. The quantity $\sum_{i=1}^{M1} ef_i > 0$ is the total amount of excess flow at all excess nodes. The flow along an edge (i,j) may be fully choked because of edge failure ($f'(i,j) = 0$) or partially choked ($0 < f'(i,j) < f(i,j)$).

The theorem states that the maximum possible throughput flow in the network, after choking the edge flows, is equal to $Q_{\max} - \sum_{i=1}^{M1} ef_i + q_d^{\max}$, where $M1$ is the number of excess nodes due to the choking of the edge flows.

The next theorem, which is an analogue of Theorem 5.2, provides the basis of an algorithm for determining the edge flows which correspond to the maximum throughput flow in the network.

Theorem 5.4 *The edge flows which correspond to the maximum throughput flow after choking the flow along several edges can be obtained by a two-stage procedure which consists of: (i) augmenting with flow the dual network until no more augmentable paths can be found, followed by (ii) augmenting with flow the dual circulation network until no more augmentable paths can be found.*

The method specified by Theorem 5.4, for determining the edge flows which correspond to the maximum throughput flow, will be illustrated by the network in Figure 5.6, which has a maximum throughput flow of 35 units. Because the throughput flow in the network has been maximised, the network flow is feasible and there is no excess or deficit flow at any internal node. Suppose that edges (4,7) and (5,6) have

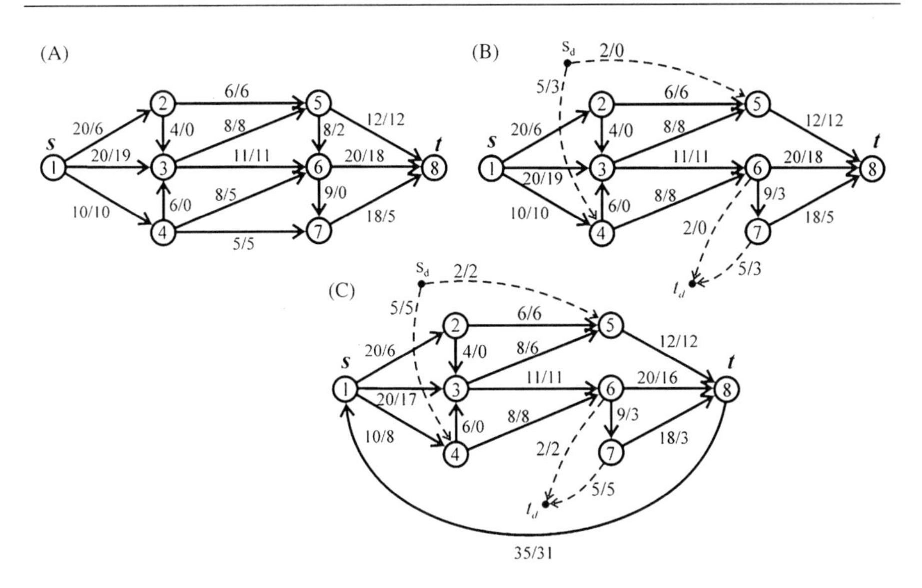

Figure 5.6 Reoptimising the throughput flow in the network after the failure of edges (5,6) and (4,7) (Todinov, 2012a).

failed (their flows have been fully choked). The failure of these edges creates two excess nodes: node 5, with excess flow of 2 units (equal to the flow of 2 units through the failed edge (5,6)), and node 4, with excess flow of 5 units (equal to the flow of 5 units through the failed edge (4,7)). The edge failures also create two deficit nodes: node 6, with a deficit flow of 2 units, and node 7, with a deficit flow of 5 units.

Next, a new source s_d is introduced, connected with the excess nodes through edges whose capacities are equal to the excess flow at the excess nodes. Similarly, a new sink t_d is introduced, connected with the deficit nodes, through edges whose capacities are equal to the absolute values of the deficit flow at the nodes.

The throughput flow in the dual network can be augmented by only $q_d^{\max} = 3$ units of flow, through the shortest augmentable path (s_d,4,6,7,t_d) (Figure 5.6B). There are no other augmentable $s_d - t_d$ paths in the dual network. Consequently, according to Theorem 5.3, the maximum throughput flow in the network is equal to

$$Q_{\max} - \sum_{i=1}^{2} f(e_i, d_i) + q_d^{\max} = 35 - (2 + 5) + 3 = 31$$

If only the maximum throughput flow after the edge failures is needed (as it is the case for discrete-event solvers of repairable flow networks), no further steps are necessary. The proposed algorithm determines only the maximum throughput flow and does not determine the corresponding edge flows, which is a big advantage. This is another reason for the significantly higher computational speed of the fast reoptimisation algorithm, compared to conventional algorithms starting from a network with empty edges.

If the edge flows corresponding to the maximum throughput flow are necessary, a circulation edge (t,s) is included, fully saturated with flow, with flow capacity 35 flow units, equal to the maximum throughput flow in the network, before the edge failures (Figure 5.6C).

After augmenting the shortest $s_d - t_d$ path $(s_d,4,1,t,6,t_d)$ with 2 units of flow and the path $(s_d,5,3,1,t,7,t_d)$ also with 2 units of flow, no more augmentable paths exist in the dual circulation network. According to Theorem 5.4, this means that the new, maximum throughput flow in the optimised network has been set, and the correct edge flows, corresponding to the maximum throughput flow, have also been set. The edge flows, which correspond to the maximum throughput flow after the failure of the two edges, are shown in Figure 5.6C.

Maximising the throughput flow in a repairable flow network, although important, does not reveal the variation of the total transmitted flow caused by component failures. This variation can be revealed by building discrete-event solvers of the flow network. The purpose is to determine the risk that, due to component failures, the throughput flow on demand, or the total throughput flow during a specified period of time, will fall below a required level. The proposed method for reoptimising the flow after an edge failure is particularly suitable for designing fast discrete-event solvers of production systems. In order to track correctly the variation of the total throughput flow during a time interval, the maximum throughput flow needs to be calculated many times after each edge failure. A typical failure history may include thousands of failures during the lifecycle of the network. Hence, the algorithm for reoptimising the throughput flow after a component failure must be very fast.

Simulating the performance of repairable flow networks during a time interval includes two types of events: *failure*, and *return from repair*. The event 'return from repair', consists of returning failed components to operation in *as good as new* condition. A return from repair is equivalent to the appearance of extra capacity between a pair of nodes in the network. Accordingly, the question here is how to achieve the new maximum throughput flow, after the return from repair.

Answering this question is easier when compared to determining the maximum throughput flow after an edge failure or after choking the flow along an edge. If, after repairing a failed edge, there are no more failed or choked edges in the network, the network flows in the repaired network are restored simply by copying the edge flows existing before any component failure or choking an edge flow.

If, after the repair of a failed component, there are still failed components in the network, the new maximum throughput flow is determined by running the Edmonds and Karp shortest-path augmentation algorithm, from the source s to the sink t of the optimised network. The edge flows corresponding to the maximum throughput flow before the failed component is returned to operation, serve as initial edge flows for the flow augmentation process. As a result, the shortest-path augmentation algorithm does not start from a network with empty edges and runs much faster. Usually, the new maximum throughput flow is found after augmenting a single path or few paths.

Initially, the maximum throughput flow in the network is determined, which corresponds to a state where there are no failed edges. Two functions *max_flow_after_failure()* and *max_flow_after_repair()* are also defined. The function *max_flow_after_failure()* determines the maximum throughput flow after an edge failure. It implements the fast reoptimisation algorithm introduced earlier.

The function *max_flow_after_repair()* determines the maximum throughput flow in the network after a return from repair. If there are no failed components in the network after returning the failed component from repair, the procedure simply copies the original flows existing in the network before any component failures. If the converse is true, the function implements the Edmonds and Karp shortest-path augmentation algorithm.

Times to events (failure or repair) are placed in a queue of events, in ascending order. The current event is the event with the smallest time. It is retrieved from the head of the queue. If the current event is an edge failure, the capacity of the failed edge is reduced to zero. The maximum throughput flow in the network is then calculated efficiently by calling the function *max_flow_after_failure()*. After a delay determined by the downtime for repair of the edge, a new life is generated for the edge and placed in the queue of events.

In the case where the event is a 'return from repair', the network topology is updated by restoring the capacity of the repaired edge to its original value. The maximum throughput flow in the network is then calculated by calling the function *max_flow_after_repair()*. Continuing this process until the end of the specified time interval and taking into account the between-event time intervals yields the total throughput flow related to the current failure–repair history. Dividing the average total throughput flow from hundreds of thousands of failure–repair histories and the total throughput flow in the network in the absence of any failures yields the production availability of the network.

It must be pointed out that the algorithm for fast reoptimisation of the flow after edge failures *does not necessarily require the throughput flow in the network to be initially maximised.* The algorithm works with *any* feasible network flow. After edge failures, the algorithm redistributes the flow in the network by eliminating the excess and deficit flow at the nodes to which the failed edges have been connected. As a result, a new feasible network flow is established.

5.3 An Algorithm for Maximising the Throughput Flow of Oil After a Component Failure in Multi-Commodity Oil Production Networks

Similar to the single-commodity case, after a component failure, an efficient maximisation of the oil flow in oil production networks can be implemented. Consider again the oil production network from Chapter 4, which contains three wells ($s1$, $s2$ and $s3$), each of which is characterised by a maximum production capacity of 300 mbd. The production fluid is a mixture of oil and water and the oil fractions characterising the separate wells are $\alpha_1 = 0.8$, $\alpha_2 = 0.4$ and $\alpha_3 = 0.6$, correspondingly.

Upon failure of a component (edge), e.g. edge (7,11), an excess and deficit flow appears at the ends of the failed edge. The dual network is shown in Figure 5.7B. The new source s_d and the new sink t_d have been connected to the network by edges with capacity equal to the actual flow (150 mbd) through the failed edge (7,11). As can be verified from Figure 5.7B, no augmentable path exists in the dual network, starting at the new source s_d and ending at the new sink t_d.

The second stage involves adding the circulation edge (t,s), with flow capacity 300 mbd, equal to the maximum throughput flow in the network before the failure of edge (7,11). The selected augmentation path is $(s_d,7,5,s2,s,12,11,t_d)$ because it passes through the well with the smallest oil fraction. This selection is essential, because augmenting a backward edge coming out of a well leads to a decrease in production from the well. In this case, in contrast to the path augmentation from Chapter 4, the wells with the smallest oil fraction are selected preferentially. The path $(s_d,7,5,s2,s,12,11,t_d)$ is augmented with the maximum possible flow of 50 mbd. The next selected augmentable path is the path $(s_d,7,8,10,5,s2,s,12,11,t_d)$, which also passes through the well with the smallest fraction of oil. This path is also augmented with the maximum possible flow of 50 mbd and, as a result, well $s2$ gets closed. There is still 50 mbd excess flow at node 7. The next path to be augmented with 50 units flow is the path $(s_d,7,8,10,6,s3,s,12,11,t_d)$, which now passes through well $s3$ which is now the well with the smallest fraction of oil. The resultant edge flows after the path augmentation are shown in Figure 5.7C.

After removing the connecting edges, the network flow is feasible everywhere and the oil flow to the sink is the maximum possible flow.

Note that the path augmentation will never encounter a situation where the amount of flow by which the production of a well is reduced, exceeds the production of the well. This is because the network flow, initially established in the network *is a feasible network flow*.

Theorem 5.5 *The maximum throughput flow in a multi-commodity production network after the failure of a component can be obtained by a two-stage procedure. The first stage consists of maximising the throughput flow in the dual network, by preferentially augmenting the shortest paths which choke the wells with the smallest oil content and increase production from the wells characterised by the largest oil content. The second stage consists of path augmentation in the dual circulation network, by preferentially selecting the shortest paths which choke the wells with the smallest oil content.*

The maximum throughput flow of oil to the sink is guaranteed by (i) the maximum throughput flow of mixture to the sink and (ii) the maximum oil content in the mixture.

Following the procedure outlined in Chapter 4, before an edge failure, the maximum throughput flow of oil to the sink is guaranteed. After an edge failure, the maximum throughput flow of mixture to the sink is guaranteed according to Theorems 5.1 and 5.2. After the edge failure, the maximum oil fraction in the mixture is guaranteed according to Theorem 4.2, because of the preferential

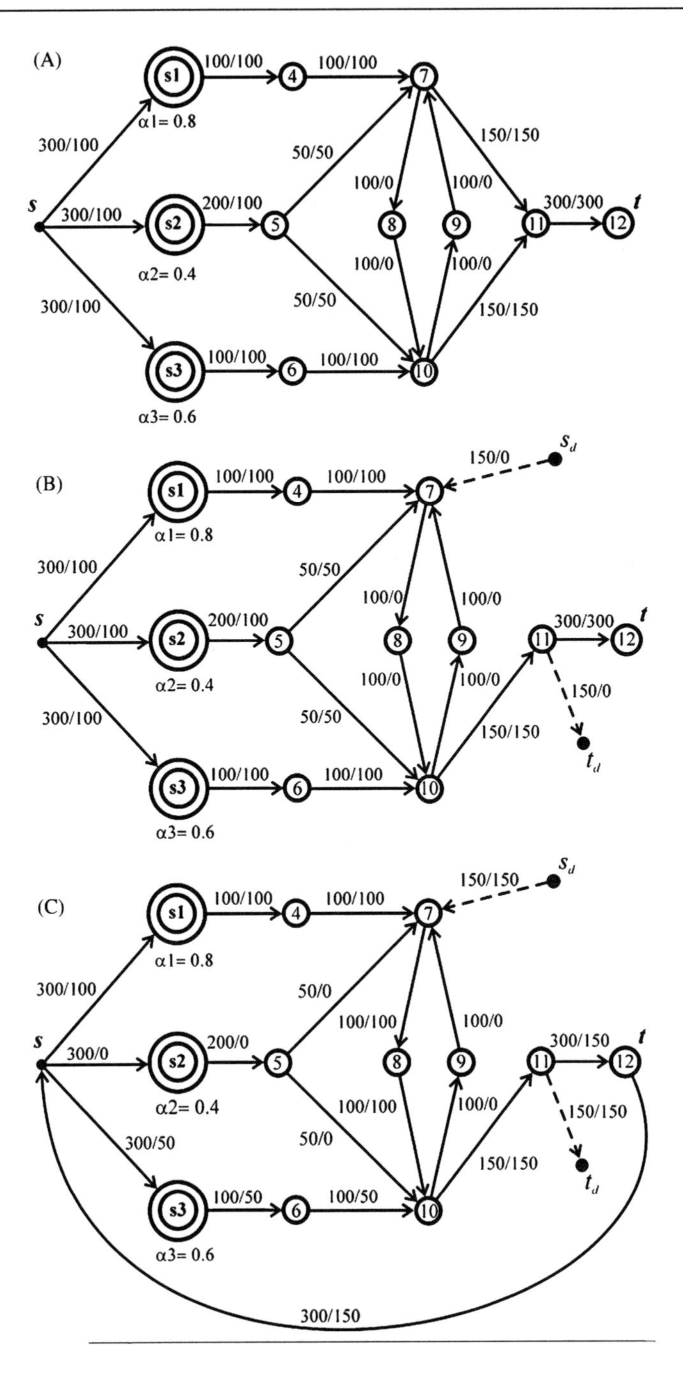

Figure 5.7 Optimising an oil production network after the failure of component (7,11).

augmentation of paths starting from the richest in oil wells and choking the wells with the smallest oil fraction.

5.4 A High-Speed Control of Large and Complex Active Power Distribution Networks

Power grids are often affected by random failures caused by lightning, fire, extreme weather conditions, failures in generators, malicious acts, etc. Furthermore, the power demand from consumers often varies drastically. Because these events are unavoidable, an optimal management of the power grids, in real time, is required to guarantee an immediate response and a stable operation even in the case of simultaneous occurrences of a number of such events.

The existing passive power distribution systems, however, cannot deliver an optimal management of the power flows. In the existing power distribution systems, if a power line is lost, the remaining electric power redistributes by following the path with the lowest impedance, which results in congestion and overloading of power lines. Sudden increase in power demand from consumers also leads to overloading of power lines.

A power distribution system with radial, tree-like structure is simple but also highly unreliable – a fault along any branch cuts off the power supply to the entire downstream part. An interconnected power grid is significantly more reliable because congestion and faults can be isolated without affecting other sections of the grid. The significant reliability increase, however, comes at a price. In a passive, interconnected grid, the flow of current along particular power lines cannot be controlled and the grid capacity is limited by the lowest-capacity line (Divan and Johal, 2006). The first line to reach its thermal capacity, limits the capacity of the entire grid, even if there is sufficient flow capacity in the grid to accommodate the required power transfer and avoid congestion and overloading. Because of this problem, the capacity utilisation of interconnected grids is only a fraction of their theoretical capacity. Furthermore, the rapid growth of electrical vehicles poses other challenges. The vehicles' batteries have significant charging rates, which requires the optimal use of the transmission capacity of the power distribution networks.

The answer to these problems are grids with active control of the power flows along the power lines and the amount of generated power from the distributed generators. Power grids possessing power flow control are increasingly considered to be the power networks of the future and the solution to the optimal management of power flows (Nguyen et al., 2009; Overbeeke, 2002). The power flow control and the interconnected topology provide extra power flow capacity, without having to build new connections. Building new connections is undesirable because of environmental, social and political considerations (Van Hertem et al., 2005). Power grids with active control and interconnected topology are both reliable and provide extra power flow capacity thereby relieving congestion and enhancing the operational envelope.

The major driver for the implementation of power distribution networks with active control is the increase in the volume and variety of distributed generation sources and renewable generation. This is in line with the trend towards increasing the flexibility of power supply and decreasing transportation losses. Hardware solutions for controlling the power flows along individual transmission lines already exist (Divan and Johal 2006; Mutale and Strbac, 2000; Van Hertem et al., 2005), some of which are cheap and accessible (Divan and Johal, 2006).

In short, algorithms for reoptimising the power flows in real time, after congestion, component failure or a sudden change in power demand are critically important for power grids. Reoptimising quickly the power flows by redistributing the flows and reducing the power generation from particular distributed generators in an optimal way: *(i) eliminates the overloading of the power lines, (ii) maximises the throughput power and (iii) matches fluctuating demand to generation and improves the grid performance.*

We need to point out that, for power distribution grids, where no cascading failures are present, the steady-state modelling approach is entirely sufficient and considering the transient electric processes is not critical to developing the fast algorithms for managing the power flows.

5.4.1 Optimal Control of the Power Flows in Power Distribution Networks with Active Control, After a Contingency Event

To reduce the risk of power outages, the power networks with active control need control algorithms that will provide a fast recovery from overloading. If overloading is present, a certain power line has reached its thermal limits, and the reliable operation of the grid is no longer guaranteed if a random failure occurs. In a meshed grid, there are always alternative, not fully loaded paths. The optimal power flow control reduces appropriately the power generation from the sources, redistributes the power flows through alternative transfer paths and relieves the overloaded lines. The result is not only an increased power flow capacity of the grid but also increased fault tolerance and reliability. A power grid without congested power lines can tolerate not only a single component failure but also simultaneous component failures.

There have been proposals for controlling the power flow based on classical algorithms for maximising the flow in a network (Amrbruster et al., 2005). According to the discussion in Chapter 4, the classical algorithms leave directed loops of parasitic flow, which waste energy and increase the congestion in the "optimised" networks. Consequently, the classical algorithms are not suitable for optimising the power flows in power networks. Furthermore, the classical algorithms are not suitable for controlling the power flows in real time. The classical algorithms for maximising the flow have a polynomial running time of order $O(m^2n)$, $O(n^2m)$, $O(n^3)$, $O(mn \log(n^2/m))$ in the size of the network, where m is the number of edges and n is the number of nodes in the network. For a power network where a component has failed, the algorithm for optimising the power flows must find a solution in real time, within the time range of milliseconds, in order to guarantee an optimal power network management.

Large power networks with active control may include thousands of nodes and power lines. Recent estimates conducted by the author showed that a computer with a processor *Intel(R) Core(TM) 2 Duo CPU T9900 @ 3.06 GHz* expends on an edge, on average, $\Delta t = 2.5 \times 10^{-6}$ s, if the classical Edmonds and Karp shortest-path algorithm for maximising the throughput flow in the network is used, after an edge failure. One of the fastest available algorithms for reoptimising the flow has an average running time proportional to m^2, where m is the number of edges in the network (see Chapter 4). For a network with 10,000 edges, the average running time of this algorithm is proportional to $(10{,}000 \times 10{,}000) \times \Delta t$, which means an average running time of 25 s! Such a big running time is unacceptable for the operation of an active power network, in real time.

If the algorithm proposed in Section 5.2 is used for reoptimising the flow after a component failure, in a network with 10,000 edges, the average running time will be proportional to the number of edges m in the network. The running time will be proportional to $10{,}000 \times \Delta t$ s, which means a running time of only 2.5 ms!

The proposed algorithm with a linear average running time has another important application. It can also be used to eliminate the congestion and overloading of transmission lines in power distribution networks. This problem is very important and arises frequently in computer networks, transportation networks and power distribution networks.

The ever-increasing demand on the existing computer networks, communication networks, transportation networks and power distribution networks requires faster power flow management, if the number of dropped calls, the delays caused by congestion and the power supply fluctuations are to be minimised.

Consider the power network in Figure 5.8A, where sections (1,2) and (2,5) have been congested. To relieve the extra load/congestion, the power flow along these edges should, for example, be reduced by 5 MW. The first step is to choke the

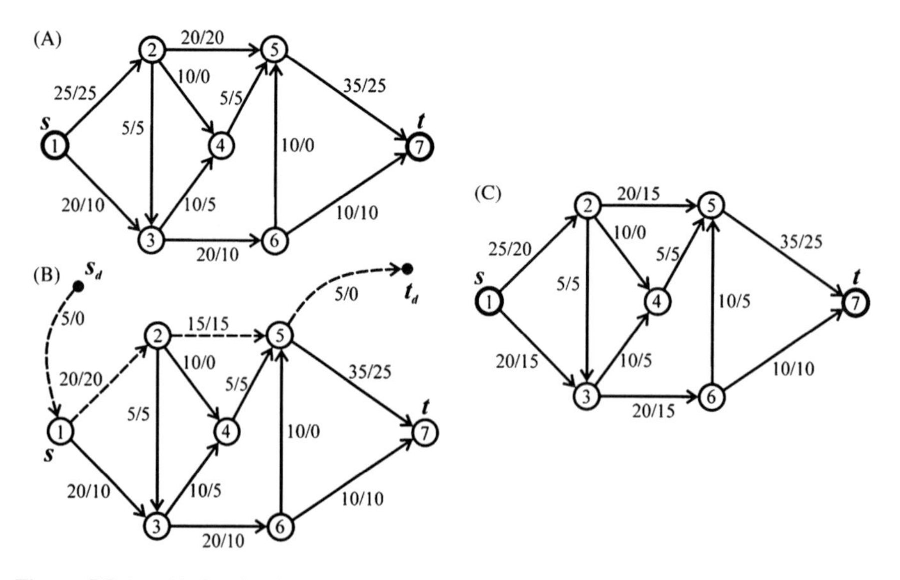

Figure 5.8 Load balancing in a network where edges (1,2) and (2,5) have been overloaded.

flows along the congested lines by limiting the capacities of these edges to the desired amount of flow through them – to 20 MW for power line (1,2) and to 15 MW for power line (2,5). As a result, excess and deficit flows of 5 MW appear at the beginning and at the end of the power lines whose flow has been choked. Node 2, however, remains a balanced node, because the excess power flow of 5 MW from choking the flow along power line (2,5) is cancelled by the deficit power flow of 5 MW from choking the flow along power line (1,2).

Additional source s_d is then added, connected to the excess node '1' by edge $(s_d,1)$, with capacity 5 MW. Similarly, additional sink t_d is also added, connected to the deficit node 5, by an edge $(5,t_d)$, with capacity 5 MW (Figure 5.8B). The algorithm for redistributing the flow in the resultant network then proceeds according to Theorem 5.4. The shortest augmentable path $(s_d,1,3,6,5,t_d)$ is augmented with 5 MW, which results in the network flows in Figure 5.8C. The power flow is feasible; the throughput flow (35 MW) is equal to the throughput flow before the redistribution. Power lines (1,2) and (2,5), however, are no longer congested.

Note that for a long directed flow path, where the choked flow along each edge is the same, only two imbalanced nodes will appear after the choking. The start node of the directed path will appear as an excess node and the end node will appear as a deficit node. In any other node i of the directed path, the deficit from choking edge $(i-1,i)$ will be counterbalanced by the excess from choking node $(i,i+1)$. The result will be a neutral node i.

In summary, the fast reoptimisation algorithm can be used with success for relieving the congestion along sections of active power networks in the absence of any failures.

In the case of preventing power lines from congestion and overloading, the optimisation has been achieved without decreasing generation from the source (without generation shedding). Generation shedding means that the generation from the source is reduced or the generation source is disconnected during the duration of the contingency event. In some cases, generation shedding is necessary in order to avoid exceeding the power line capacities.

Such is the case in Figure 5.9, featuring a simplified active power distribution network, where the power line (2,3) suddenly fails. The network includes only a single generator (source) of flow s and a single load (sink) t.

The parameters of each edge (transmission line) are given in the format 'c/f', where c is the capacity of the edge (transmission line) in MW and f is the actual power flow through the edge in MW. The maximum throughput power flow for the network in Figure 5.9A is 16 MW. In the case of failure of power line (2,3), carrying power flow of magnitude 3 MW, the overloading caused by the passive redistribution of the excess power can be avoided by an appropriate combination of altering the magnitudes of the power flows along certain power lines and generation shedding. Following the maximum flow after edge failure algorithm, the shortest s_d-t_d paths in the dual network in Figure 5.9B are augmented first. Because no augmentation of s_d-t_d paths in the dual network is possible (Figure 5.9B), a circulation edge (t,s) with capacity equal to the maximum throughput flow of 16 MW is introduced (Figure 5.9C). In the dual circulation network (Figure 5.9C),

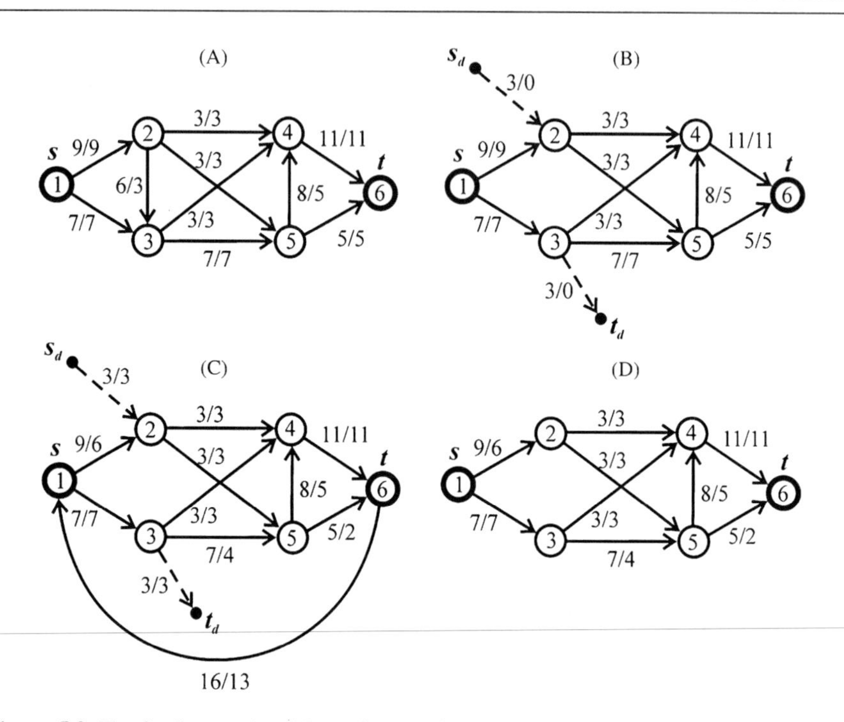

Figure 5.9 The fundamental problem of power flow maximisation in an active power network after a failure of a power line.

the shortest augmentable $s_d - t_d$ path (s_d,2,1,6,5,3,t_d) has been augmented with 3 MW flow, after which, further $s_d - t_d$ path augmentations are no longer possible.

As it can be seen in Figure 5.9D, the production from the generator has been reduced from 16 to 13 MW, to accommodate the new maximum throughput flow after the failure of power line (2,3) (Figure 5.9D).

A very important application of the decongestion algorithm is in handling variable demand from consumers and variable production from sources. In case of a sudden reduction of demand for example, an appropriate power shedding and power flow redistribution need to be done to match the power generation from the sources with the power consumption from the consumers with reduced demand, without reducing the power supply to the rest of the consumers. Suppose that there are two consumers c1 and c2, with power consumption 6 and 10 MW, respectively (Figure 5.10A). Suppose that the power consumption from consumer c2 suddenly drops from 10 to 5 MW.

The power network with multiple consumers is first transformed into an $s - t$ network (with a single source and a single sink) by introducing a super-source s and a super-sink t, as it is shown in Figure 5.10B. The flow capacities of edges (4,6) and (5,6), connecting the consumers (nodes 4 and 5) with the super-sink t, are equal to the flow consumption from the consumers. As a result, the consumers disappear and throughput edges appear instead (Figure 5.10B).

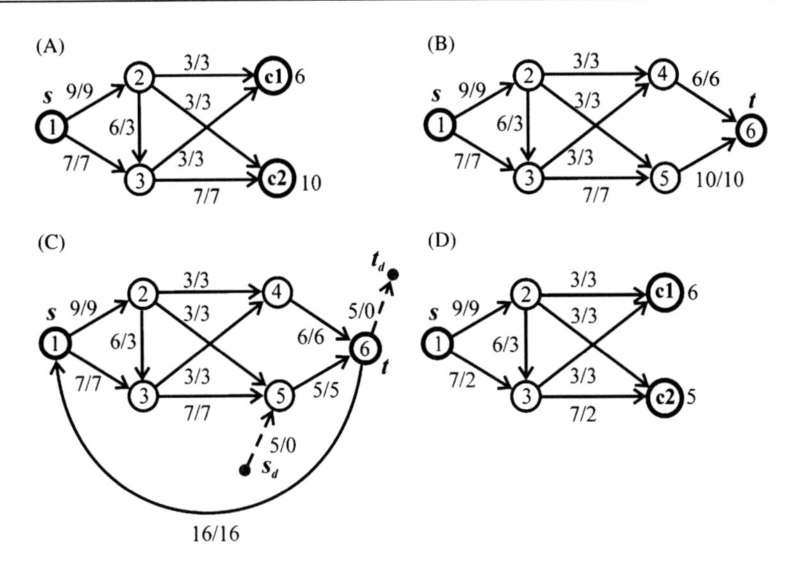

Figure 5.10 Reoptimisation algorithm for variable power consumption.

Because, the demand from consumer $c2$ has suddenly decreased from 10 to only 5 MW, the capacity of edge (5,6) is reduced to 5 MW. As a result, excess flow of 5 MW appears at node 5 and deficit flow of 5 MW appears at node 6. Augmenting $s_d - t_d$ paths in the dual network is not possible; this is why the circulation edge (t,s) is introduced, with capacity 16 MW, equal to the throughput flow in the network before the reduction in demand from consumer $c2$. The $s_d - t_d$ path $(s_d,5,3,1,6,t_d)$ can be augmented with 5 MW flow. After the removal of the connecting edges, the excess and deficit flows disappear in the original network. The network has been reoptimised following the decrease in power consumption from consumer $c2$, by generation shedding from the source s.

In a similar fashion, the network can also be reoptimised in the case of a variation of the power generation from some of the sources. The high computational speed of the proposed approach is due to the circumstance that the reoptimisation after an edge failure is done by altering the flows along a relatively small number of edges. In contrast, the reoptimisation by using classical algorithms starts from empty edges and is conducted by altering the flows along most of the edges in the network.

5.4.2 Transforming an Active Power Network, with Distributed Sources of Generation and Loads, into a Single-Source Single-Sink Network

In reality, a number of sources of generation and loads (consumers) are present in the same distribution network. This is for example the case of power grids with a large number of distributed renewable sources of generation. The power demand from the loads can also vary suddenly, in unpredictable manner. Furthermore, the sources have a limited flow generation capacity and the consumers have also a limited flow consumption capacity.

To reduce the network to a single-source, single-sink network, the technique discussed in Chapter 2 can be applied. A super-source s, with unlimited flow generation capacity, is introduced, connected with the sources of generation through empty edges with capacities equal to the flow generation capacities of the sources. Similarly, a super-sink t, with unlimited flow consumption capacity is also introduced. The super-sink is connected with the loads (consumers) through empty edges with capacities equal to the flow consumption capacities characterising the loads.

The network in Figure 5.11 features three sources of power $s1$, $s2$ and $s3$ and two loads (consumers) $L1$ and $L2$. The power generation capacities of the sources are $s1$ (15 MW), $s2$ (20 MW) and $s3$ (30 MW). The power consumption capacities of the loads are $L1$ (15 MW) and $L2$ (35 MW).

The possible transmission paths and their maximum allowable power flow capacities have also been specified.

The network in Figure 5.11A characterised by multiple sources and loads can be transformed into an $s-t$ network (with a single source and a single sink) by introducing a super-source s and a super-sink t, as it is shown in Figure 5.11B. The flow capacities of edges $(s,s1)$, $(s,s2)$ and $(s,s3)$ connecting the super-source s with the sources of generation (nodes $s1$, $s2$ and $s3$) are equal to the power flow generation capacities of the sources. Similarly, the flow capacities of edges $(L1,t)$ and $(L2,t)$ connecting the loads (nodes $L1$ and $L2$) with the super-sink t are equal to the flow consumption capacity of the loads. As a result, all sources and loads disappear and throughput edges appear instead (Figure 5.11B).

The methods proposed in this chapter constitute a novel idea in the real-time control of active power networks and power supply reliability. The proposed algorithms are essentially a new generation of very fast algorithms for reoptimising power flows in real time, upon contingency events. The algorithms for reoptimising the flow after contingency events are also very important in applications such as oil and gas production, telecommunications, transportation, water distribution, manufacturing, electronics, supply logistics and chemical production. Consequently, these industrial sectors could benefit significantly from the developed high-speed algorithms for real-time

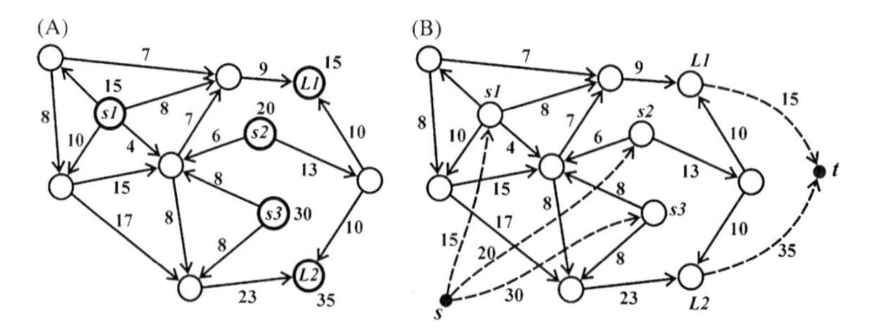

Figure 5.11 Transforming an active power network with multiple sources and loads into a single-source/single-sink network.

control. The possibility for reoptimising the edge flows locally, by independent agents with no knowledge of the entire network topology, and still achieve a global maximum of the throughput flow, is a major advantage of the proposed algorithm. The algorithm essentially *provides a solution to the long-standing problem of decentralised optimal control of large and complex power distribution networks.*

Appendix A

Proof of Theorem 5.3

After the simultaneous choking of the flow along n edges (x_i,y_i), $i = 1, \ldots, n$, the network flow is disturbed only locally, at nodes x_i and y_i to which the affected edges have been connected. As a result of the choking, the flows along the edges have been reduced from $f(x_i,y_i)$ to $f'(x_i,y_i)$, where $0 \le f'(x_i,y_i) < f(x_i,y_i)$, $i = 1,2, \ldots, n$.

Suppose that choking the flow created $M1$ excess nodes e_i with excess flow ef_i ($i = 1, \ldots, M1$) and $M2$ deficit nodes d_j with deficit flows df_j ($j = 1, \ldots, M2$) (Figure 5.A1B). Note that after choking the edge flows, the sum of the excess flow at the excess nodes is always equal to the sum of the deficit flow at the deficit nodes, $\sum_{i=1}^{M1} ef_i = \sum_{j=1}^{M2} df_i$.

To create the dual network, a new source s_d is introduced, connected by empty edges with the excess nodes. The capacities of the connecting edges are equal to the excess flows ef_i at the excess nodes e_i, $i = 1, \ldots, M1$. Similarly, a new sink is introduced, connected by empty edges with the deficit nodes d_j, $j = 1, \ldots, M2$. The

(A)

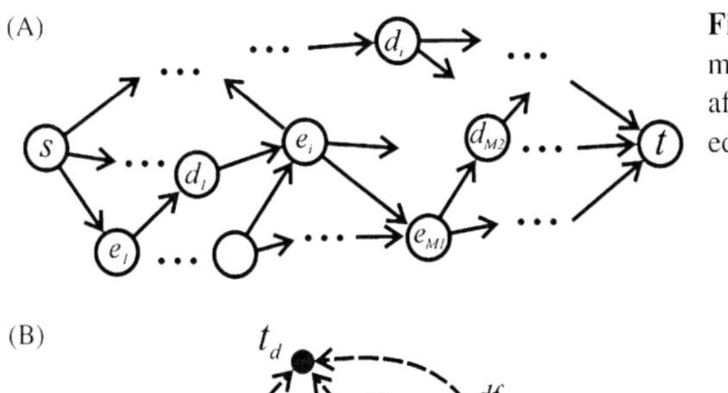

(B)

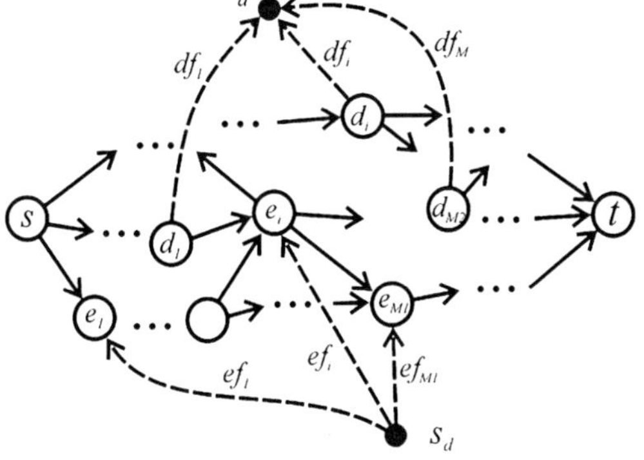

Figure 5.A1 An algorithm for maximising the flow in the network after choking the flow along several edges.

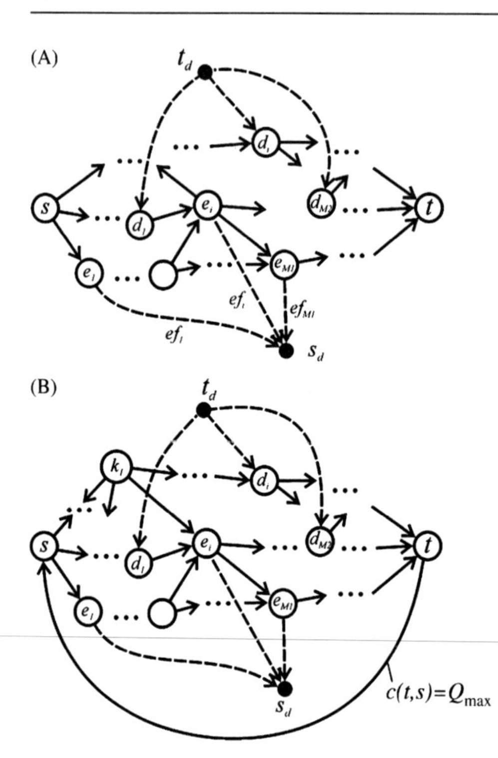

Figure 5.A2 (A) Equivalent process of augmenting with flow the dual network. (B) Determining directed $s_d - t_d$ augmentation paths consisting of backward edges only.

capacities of the connecting edges are equal to the deficit flows df_i at the deficit nodes.

Instead of augmenting the empty edges connecting the new source s_d with the excess nodes and the new sink t_d with the deficit nodes (Figure 5.A1B), an equivalent operation is introduced, by reversing the direction of the empty edges at the new source and the new sink and saturating them with flow (Figure 5.A2A). As a result, the new source s_d becomes an excess node and the new sink t_d becomes a deficit node.

Now the equivalent operation to augmenting the dual network in Figure 5.A1B is to purge flow from edges (s_d,e_i), $i = 1, \ldots ,M1$, and edges (d_j,t_d), $j = 1,\ldots,M2$, in Figure 5.A2A, by augmenting the shortest $s_d - t_d$ paths, until all of the connecting edges become empty. The advantage of the equivalent operation is that it makes excess and deficit nodes disappear at the beginning and establishes a feasible flow in all parts of the dual network.

There are two mutually exclusive possibilities for the network in Figure 5.A2A, the first of which is: (i) the amount of purged flow $q_d^{\max}$ from the backward edges (s_d,e_i) $(i = 1, \ldots ,M1)$ in the dual network is equal to the sum of their flow capacities and is also equal to the sum of the flow capacities of edges (d_j,t_d) $(j = 1, \ldots , M2)$: $q_d^{\max} = \sum_{i=1}^{M1} ef_i = \sum_{j=1}^{M2} df_j$.

After the $s_d - t_d$ path augmentation, edges (s_d,e_i) and (d_j,t_d), connected to the new source s_d and the new sink t_d will be empty, which essentially eliminates the excess flow at the excess nodes e_i ($i = 1, \ldots, M1$) and the deficit flow at the deficit nodes d_j ($j = 1, \ldots, M2$) in the optimised network. In other words, the flow conservation law at the nodes and the capacity constraints will be fulfilled everywhere in the optimised network. (During a Ford–Fulkerson type of augmentation, the feasibility of the network flow is preserved.)

Indeed, according to the first possibility, eliminating the excess flow at the excess nodes e_i ($i = 1, \ldots, M1$) and the deficit flow at the deficit nodes d_j ($j = 1, \ldots, M2$) has essentially been done along the shortest paths, starting from the new source s_d and ending at the new sink t_d. Suppose that neither any of the excess nodes e_i nor any of the deficit nodes d_j coincides with the original sink t. Then, augmenting a path starting at node s_d and ending at node t_d in the dual network (Figure 5.A2A) implies that each time the original sink t is visited, two edges entering the sink t will necessarily be part of the augmented path. One of the edges will be a forward edge and the other will be a backward edge.

According to the Ford–Fulkerson type of augmentation, the augmented flow which is added to the forward edge incident to the sink t is subtracted from the backward edge incident to the sink t and, as a result, the throughput flow in the original network after augmenting the dual network will not be altered. Now, assume that one of the nodes (e.g. node d_j) coincides with the original sink t. Because all edges incident to the sink are ingoing edges, node d_j, coinciding with the sink t, will be the deficit node and node e_i will be the excess node. Failure of edge (e_i,d_i) will remove flow $f(e_i,d_i)$ entering the sink t, which will later be added back after augmenting another edge entering the sink. This is because the augmented $s_d - t_d$ path must necessarily pass through the original sink t. As a result, the throughput flow after augmenting paths in the dual network will not be altered.

Now, consider the second possibility: (ii) the maximum purged (augmented) flow $q_d^{\max}$ from edges (s_d,e_i), $i = 1, \ldots, M1$, in the dual network is smaller than the sum of the excess flow from the excess nodes ($q_d^{\max} < \sum_{i=1}^{M1} ef_i$). As a result, the edges connected to the new source and the new sink will not be empty at the end of the first stage, consisting of augmenting $s_d - t_d$ paths in the dual network.

Let the original sink t be connected with the original source s through a back (circulation) edge (t,s) with capacity $Q_{\max}$ (Figure 5.A2B), which is equal to the maximum throughput flow in the original network, before choking the flow along edges (x_i,y_i), $i = 1, \ldots, n$. The network in Figure 5.A2B, obtained from the dual network from Figure 5.A2A by adding the circulation edge (t,s), will be referred to as the *dual circulation network*. Before proving Theorem 5.3, two lemmas will be proved.

Lemma 5.A1 *If there are no augmentable $s_d - t_d$ paths in the dual network, the excess flow at the excess nodes e_i and the deficit flow at the deficit nodes d_i can always be reduced by augmenting an $s_d - t_d$ path in the dual circulation network, where the circulation edge (t,s) belongs to the augmented path.*

Proof This lemma will be proved by showing that the remaining excess and deficit flow at nodes e_i and d_j ($i = 1, \ldots, M1$; $j = 1, \ldots, M2$) can always be reduced by selecting, for example, $s_d - t_d$ paths consisting of backward edges only. The original sink t is connected with the original source s, through a back edge (t,s) with capacity Q_{max} (Figure 5.A2B), which is the maximum throughput flow in the original network before choking the flow along edges (x_i,y_i), $i = 1, \ldots, n$.

The network flow in the dual circulation network is feasible because the flow conservation law is honoured at each node and the edge capacity constraints are not violated. There are no excess and deficit nodes in the network, except at the new source s_d and at the new sink t_d (Figure 5.A2B). Let us start from the new source s_d, by selecting a non-empty backward edge (s_d,e_i), and consider the edges going into node e_i. Because node e_i is now a balanced node, the sum of the flows going into node e_i is equal to the sum of the flows going out of node e_i. Consequently, if edge (e_i,s_d) carries some flow out of node e_i, there must be a directed edge (k_1,e_i) going into node e_i and carrying flow greater than zero. Let us consider the start node k_1 of this edge (Figure 5.A2B). The reasoning which has been applied to node e_i can now be applied to node k_1 and so on, until either a visited node is reached or the new sink t_d is reached. Suppose that a visited node v has been reached before reaching the new sink t_d. This means that a cyclic path has been encountered, consisting of backward edges only. The edge (s_d,e_i), however, cannot be part of this cyclic path, because all edges incident to node s_d are backward edges. Next, the edge with the smallest amount of flow belonging to the encountered cyclic path is identified, and its flow is subtracted from all edges of the cyclic path. During this operation, the edge carrying the smallest amount of flow becomes empty and the flow conservation law at the repeated node v, will not be violated.

Because the edge through which the repeated node v has been first reached is not part of the cycle, we continue from node v and the same process is repeated until no more directed cyclic paths are encountered. After each flow subtraction from the edges of the encountered directed cyclic path, at least one edge from the network becomes empty and is *never filled with flow again* because only backward edges are selected for the $s_d - t_d$ path. Reaching the new sink t_d is guaranteed after at most m repetitions of this process where m is the number of edges in the network. Reaching the new sink is always guaranteed because all nodes in the network are balanced, except nodes s_d and t_d. During the process of flow removal from a backward edge coming out of a node and from another backward edge going into the node, the flow conservation law at the node is always preserved. As long as the flow along a backward edge (k_ik_{i+1}) is nonzero, there must be a backward edge $(k_{i+1}k_{i+2})$ going into node k_{i+1} and carrying nonzero flow, and so on, until the new sink t_d is reached (Figure 5.A2B).

In short, an augmentable path from the new source s_d to the new sink t_d always exists. The bottleneck edge from this path, carrying the smallest amount of flow is then found and the edge flows along the path are augmented with this bottleneck flow. As a result, the edges carrying the bottleneck flow become empty, and both the flow along edge (s_d,e_i), and the flow along edge (d_i,t_d) decrease.

Note that during this process, the $s_d - t_d$ path must necessarily include the circulation edge (t,s). Otherwise, it will follow that there is an augmentable $s_d - t_d$ path in the dual network, which contradicts the condition of the lemma.

Here, we need to point out that Lemma 5.A1 does not exclude the possibility of augmenting $s_d - t_d$ paths which contain *both* backward and forward edges. It only states that an augmentation of $s_d - t_d$ paths in the dual circulation network is *always possible.* □

The next lemma guarantees the absence of augmentable s–t paths in the optimised network.

Lemma 5.A2 *If no augmentable $s_d - t_d$ path exists in the dual network, an augmentation of an $s_d - t_d$ path in the dual circulation network, results in the absence of augmentable s–t paths in the optimised network.*

Proof Consider the dual network in which there are no augmentable $s_d - t_d$ paths, e.g. the dual network, immediately after the first stage (Figure 5.A3).

In the dual network, define a set of nodes A that can be reached through augmentable paths from node s_d. An augmentable path is a path along which there are no fully saturated forward edges or empty backward edges.

The set $\overline{A}$ includes all nodes that cannot be reached from s_d through augmentable paths ($A \cup \overline{A} = V$ and $A \cap \overline{A} = \emptyset$). The new sink t_d does not belong to set A because, according to the condition of the lemma, there is no augmentable $s_d - t_d$ path in the dual network. The original source s belongs to set A because, according to Lemma 5.A1, there is always an augmentable path in the dual circulation network, therefore, there is always an augmentable path from the new source s_d to the original source s.

However, the original sink t does not belong to set A, because, Lemma 5.A1 states that there is always an augmentable path in the dual circulation network from the new source s_d to the new sink t_d, which includes the circulation edge (t,s). This means that there is always an augmentable path from the original sink t to the new sink t_d in the dual network. If the original sink t was reachable from s_d through an augmentable path (if the sink t belonged to set A), the new sink t_d would also be reachable from s_d and this would contradict the condition of the lemma that there is no augmentable $s_d - t_d$ path in the dual network. Consequently, the original sink t belongs to the set $\overline{A}$.

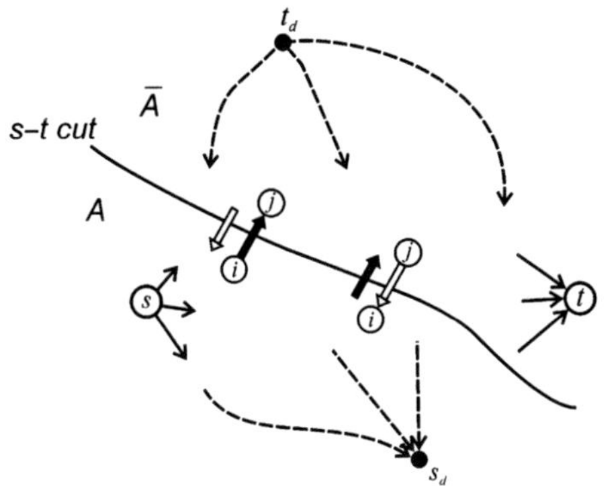

Figure 5.A3 An $s - t$ cut, defined by the reachability of nodes from the new sink s_d.

Because a node in the network can either belong to set A or to set $\overline{A}$ and no node can simultaneously belong to both sets, $A \cap \overline{A} = \varnothing$. In addition, $A \cup \overline{A} = V$ holds, where V is the set of all nodes in the network, and as it was demonstrated earlier, $s \in A$ and $t \in \overline{A}$. As a result, the sets A and $\overline{A}$ define an $s-t$ cut $A - \overline{A}$ in the optimised network.

Edges which cross the cut from set A to set $\overline{A}$ are fully saturated with flow while edges which cross the s-t cut in the opposite direction, from set $\overline{A}$ to set A, are empty (Figure 5.A3). Indeed, if a forward edge (i,j) is not fully saturated with flow or a backward edge (i,j) is not fully empty, this will make edge (i,j) augmentable and node j will be reachable from s_d, because node i belongs to set A and is therefore reachable from s_d through an augmentable path. Because, node j is in set $\overline{A}$, it is not reachable through an augmentable path from the new source s_d and therefore all forward edges crossing the $s - t$ cut must be fully saturated and all backward edges must be empty (Figure 5.A3).

Next, note that after each augmentation of an $s_d - t_d$ path in the dual circulation network, the augmented path cannot possibly cross the $s - t$ cut through a directed edge from a node in set A towards a node in set $\overline{A}$, because all forward edges crossing the $s - t$ cut are fully saturated and all backward edges are empty – therefore, none of these edges can be augmented.

The augmented $s_d - t_d$ path however may cross the $s - t$ cut through an edge whose starting node j is in set $\overline{A}$ and whose end node i is in set A (Figure 5.A4). To do so, the augmented path must either enter the A set through a fully saturated backward edge, decreasing its flow, or through an empty forward edge, increasing its flow (Figure 5.A4). This is because there are no other kind of edges crossing the $s - t$ cut. After entering the set A, in order to reach the new sink t_d, the augmented $s_d - t_d$ path must come back and cross the $s - t$ cut again. Now, except the edge through which the $s_d - t_d$ path entered the A set, there are no other augmentable edges crossing the $s - t$ cut along which the $s_d - t_d$ path can return to set $\overline{A}$. The only possibility for the $s_d - t_d$ path *to return to set* $\overline{A}$ *is through the same edge through which it has entered set A*. This means, however, that the augmented

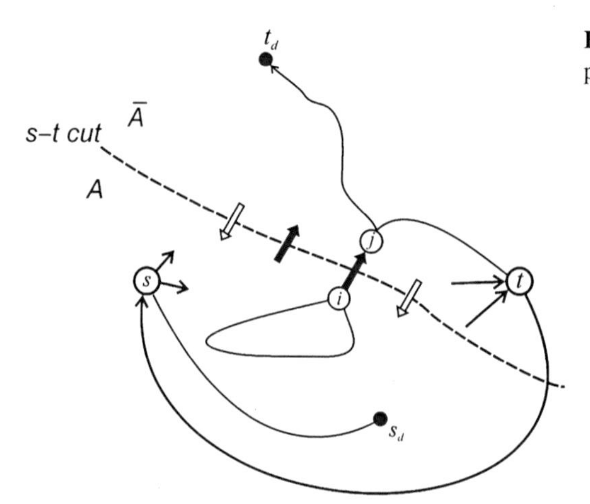

Figure 5.A4 An augmented $s_d - t_d$ path in the dual circulation network.

$s_d - t_d$ path will restore the state of the edge through which it entered set A and leave it in the way it was before entering the set A (Figure 5.A4). This is because the bottleneck flow with which the path is augmented is the same for all edges along the path. As a result, if edge (j,i), through which the path entered set A, was a fully saturated backward edge, it will be left a fully saturated backward edge after the augmented path leaves set A and enters set $\overline{A}$. If edge (j,i), through which the path entered set A, was an empty forward edge, it will be left empty after the augmented path leaves set A and enters set $\overline{A}$. The augmented $s_d - t_d$ path must certainly enter the $\overline{A}$ set because the new sink t_d belongs to the $\overline{A}$ set. As a result, the number of times the path may cross the $s - t$ cut from $\overline{A}$ to A and from A to $\overline{A}$ is an *even number*. The state of the edges crossing the $s - t$ cut from set A to set $\overline{A}$ will remain exactly the same as it was before the augmentation of the $s_d - t_d$ path in the dual circulation network – fully saturated forward edges and fully empty backward edges. Consequently, there will be no augmentable $s - t$ path in the original network, after the augmentation of the $s_d - t_d$ path in the dual circulation network. This proves the lemma. □

Now Theorem 5.3 can be proved easily. The first stage of augmentation of $s_d - t_d$ paths in the dual network ends when no more augmentable $s_d - t_d$ paths can be found. Let the maximum throughput flow, with which the dual network has been augmented at the end of the first stage, be $q_d^{\max}$. The remaining flow to be purged from the edges connecting the new source with the excess nodes is given by

$$q_{\text{rem}} = \sum_{i=1}^{M1} ef_i - q_d^{\max} \tag{5.A1}$$

According to Lemma 5.A1, at this stage, there is always an augmentable $s_d - t_d$ path in the dual circulation network, which reduces the remaining flow q_{rem}, given by Eq. (5.A1). Because the augmentable $s_d - t_d$ path in the dual circulation network always includes the circulation edge (t,s), each augmentation of an $s_d - t_d$ path subtracts an equal amount of flow from the edges going out of the source s and from the edges going into the sink t. According to Lemma 5.A2, after each augmentation of an $s_d - t_d$ path in the dual circulation network, there will be no augmentable $s - t$ path in the original network. It is guaranteed that by augmenting $s_d - t_d$ paths in the dual circulation network, the flow q_{rem} which has remained in the edges (s_d, e_i), $i = 1, \ldots, M1$, connecting the new source s_d with the excess nodes will disappear.

The augmentation of $s_d - t_d$ paths is repeated until the entire excess flow q_{rem} disappears. Consequently, the new throughput flow $Q'_{\max}$ after choking the flows along the edges, will be

$$Q'_{\max} = Q_{\max} - \sum_{i=1}^{M1} ef_i + q_d^{\max} \tag{5.A2}$$

where $Q_{\max}$ is the maximum throughput flow before the choking of the edges. The flow given by equation 5.A2 will also be the new maximum throughput flow, because, according to Lemma 5.A2, there will be no augmentable $s - t$ path in the network. This finally proves the theorem. □

6 The Dual Network Theorem for Static Flow Networks and Its Application for Maximising the Throughput Flow

6.1 Analysis of the Draining Algorithm for Maximising the Throughput Flow in Static Networks

A large number of algorithms have been proposed for solving the maximum flow problem, by starting from an empty network and gradually saturating the edges with flow, until the whole network is fully saturated with flow (Dinic, 1970; Edmonds and Karp, 1972; Elias et al., 1956; Ford and Fulkerson, 1956; Goldberg and Tarjan, 1988; Hochbaum, 2008; Karzanov, 1974; Sleator and Tarjan, 1980). These algorithms have been comprehensively reviewed in Hu (1969), Tarjan (1983), Ahuja et al. (1993) and Asano and Asano (2000).

Recently, a different concept for maximising the flow in static flow networks was proposed by Dong et al. (2009), by starting from a network with fully saturated edges. A backward edge connecting the sink with the source is also introduced. The algorithm drains flow from paths connecting deficit nodes and excess nodes, until all nodes become balanced and a maximum throughput flow is attained. The node balancing is conducted along the shortest augmentable paths between excess and deficit nodes. The obtained drained network flow is subsequently subtracted from the flows of the fully saturated edges in the original network, to obtain the maximum throughput flow in the original network (Dong et al., 2009).

The definition of excess and deficit nodes given by Dong et al. (2009) is as follows. For a network with edges fully saturated with flow, if the sum of capacities of all edges going into a node e (different from the source and the sink) is greater than the sum of capacities of all outgoing edges, the node is said to be an *excess node*. The amount of excess flow ef at an excess node e is given by

$$ef = \sum_{i \in \delta +} c(i,e) - \sum_{j \in \delta -} c(e,j) > 0 \tag{6.1}$$

Conversely, if the sum of capacities of all edges going into a node d (different from the source and the sink) is smaller than the sum of capacities of all outgoing

Flow Networks. DOI: http://dx.doi.org/10.1016/B978-0-12-398396-1.00006-4

edges, the node is said to be a *deficit node*. The amount of deficit flow df in the deficit node is

$$df = \sum_{i \in \delta +} c(i, d) - \sum_{j \in \delta -} c(d, j) < 0 \qquad (6.2)$$

Finally, if the sum of capacities of all edges going into a particular node is equal to the sum of capacities of all edges going out of the node, the node is referred to as a *balanced node*. The amount of excess/deficit flow at a balanced node is zero.

Unfortunately, the algorithm proposed in Dong et al. (2009) *is fundamentally flawed* and, as a result, it yields suboptimal solutions. This can be demonstrated by the simple counterexample network from Figure 6.1A, which contains one deficit node (node 2) and one excess node (node 1). The labels on the edges of the network in Figure 6.1A are their flow capacities. According to Dong et al. (2009), the flooding network for the network in Figure 6.1A is the one in Figure 6.1B. Following the draining algorithm, after augmenting the flooding network in Figure 6.1B with 10 units flow, along the shortest path $(s',2,6,t,s,1,t')$ and subtracting the network flows from the flows of the fully saturated edges in Figure 6.1A, an incorrect value for the maximum throughput flow is obtained, equal to 90 units (Figure 6.1C). However, the maximum throughput flow in the network from Figure 6.1A is 100 units, and the correct edge flows, yielding this throughput flow are shown in Figure 6.1D.

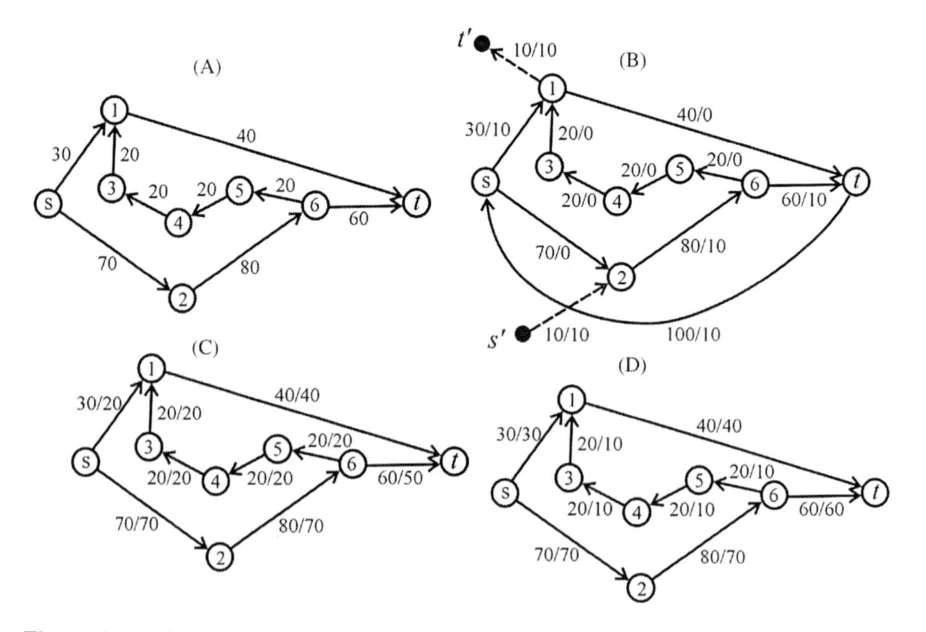

Figure 6.1 A simple counterexample network, for which the draining algorithm in Dong et al. (2009) produces an incorrect result.

This simple counterexample shows that there are networks for which the draining algorithm proposed in Dong et al. (2009) fails to determine the correct maximum throughput flow. In what follows, we show that a *correct algorithm for determining the maximum throughput flow in a network whose edges are fully saturated with flow* must initially include a node balancing stage which *does not include the circulation edge (t,s)*. The purpose is to make the maximum possible redistribution of excess and deficit flows between the excess and deficit nodes, *before* draining the excess flow from the network. However, the draining algorithm proposed in Dong et al. (2009) works on a network which always includes the circulation edge (*t*,*s*). This is the reason why, for a number of networks, this algorithm yields suboptimal solutions.

Another drawback of the algorithm proposed in Dong et al. (2009) is the increased tendency of leaving directed loops with parasitic flow in the optimised network, which is *highly undesirable*. This drawback can be illustrated with the network in Figure 6.2A, where the labels stand for edge capacities.

According to the method proposed by Dong et al. (2009), in the network from Figure 6.2A, there are no excess and deficit nodes, so that no flow is removed from any edge in the fully saturated with flow network. The edge flows maximising the throughput flow are according to Figure 6.2B. The first number 'c' of the labels 'c/f' on the edges stands for the edge capacity and the second number 'f' stands for the actual flow, along the edge. As a result, the maximum throughput flow of 30 units (Figure 6.2B) is obtained correctly, but the directed cyclic path (3,2,4,3) with parasitic flow of 10 units, still remains.

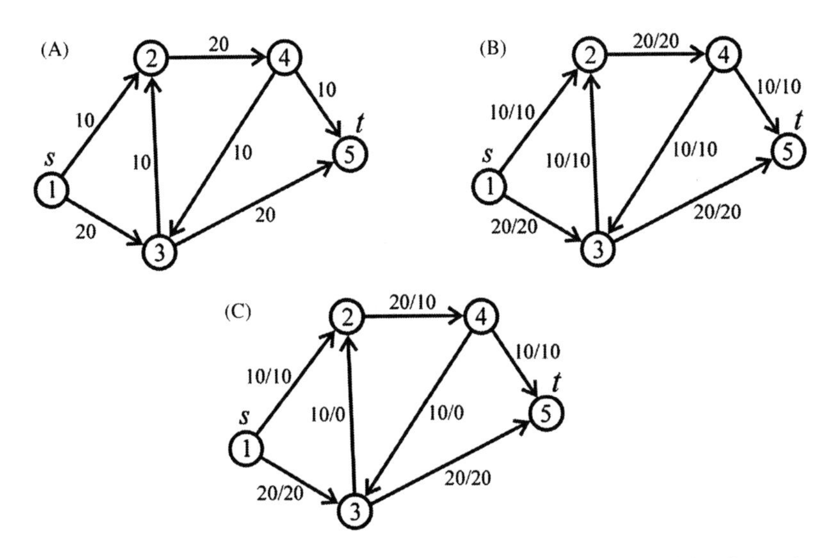

Figure 6.2 (A) Initial network; (B) A network with a parasitic loop (3,2,4,3) of 10 units flow, after the throughput flow maximisation, by the method proposed in Dong et al. (2009); (C) The same network can be optimised to have the same maximum throughput flow of 30 units, with no parasitic loops of flow.

If the classical Edmonds and Karp shortest-path algorithm (1972) was used for maximising the throughput flow, by starting from a network with empty edges (Figure 6.2A), the shortest path (1,3,5) will be augmented first, with 20 units of flow. This path augmentation will be followed by augmenting the path (1,2,4,5) with 10 units of flow. The result is the network in Figure 6.2C, carrying the maximum throughput flow of 30 units. No parasitic loops of flow exist in the optimised network from Figure 6.2C.

6.2 The Dual Network Theorem for Static Flow Networks

The analysis in Section 6.1 poses the natural question about the correct algorithm which exploits the concept of starting from a network with fully saturated edges.

Here, by using Theorem 5.3, a correct algorithm can be developed for determining the maximum throughput flow in a static network, whose edges are initially fully saturated with flow. Theorem 5.3 applies to a network where the maximum throughput flow Q_{max} has been set (e.g. by any of the classical algorithms for maximising the flow in a network) and in which several edges ($i = 1,2, \ldots ,K$) fail simultaneously. The failed edge i is directed from node e_i to node d_i. The flows through the edges before their failures are $f(e_i,d_i)$, $i = 1,2, \ldots ,K$.

It is important to note that the definition of excess nodes and deficit nodes given in Chapter 5 with Eqs. (5.1) and (5.2) is *different* from the definition of excess and deficit nodes given by Dong et al. (2009) with Eqs. (6.1) and (6.2). The definition given by Eqs. (6.1) and (6.2) is based on the capacities of the ingoing and outgoing edges, while the definition by Eqs. (5.1) and (5.2) is based on the actual ingoing and outgoing flows of a node. The definition by Eqs. (5.1) and (5.2) is related to networks with an initial *feasible network flow*, where several edges fail simultaneously.

Consider now the static network in Figure 6.3, whose edges are fully saturated with flow and where excess and deficit nodes will normally exist, in the sense of the definition given by Eqs. (6.1) and (6.2). Suppose that nodes e_i ($i = 1, \ldots ,M$)

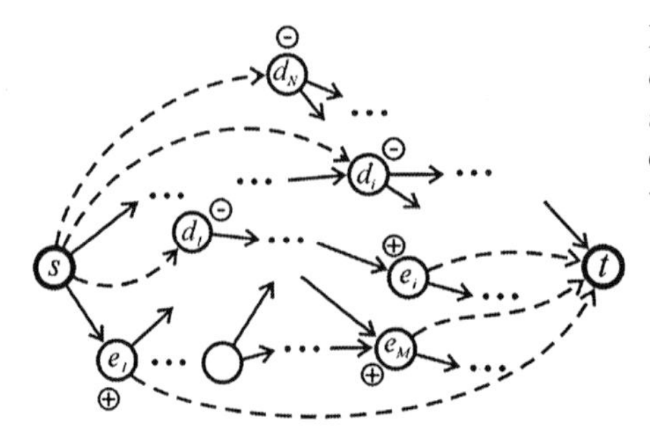

Figure 6.3 By using ghost edges (the dashed lines), the excess and deficit nodes in any static flow network with fully saturated edges can be transformed into a network where all internal nodes are balanced.

are excess nodes and nodes d_i ($i = 1, \ldots ,N$) are deficit nodes (the rest of the internal nodes are balanced nodes). If imbalanced nodes are present, the network flow *is not feasible at the start*, and the purpose is to make it feasible and maximise the throughput flow, by an appropriate flow redistribution between excess and deficit nodes and by draining flow from the network.

Now let us connect all excess nodes with the sink t, by *ghost edges* directed to the sink, with flow capacities equal to the amount of excess at the excess nodes (Figure 6.3). Simultaneously, let us also connect the source s with all deficit nodes by ghost edges directed towards the deficit nodes, with flow capacities equal to the deficit flow at the deficit nodes. This operation will transform the original network into a network where all internal nodes are balanced (Figure 6.3). The ghost edges are drawn by dashed lines.

In the network in Figure 6.1A, there is a single excess node (node 1) with excess flow equal to 10 units and a single deficit node (node 2), with a deficit flow equal to 10 units. Node 1 can be connected to the sink t by a ghost edge directed towards the sink, with a flow capacity of 10 flow units. The source s can also be connected to the deficit node 2, by a directed ghost edge with a flow capacity of 10 units. These operations transform the network in Figure 6.1A into the network in Figure 6.4A, where all internal nodes are balanced. The ghost edges in Figure 6.4A are drawn by dashed lines.

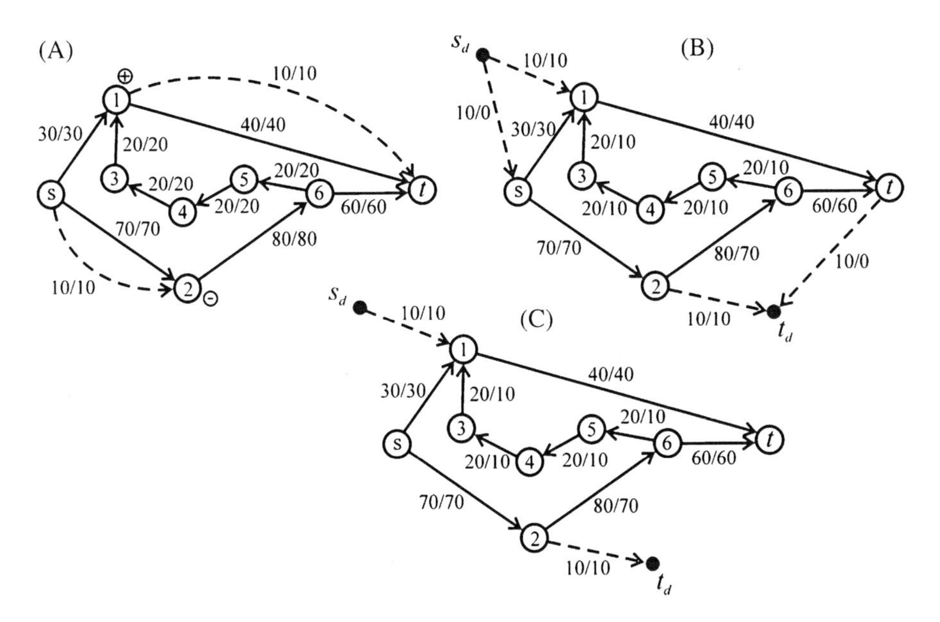

Figure 6.4 (A) The network from Figure 6.1A, transformed into a network with balanced internal nodes, by introducing ghost edges; (B) A dual network of the network in Figure 6.4A; (C) A dual network of the network in Figure 6.4A, after removing (cancelling) edges (s_d,s) and (t,t_d), characterised by the same capacity.

After the introduction of ghost edges, the network flow becomes feasible everywhere. In other words, the flow conservation at the nodes and the capacity constraints of the edges are honoured in the network. The maximum throughput flow $Q_{max,g}$ in the network, which includes all ghost edges, can be determined immediately – it is equal to the sum of capacities of all edges going into the sink.

$$Q_{\mathrm{max,\,g}} = \sum_i c(i,t) + \sum_k c_{\mathrm{g}}(k,t) \tag{6.3}$$

In Eq. (6.3), $\sum_{\mathrm{i}} c(i,t)$ is the sum of capacities of all real edges going into the sink t and $\sum_{\mathrm{k}} c_g(k,t)$ is the sum of capacities of all ghost edges going into the sink. For the network in Figure 6.4A, e.g., $Q_{\mathrm{max,g}} = 110$ units of flow.

Considering the flow conservation in the network, the maximum throughput flow can also be presented as a sum of the capacities of all edges coming out of the source,

$$Q_{\mathrm{max},\,g} = \sum_i c(s,i) + \sum_m c_{\mathrm{g}}(s,m) \tag{6.4}$$

where $\sum_{\mathrm{i}} c(s,i)$ is the sum of capacities of all real edges coming out of the source s and $\sum_{\mathrm{m}} c_{\mathrm{g}}(s,m)$ is the sum of capacities of all ghost edges coming out of the source s.

Now, suppose that all ghost edges in the network from Figure 6.4A 'fail' simultaneously. Because, by saturating all edges with flow, the throughput flow has essentially been maximised, the problem is now reduced to a problem related to Theorem 5.3, dealing with a network with feasible edge flows and a maximised throughput flow, in which several edges (the ghost edges) have been fully choked. Because the conditions of Theorem 5.3 are fulfilled, it can be applied for determining the new maximum throughput flow in the network, after the 'failure' of all ghost edges. Because, removing ghost edges essentially transforms the network in Figure 6.4A into the original network from Figure 6.1A, applying Theorem 5.3 essentially determines the new maximum throughput flow in the original network from Figure 6.1A.

Theorem 5.3 is related to the dual network. Accordingly, a new source s_d is introduced, connected to the excess nodes, and a new sink t_d is introduced connected with the deficit nodes (Figure 6.4B). Because the original source s is connected to all deficit nodes, there is a single edge (s_d,s) connecting the new source s_d with the original source s, whose flow capacity is equal to the sum of the deficit flows at all deficit nodes. Because the original sink t is connected to all excess nodes, there is also a single edge (t,t_d) connecting the original sink t with the new sink t_d, whose flow capacity is equal to the sum of the excess flows at all excess nodes.

Paths connecting the new source s_d with the new sink t_d are augmented, until no more augmentable paths can be found. At this point, suppose that the sum of the flows along the edges connecting the new source s_d with the excess nodes is q_d^{max}. The quantity q_d^{max} is essentially the maximum throughput flow in the dual network. According to Theorem 5.3, the new maximum throughput flow after the failure of the ghost edges is equal to the maximum throughput flow $\sum_i c(i,t) + \sum_k c_g(k,t)$

before the failure of the ghost edges minus the sum of the flows $\sum_i c_{gi}$ through the ghost edges plus the maximum throughput flow $q_d^{\max}$ in the dual network. Because the ghost edges do not share common nodes, the total excess flow at the excess nodes, after the failure of the ghost edges, is equal to the sum of the flows through the ghost edges. The maximum throughput flow in the original network (without ghost edges) becomes

$$Q_{\max} = \sum_i c(i,t) + \sum_k c_g(k,t) - \sum_i c_{gi} + q_d^{\max} \tag{6.5}$$

where $\sum_i c_{gi}$ is the sum of capacities of all ghost edges. The term $\sum_i c_{gi}$, however, can also be presented as

$$\sum_i c_{gi} = \sum_k c_g(k,t) + \sum_m c_g(s,m) \tag{6.6}$$

where $\sum_k c_g(k,t)$ is the sum of capacities of all ghost edges going to the sink t and $\sum_m c_g(s,m)$ is the sum of capacities of all ghost edges coming out of the source s. Expression (6.6) holds because each ghost edge is either coming out of the source s or going into the sink t and no ghost edge connects directly the source s and the sink t. Substituting Eq. (6.6) into Eq. (6.5) yields

$$Q_{\max} = \sum_i c(i,t) - \sum_m c_g(s,m) + q_d^{\max} \tag{6.7}$$

for the maximum throughput flow in the original network.

Equation (6.7) has been derived from Theorem 5.3 and is, in effect, a new fundamental result for the maximum throughput flow in a static flow network.

If Eq. (6.4) was used as an expression for the maximum throughput flow in the network with ghost edges, the equation

$$Q_{\max} = \sum_i c(s,i) + \sum_m c_g(s,m) - \sum_i c_{gi} + q_d^{\max} \tag{6.8}$$

will be obtained from Theorem 5.3 for the maximum throughput flow in the original network. After the substitution of Eq. (6.6), Eq. (6.8) yields

$$Q_{\max} = \sum_i c(s,i) - \sum_k c_g(k,t) + q_d^{\max} \tag{6.9}$$

for the maximum throughput flow $Q_{\max}$.

Equation (6.9) is an alternative equation for the new maximum throughput flow. In Eq. (6.9), $\sum_k c_g(k,t)$ is the sum of the excess flow at all excess nodes. In Eq. (6.7), $\sum_m c_g(s,m)$ is the sum of the deficit flow at all deficit nodes. The results Eqs. (6.9) and (6.7) can then be summarised in the next two theorems.

Theorem 6.1 (Dual network theorem for static flow networks) *The maximum throughput flow in any static flow network is equal to the sum of capacities of all edges coming out of the source minus the total excess flow at the excess nodes plus the maximum throughput flow in the dual network (Todinov 2012a).*

Theorem 6.2 (Dual network theorem, alternative formulation) *The maximum throughput flow in any static flow network is equal to the sum of the capacities of all edges going into the sink minus the total deficit flow at the deficit nodes plus the maximum throughput flow in the dual network (Todinov 2012a).*

The dual network theorem establishes a very important link between the maximum throughput flow in a static flow network and the maximum throughput flow in its dual network. *The dual network theorem replaces the task of determining the maximum throughput flow in the original network with the task of determining the maximum throughput flow in its dual network.* In the case of few unbalanced nodes in the original network, determining the maximum throughput flow in the dual network is a task which is significantly easier than the task of determining the maximum throughput flow in the original network. The main reason for this important tradeoff is that *the dual network is already saturated with flow*. As a result, in case of fewer imbalanced nodes, fewer path augmentations are normally needed for maximising the throughput flow in the dual network. Equations (6.7) and (6.9) can be checked easily. Excluding $q_d^{\max}$ from the system of Eqs. (6.7) and (6.9) results in

$$\sum_i c(s,i) - \sum_k c_g(k,t) = \sum_i c(i,t) - \sum_m c_g(s,m) \tag{6.10}$$

which is the flow conservation law in static networks: the sum of capacities of all edges coming out of the source minus the total excess flow at the excess nodes is always equal to the sum of capacities of all edges going into the sink minus the absolute value of the total deficit flow at all deficit nodes.

Applying Theorem 6.1 and Eq. (6.9) to the network in Figure 6.4A yields $Q_{\max} = (70 + 30) - 10 + 10 = 100$ flow units because the maximum throughput flow in the dual network is $q_d^{\max} = 10$. Applying Theorem 6.2 and Eq. (6.7) to the network in Figure 6.4A also yields $Q_{\max} = (60 + 40) - 10 + 10 = 100$ for the maximum throughput flow.

For static flow networks, Theorem 6.1 or Theorem 6.2 should be used for maximising the throughput flow in cases where the maximum throughput flow is of interest *and not the actual edge flows*. This is indeed the case in building fast discrete-event simulators for determining the performance of repairable flow networks, where the maximum throughput flow after an edge failure and not the actual edge flows is important.

In future research, Theorems 6.1 and 6.2 could be instrumental in proving new fundamental results about static flow networks.

The edge flows corresponding to the maximum throughput flow in a network can be determined by the two-stage procedure outlined in Chapter 5. The first stage consists of maximising the throughput flow in the dual network with a new source s_d and new sink t_d, by augmenting the shortest augmentable paths, until no more augmentable paths can be found.

The second stage follows the first stage and consists of path augmentation in the *dual circulation network*. This is obtained by including a circulation edge (t,s) in the dual network, with capacity $c(t,s)$, equal to the sum of capacities of the edges going out of the original source s or the sum of capacities of the edges going into the sink t, whichever is smaller:

$$c(t,s) = \min\left\{\sum_i c(s,i); \quad \sum_k c(k,t)\right\}$$

The initial edge flows, with which the augmentation of the dual circulation network starts, are the edge flows obtained after maximising the throughput flow in the dual network.

The throughput flow in the as-defined dual circulation network is maximised by augmenting the shortest feasible $s_d - t_d$ paths, until all outgoing edges from the new source s_d are fully saturated with flow. This is always guaranteed, and a proof of this fact has already been given in the appendix to Chapter 5.

Going back to Figure 6.4C, because the deficit and excess flow has been eliminated during the flow redistribution stage, all edges coming out of the new source s_d are fully saturated with flow and no flow augmentation in the dual circulation network is needed. The edge flows which correspond to the new maximum throughput flow are shown in Figure 6.4C.

6.3 Improving the Average Running Time of Maximising the Throughput Flow in the Dual Network

Theorems 6.1 and 6.2 suggest an algorithm for maximising the throughput flow in a static flow network. The average running time of maximising the flow in the dual network can be improved significantly, if the two-stage algorithm discussed in Chapter 4 is used. The essence of this algorithm is to augment the shortest $s_d - t_d$ paths in the dual network which contain only backward edges, except the first edge which comes out of the source s_d and the last edge which goes to the new sink t_d. After no more augmentable paths with backward edges belonging to the original network can be found, the second stage of the algorithm includes augmenting the shortest $s_d - t_d$ paths that may include both forward and backward edges belonging to the original network.

The augmentation of the shortest $s_d - t_d$ paths is the standard Ford–Fulkerson type of augmentation.

Very often, the throughput flow in the dual network is maximised during the first stage of $s_d - t_d$ paths augmentation, which includes only $s_d - t_d$ paths with backward common edges with the original network. Often, there is either no need of a second stage or there is a need of augmentations of $s_d - t_d$ paths, which include both backward and forward common edges with the original network, but their number does not significantly exceed the number of augmentations during the first stage.

Suppose that the maximum throughput flow in the dual network is found during the first stage only (augmenting paths with backward common edges only with the original network), or during the first stage and after augmenting a comparable number of $s_d - t_d$ paths which have both backward and forward common edges with the original network. In this case, according to Chapter 4, the average running time is $O(m^2)$, where m is the number of edges in the original network.

Here, we must point out that an essential part of the proposed method is that *the shortest $s_d - t_d$ paths in the dual network must be augmented.* Otherwise, for some networks, the algorithm could have an unacceptably low running time. This can be readily demonstrated on the network from Figure 6.5, where there are four internal imbalanced nodes: the excess node 5, with excess flow equal to $2 \times 10^9 - 1$, the excess node 4, with excess flow equal to 1, the deficit node 2, with deficit flow equal to $-(2 \times 10^9 - 1)$, and the deficit node 3, with deficit flow equal to -1.

Initially, all edges of the network in Figure 6.5 are fully saturated with flow. The first identified feasible path $(s_d,4,3,t_d)$ in the dual network can be augmented with 1 unit of flow. As a result, edge (3,4) will become empty. Suppose that the next selected augmentable path in the dual network is the augmentable path $(s_d,5,3,4,2,t_d)$. This path can be augmented with the maximum possible flow of 1 unit. As a result, edge (3,4) will be fully saturated again. Now suppose that the next selected augmentable path is $(s_d,5,4,3,2,t_d)$, which is augmented with the maximum possible flow of 1 unit. If the sequence of augmentable paths $(s_d,5,3,4,2,t_d)$ and $(s_d,5,4,3,2,t_d)$ is repeated, there will be 2×10^9 path augmentations before the dual network is fully saturated! However, if the shortest augmentable paths were used, after the initial augmentation of path $(s_d,4,3,t_d)$, the maximum throughput flow could be found after only two path augmentations. The shortest path $(s_d,5,3,2, t_d)$ is augmented with the maximum possible 10^9 units flow, followed by augmenting the shortest path $(s_d,5,4,2,t_d)$ with the maximum possible flow of $10^9 - 1$ units. After the saturation of the dual network, a maximum throughput flow equal to 1 will be set in the original network.

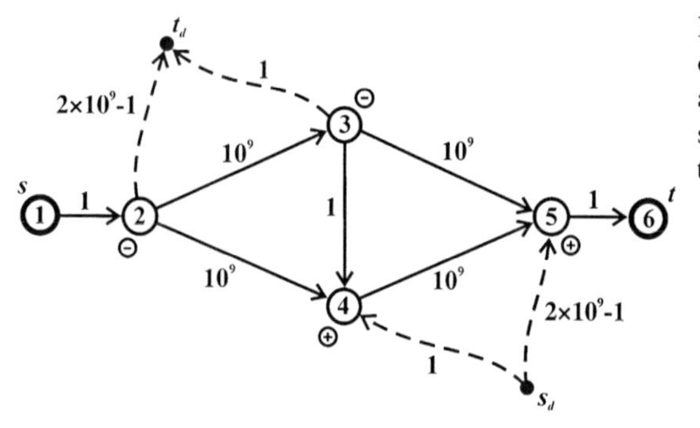

Figure 6.5 A slow convergence of the solution could occur if arbitrary paths and not the shortest paths are augmented in the dual network.

6.4 Application of the Dual Network Theorem for Determining the Maximum Throughput Flow in a Static Flow Network

The algorithm for determining the maximum throughput flow in a static network by 'failure' of ghost edges can be simplified significantly by noticing that during the flow redistribution stage (maximising the throughput flow in the dual network), the source s or sink t in the original network cannot be visited by the augmented path (Figure 6.6A).

Indeed, suppose that such a possibility exists. Without loss of generality, let us examine *the first* augmented path that starts at the new source s_d and at some point visits the original source s.

In order to reach the new sink t_d, the augmented path must get out of the original source s. Because, according to our assumption, this is *the first* augmentation path that visits the source s, all outgoing edges from the source s, except the one from which the source s has been visited by the augmented path, must be fully saturated with flow. There is no possibility to augment a path passing through the original source s, because no flow can be increased along any of the edges coming out of the source. Similar considerations apply to the first augmented path visiting the original sink t.

Consequently, during the flow redistribution stage in the dual network (Figure 6.6A), no augmentable path can possibly visit the source s or the sink t. As a result, at the end of the first stage of the algorithm, the edge connecting the new source s_d and the original source s and the edge connecting the original sink t and the new sink t_d will always remain empty. Therefore, the dual network can be simplified significantly, by ignoring the edge connecting the new source s_d and the original source s and the edge connecting the original sink t and the new sink t_d (Figure 6.6A). These are included only during the second stage – the draining stage of the dual circulation network (Figure 6.6B).

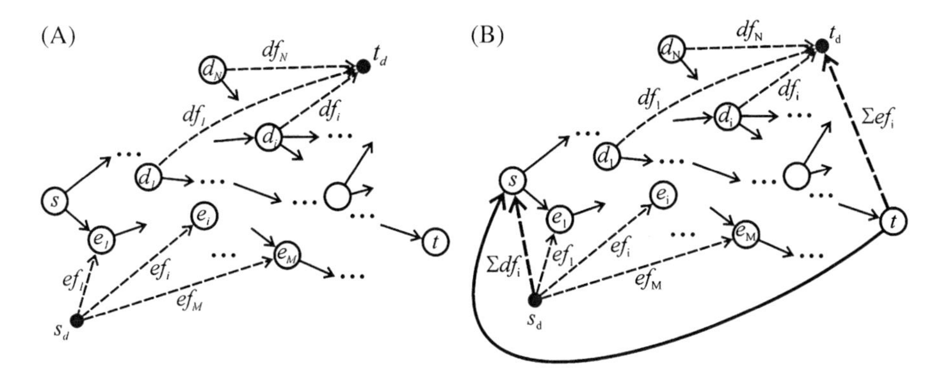

Figure 6.6 (A) Flow re-distribution stage in the dual network and (B) Flow draining stage in the dual circulation network.

During the draining stage, a circulation edge (t,s) is introduced (Figure 6.6B), with capacity equal to either the sum of capacities of the edges coming out of the source s or the sum of capacities of all edges going into the sink t, whichever is smaller. By using the circulation edge (t,s), the flow can be augmented directly through the path (s_d,s,t,t_d). If there are two edges (s_d,s) and (t,t_d) connecting the new source with the old source and the old sink with the new sink, the edge with the smaller flow capacity, either edge (s_d,s) or edge (t,t_d), can then be removed from the network without affecting the procedure. The capacity of the other edge is corrected by subtracting from it the smaller capacity of the removed edge. If the sum of the excess flow at the excess nodes is equal to the absolute value of the sum of the deficit flow at the deficit nodes, both edges (s_d,s) and (t,t_d) can be removed from the network. This is exactly the case for the network in Figure 6.4B. After removing edges (s_d,s) and (t,t_d), the resultant (augmented) network is shown in Figure 6.4C.

In summary, the algorithm for maximising the throughput flow in any static network has several distinct stages:

- **Flow redistribution stage.** Maximises the throughput flow in the dual network (Figure 6.6A). If all edges coming out of the new source s_d are fully saturated with flow, the new maximum throughput flow in the original network has been found. If only the maximum throughput flow is needed and not the edge flows, the algorithm ends with this stage. Determining the maximum throughput flow, without necessarily determining the corresponding edge flows is important while comparing a large number of network designs and selecting the design with the largest throughput flow. In this case, the exact edge flows corresponding to the maximum throughput flow are not necessary. If the throughput flow, as well as the edge flows are needed, another stage is necessary.

  If there is at least a single edge coming out of the new source s_d, which has not been fully saturated with flow, the algorithm continues with the next, draining stage.
- **Draining stage**. Maximises the throughput flow in the dual circulation network, until all edges coming out of the new source s_d are fully saturated with flow (Figure 6.6B).
- **Eliminating directed loops of flow**. Identifies and removes directed loops of flow in the network with maximum throughput flow, by applying Algorithm 4.1. During this operation, the maximum throughput flow remains unaffected.

  The algorithm can be detailed by the following basic steps.

Algorithm 6.1

Initial state of the network – all edges are fully saturated with flow, to their full capacity.
All deficit nodes and excess nodes in the network are known before the start of the algorithm.

1. **Flow redistribution stage** (maximising the throughput flow in the dual network).
   1.1. Create a dual network by linking the new source s_d with all excess nodes, through edges with capacities equal to the amount of excess flows at the excess nodes. Similarly, all deficit nodes are linked with the new sink through edges with capacities equal to the absolute value of the deficit flows at the deficit nodes.

   1.2. **If** (there are no deficit and excess nodes) **then**

The maximum throughput flow in the network is equal to the sum of capacities of the edges coming out of the source s. **Go to step 3.**

1.3. **If** (there are both deficit and excess nodes) **then**

Augment the shortest $s_d - t_d$ paths in the dual network until no more augmentable $s_d - t_d$ paths can be found. Record the total amount of throughput flow $q_d^{\max}$ in the dual network.

1.4. The maximum throughput flow in the original network is equal to the sum of capacities of all edges coming out of the source s, minus the total amount of excess flow at all excess nodes plus the maximum throughput flow $q_d^{\max}$ in the dual network.

If (only the maximum throughput flow is required and not the edge flows) **then Stop.**

If (all edges coming out of the new source are fully saturated) **then Stop**. There is no need of a flow draining stage.

2. Draining stage. (maximises the throughput flow in the dual circulation network)

2.1. Introduce a $t-s$ circulation edge with capacity equal to either to the sum of capacities of the edges going into the sink t or the sum of the capacities of the edges coming out of the source s, whichever is smaller (Figure 6.6B).

Find the sum Δ of the excess flow at the excess nodes and the deficit flow (taken with minus sign) at the deficit nodes.

If the sum Δ is positive ($\Delta > 0$), connect the old sink t with the new sink t_d, by a directed edge with capacity Δ. If the sum is negative ($\Delta < 0$), connect the new source s_d with the original source s by a directed edge with capacity $|\Delta|$.

2.2. Augment the shortest $s_d - t_d$ paths in the dual circulation network, until all edges coming out of the new source are fully saturated with flow and no more augmentable $s_d - t_d$ paths can be found. (This is guaranteed, according to Theorem 5.3.)

If the edge flows corresponding to the maximum throughput flow are of interest, then an extra stage, eliminating loops of parasitic flow must be executed.

3. Extra stage. Eliminating the directed loops of flow.

This stage proceeds according to the detailed algorithm described in Chapter 4.

Note that if no deficit and excess nodes exist simultaneously, the flow redistribution stage consisting of augmenting the shortest $s_d - t_d$ path in the dual network is skipped and the algorithm continues with the flow draining stage.

The proof of Theorem 5.3 is essentially the proof of optimality for the proposed algorithm. This proof guarantees that the algorithm will always determine the maximum throughput flow and, during the draining stage, a state will be reached where all edges coming out of the new source will be fully saturated with flow and the algorithm will terminate.

6.4.1 Solved Examples

The proposed algorithm will be illustrated by the network in Figure 6.7A. In the network, there is a single excess node (node 5) and a single deficit node (node 8). The excess flow at the excess node is $ef_5 = 12$. The deficit flow at the deficit node is $df_8 = -9$. In the dual network (Figure 6.7B), edge $(s_d,5)$ connects the new source s_d with the excess node 5 and edge $(8,t_d)$ connects the deficit node 8 with the new sink t_d. The capacity of connecting edge $(s_d,5)$ is 12 flow units – equal to the excess flow at node 5. The capacity of connecting edge $(8,t_d)$ is 9 units – equal to the absolute value of the deficit flow at node 8.

The maximum throughput flow in the dual network is attained by augmenting the shortest path $(s_d,5,6,8,t_d)$ with 4 units of flow. After this operation, there are no more augmentable paths in the dual network and the maximum throughput flow in the dual network is $q_d^{\max} = 4$. (Figure 6.7C).

The total excess flow at all excess nodes is 12 units. The sum of capacities of all edges coming out of the source s is 32 units. According to Theorem 6.1, the maximum throughput flow in network 6.7A is equal to $32 - 12 + 4 = 24$ units.

The edge flows corresponding to this maximum throughput flow are obtained by following the steps of Algorithm 6.1. The dual circulation network is shown in

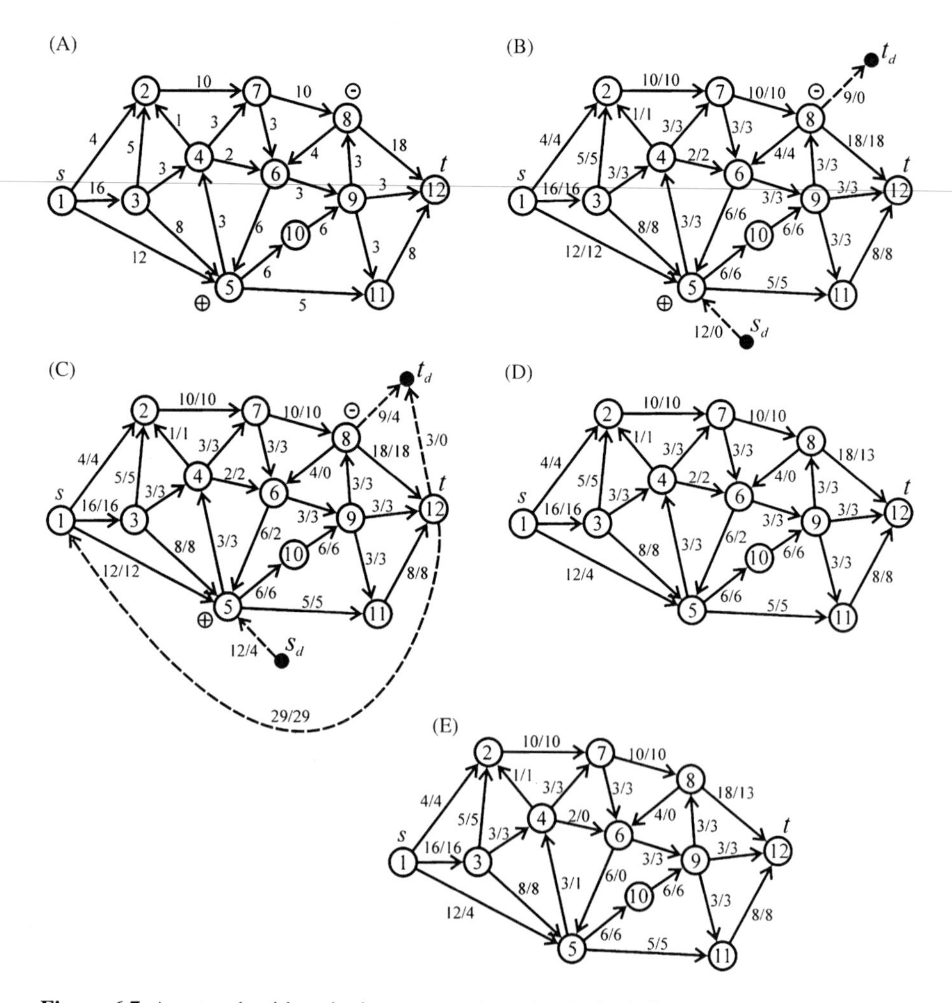

Figure 6.7 A network with a single excess node and a single deficit node.

Figure 6.7C. The shortest $s_d - t_d$ path $(s_d,5,1,12,t_d)$ is augmented with 3 units flow, while the next shortest $s_d - t_d$ path $(s_d,5,1,12,8,t_d)$ is augmented with 5 units flow. As a result, after removing the connecting edges, the excess and deficit flows disappear and the final state of the edge flows is shown in Figure 6.7D. The maximum throughput flow is indeed 24 units.

Finally, stage 3 of the algorithm identifies and removes the direct cyclic path (4,6,5,4) carrying parasitic flow of 2 units. After removing this parasitic loop of flow from the network, the final edge flows are according to Figure 6.7E. In the network from Figure 6.7E, there are no more directed loops with parasitic flow.

In the network in Figure 6.8A, there are two excess nodes '7' and '10' and a single deficit node '6'. The excess flow at the excess nodes is $ef_7 = 11$ and $ef_{10} = 6$. The total excess flow is $ef_7 + ef_{10} = 17$ units. The deficit flow at the deficit node is $df_6 = -10$. In the dual network (Figure 6.8B), two edges $(s_d,7)$ and $(s_d,10)$ connect the new source s_d with the excess nodes and edge $(6,t_d)$ connects

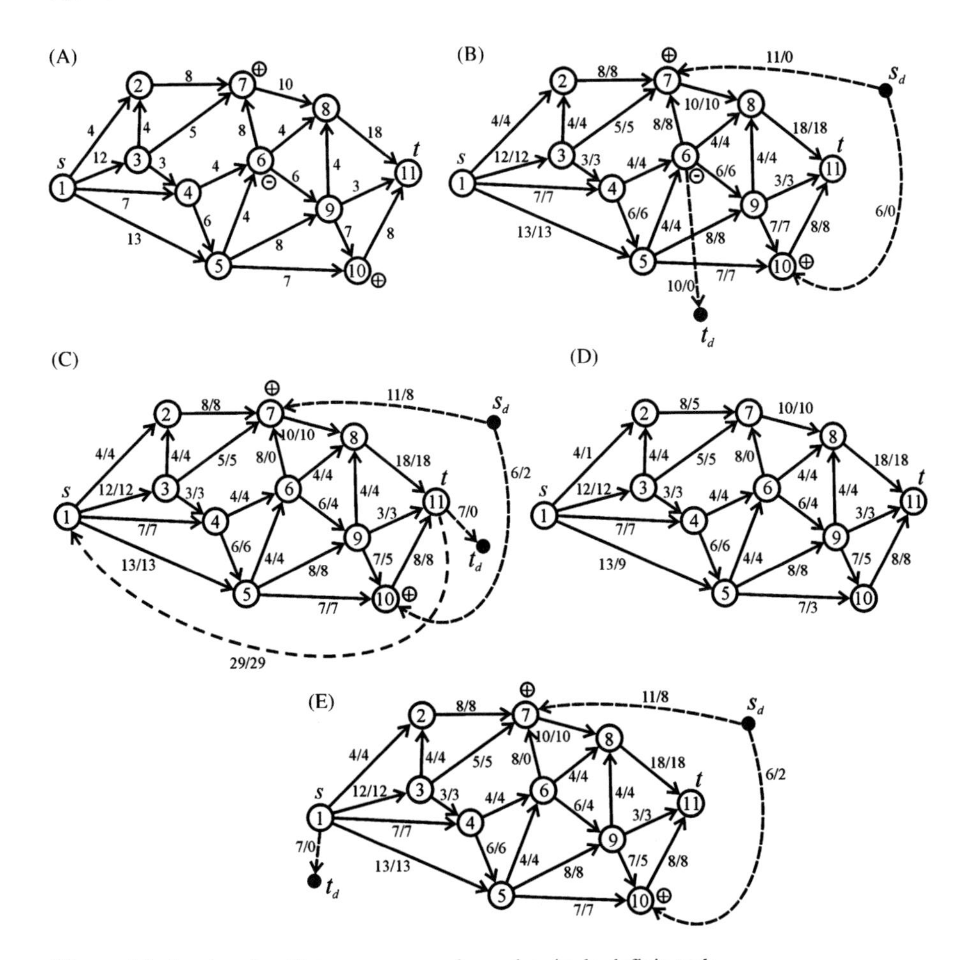

Figure 6.8 A network with two excess nodes and a single deficit node.

the deficit node '6' with the new sink t_d. The capacities of the connecting edges are equal to the excess at the excess nodes or the absolute value of the deficit flow, at the deficit node. The dual network can be augmented with $q_d^{\max} = 10$ units of flow, which is the maximum throughput flow in the dual network. The shortest $s_d - t_d$ path $(s_d,7,6,t_d)$ can be augmented with 8 units flow and the next shortest path $(s_d,10,9,6,t_d)$ can be augmented with 2 units of flow. As a result, node 6 becomes a balanced node. The sum of capacities of the edges coming out of the source s is 36 units. According to Theorem 6.1, the maximum throughput flow in the network in Figure 6.8A is equal to $36 - 17 + 10 = 29$ units.

The edge flows corresponding to the maximum flow of 29 units are obtained by following the steps of the algorithm.

The dual circulation network is created by introducing a t–s circulation edge with capacity equal to 29 units. This is because the sum of the capacities of the edges going into the sink is 29 and it is smaller than the sum of capacities (36) of the edges coming out of the source s. Edge (t,t_d), with capacity 7 units is also introduced, equal to the sum of the total excess flow at the excess nodes and the total deficit flow at the deficit nodes in the original network $(17 - 10 = 7)$.

After augmenting the shortest path $(s_d,10,5,1,11,t_d)$ with 4 units of flow and path $(s_d,7,2,1,11,t_d)$ with 3 units of flow, and after removing the connecting edges, the excess flow at the excess nodes disappears and the final state of the edge flows is shown in Figure 6.8D. The maximum throughput flow is indeed 29 units. There are no loops of parasitic flow in the network.

Because there are only excess nodes after the augmentation of the dual network, the algorithm can further be simplified by removing the t–s circulation edge and connecting the new sink t_d directly to the old source s through edge $(1,t_d)$ with capacity equal to the sum of the total excess flow at the excess nodes and the total deficit flow at the deficit nodes in the original network, (Figure 6.8E).

A similar simplification can be made if, at the end of the path augmentation in the dual network, only deficit nodes remain. In this case, the circulation edge can also be removed and the new source s_d can be directly connected to the original sink t, through an edge with capacity equal to the absolute value of the sum of the total excess flow at the excess nodes and the total deficit flow at the deficit nodes in the original network.

6.5 Area of Application of the Proposed Throughput Flow Maximisation Algorithm

The proposed algorithm, based on maximising the throughput flow in the dual network, will be very efficient in networks where most of the nodes are well-balanced nodes, and few excess and deficit nodes exist. In this case, *any known algorithm for maximising the throughput flow, by gradually filling an empty network, will be outperformed by the proposed algorithm.*

The condition of few existing excess and deficit nodes is closely matched by the design of real production networks, e.g. oil and gas production networks. Commonly, for well-designed production networks, the sum of capacities of the edges going into a node is equal to the sum of capacities of the outgoing edges. As a result, the designed networks are usually with balanced nodes.

Furthermore, for repairable flow networks, where the throughput flow has been optimised before a component failure, for each node, the sum of all ingoing flows *is always* equal to the sum of the outgoing flows. In this case, an edge failure produces only two imbalanced nodes and this circumstance has been used for developing the fast method for reoptimising the flow in repairable flow networks described in Chapter 5.

If only excess nodes are present in the fully saturated network, the maximum throughput flow is determined immediately: it is equal to the sum of capacities of all edges coming out of the source s minus the total excess flow at all excess nodes. If only deficit nodes are present in the fully saturated network, the maximum throughput flow can also be obtained immediately: it is equal to the sum of the capacities of all edges going into the sink minus the total deficit flow at all deficit nodes. Finally, if both excess nodes and deficit nodes are simultaneously present, but there is no augmentable $s_d - t_d$ path in the dual network, the maximum throughput flow is equal to the sum of capacities of all edges coming out of the source minus the total excess flow at all excess nodes. According to Theorem 6.2, the maximum throughput flow is also equal to the sum of capacities of all edges going into the sink minus the absolute value of the total deficit flow at all deficit nodes. The running time of the algorithm varies significantly.

In the case where the maximum throughput flow in the dual network is determined by augmenting a single $s_d - t_d$ path or few $s_d - t_d$ paths, the average running time depends linearly in the size of the network. In the case where most of the nodes in the network are imbalanced nodes, the proposed algorithm will have to perform a lot of work to eliminate the excess and deficit flows from the imbalanced nodes. This point has been illustrated by the extreme-case network in Figure 6.9, where there are many excess and deficit nodes and, in addition, the capacities of the edges connecting the source and the sink to the network are very small.

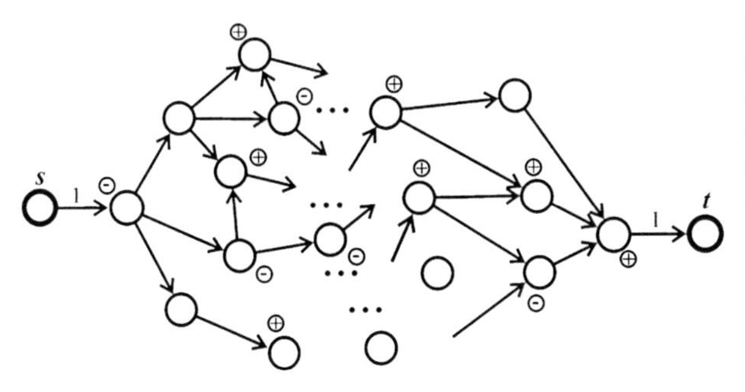

Figure 6.9 An extreme-case network with many imbalanced nodes. Edges with unit capacities connect the source s and the sink t with the rest of the network.

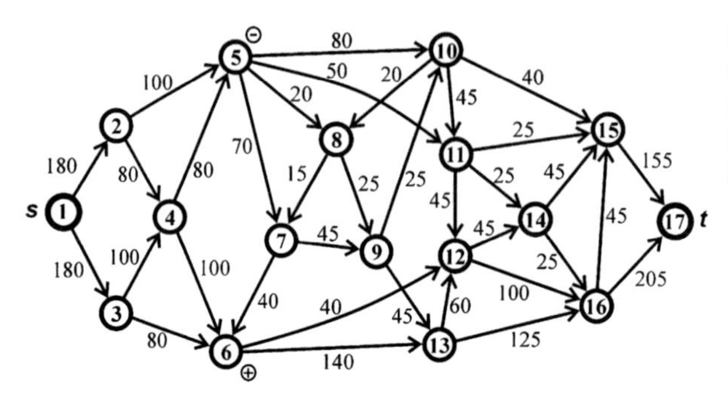

Figure 6.10 Test network demonstrating the advantage of the proposed algorithm in the case of very few imbalanced nodes.

In this extreme case, instead of augmenting a single $s-t$ path and terminating, the algorithm will shift unnecessarily all of the excess flow in the network until it determines the throughput flow of 1 unit.

The ideal application of the proposed method is in repairable flow networks, where, upon failure of an edge, exactly two imbalanced nodes are formed, in close proximity. Eliminating the excess and deficit flows from the pair of imbalanced nodes resulting from the edge failure is often performed in linear time or in constant time (Todinov 2011b, 2012a).

6.5.1 Comparing the Performance of the Proposed Algorithm with the Performance of a Conventional Algorithm

To demonstrate the clear advantages of the proposed algorithm over classical algorithms, in the case of very few imbalanced nodes, a comparison has been made. The comparison has been made only in terms of determining the value of the maximum throughput flow, not in terms of determining the edge flows leading to the maximum throughput flow. The extra stage of the proposed algorithm, related to removing parasitic loops of flow, has also been excluded.

The throughput flow of the network in Figure 6.10 was maximised by using the classical Edmonds and Karp algorithm, starting from an empty network. One million runs of the Edmonds and Karp algorithm on a computer with processor *Intel (R) Core(TM) 2 Duo CPU T9900 @ 3.06 GHz* took 27.9 s.

The maximum throughput flow in the network from Figure 6.10 was also determined by running the proposed algorithm. There are only two imbalanced nodes in the network: the excess node 6, with 40 units excess flow, and the deficit node 5, with 40 units deficit flow. The path (6,7,5) between the excess node 6 and deficit node 5 was augmented with 40 flow units. After the augmentation, the excess and deficit flows at nodes 6 and 5 disappear and the maximum throughput flow in the network is 360 units.

Running the new optimisation algorithm one million times takes only 1.06 s, 26 times faster than the conventional algorithm!

7 Reliability of the Throughput Flow: Algorithms for Determining the Throughput Flow Reliability

7.1 Probability that an Edge Will Be in a Working State on Demand

The edges in real networks are unreliable and are subject to failures. As a result, each edge (i,j) of a real flow network is characterised by *a cumulative time-to-failure distribution*

$$F_{ij}(t) = \Pr(T_f \leq t) \tag{7.1}$$

that gives the probability that the time to failure T_f of edge (i,j) will be smaller than or equal to a specified time t.

7.1.1 Hazard Rate: A General Time-to-Failure Model

Unreliable nodes and edges in a network are characterised by their hazard rates. The hazard rate $h(t)$ is *the proportion of items in service that fail per unit interval* (Barlow and Proschan, 1965, 1975; Bazovsky, 1961; Ebeling, 1997). *Given that the system or component has survived time t, $h(t)\Delta t$ gives the probability of failure within the small time interval $t,t+\Delta t$*, Figure 7.1.

The hazard rate can be presented as

$$h(t) = \frac{f(t)}{R(t)} \tag{7.2}$$

where $f(t) = \mathrm{d}F(t)/\mathrm{d}t$ is the probability density of the time to failure, $F(t)$ is the cumulative distribution of the time to failure and $R(t) = 1 - F(t)$ is the probability of surviving time t. Then, the conditional probability of failure in the infinitesimally small time interval $(t, t + \mathrm{d}t)$, given that the edge has survived time t, is given by

Flow Networks. DOI: http://dx.doi.org/10.1016/B978-0-12-398396-1.00007-6

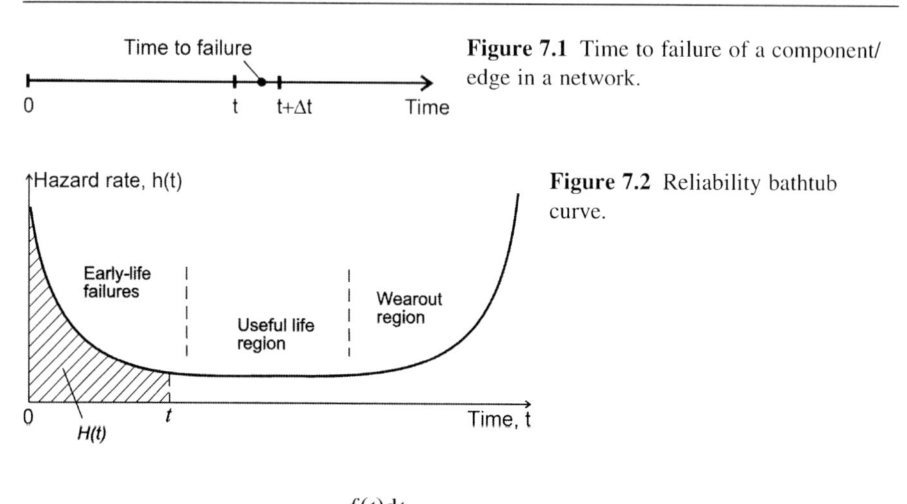

Figure 7.1 Time to failure of a component/edge in a network.

Figure 7.2 Reliability bathtub curve.

$$\Pr(t < T \leq t + \mathrm{d}t | T > t) = \frac{f(t)\mathrm{d}t}{R(t)} \tag{7.3}$$

If the hazard rate dependence $h(t)$ is a known function of the time (Figure 7.2), the time-to-failure distribution of an edge can be obtained by integration (Barlow and Proschan, 1965, 1975; Blake, 1979; Ross, 2002). Equation (7.2) can also be presented as

$$h(t) = -\frac{R'(t)}{R(t)} \tag{7.4}$$

which is a separable differential equation with initial condition $R(t = 0) = 1$. Presenting Eq. (7.4) as

$$-h(t)\mathrm{d}t = \mathrm{d}R(t)/R(t)$$

and integrating both sides from 0 to t gives

$$-\int_0^t h(\nu)\mathrm{d}\nu = \ln R(t) + C$$

From $R(t = 0) = 1$, $C = 0$ is obtained for the integration constant C. Finally, the time to failure of the edge can be presented as

$$F(t) = 1 - \exp\left[-\int_0^t h(\nu)\mathrm{d}\nu\right] \tag{7.5}$$

$H(t) = \int_0^t h(\nu)\mathrm{d}\nu$ will be referred to as *cumulative hazard rate*. It is equal to the area beneath the hazard rate curve shown in Figure 7.2 (the hatched region).

Hence, the time-to-failure distribution can be expressed as a function of the cumulative hazard rate:

$$F(t) = 1 - \exp[-H(t)] \tag{7.6}$$

Equation (7.5) is a very general equation and can be determined by integrating the time dependence of the hazard rate function, if it is known. If the amount of historic failure data is insufficient to determine the hazard rate time dependence, a constant hazard rate is assumed.

The hazard rate, characterising the components building a flow network, follows a curve with bathtub shape (Figure 7.2), with three distinct regions. Most of the failures in the first region, referred to as *early-life failure region or infant mortality region*, are quality related and result from inherent defects due to poor design, manufacturing, assembly, lack of experience in operating the equipment, etc. The second region of the bathtub curve, referred to as *useful life region*, is characterised by approximately constant hazard rate. The negative exponential distribution is the model of the times to failure in this region. Failures in this region are usually caused by external factors, usually a random overstress of the components, and are not caused by ageing, wearout or degradation.

The third region, referred to as *wearout region*, is characterised by an increasing with age hazard rate, due to accumulated wear and degradation (e.g. caused by wear, erosion, corrosion and fatigue).

7.1.2 Negative Exponential Distribution of the Time to Failure: Mean Time to Failure

Suppose that the probability of failure of a component (an edge or a node) within a short time interval is constant. In other words, the hazard rate $h(t) = \lambda = \text{const.}$ and the probability $\lambda\,\Delta t$ that a component, which has survived time t, will fail in the small time interval $t, t + \Delta t$, is constant and does not depend on the age t of the component. Under this assumption, the integral in Eq. (7.5) becomes $\int_0^t h(\nu)\mathrm{d}\nu = \lambda t$ and Eq. (7.5) becomes

$$F(t) = 1 - \exp(-\lambda t) \tag{7.7}$$

Equation (7.7) is widely known as the *negative exponential distribution* of the times to failure. The probability density function of the time to failure is obtained by differentiating the cumulative distribution function (7.7) with respect to t:

$$f(t) \equiv \frac{\mathrm{d}F(t)}{\mathrm{d}t} = \lambda \exp(-\lambda t) \tag{7.8}$$

The probability density at time t multiplied by the length of an infinitesimally small time interval, $f(t)\mathrm{d}t$, gives the probability of failure of the edge in the infinitesimal time interval $t,t+dt$. Note that this probability has not been conditioned on the survival of time t:

$$\Pr(t < T \leq t + \mathrm{d}t) = f(t)\mathrm{d}t \tag{7.9}$$

The value $\lambda \mathrm{d}t$ gives also the probability of failure in the infinitesimal time interval $t,t+\mathrm{d}t$, but this probability is conditional. It is specified on the condition that the edge has survived time t:

$$\Pr(t < T \leq t + \mathrm{d}t | T > t) = \lambda \mathrm{d}t \tag{7.10}$$

The conditional probability is constant if the probability of failure in a small time interval practically does not depend on the age of the edge. In this case, the time to failure follows the negative exponential distribution. The probability that the edge will fail within a specified time interval is the same, irrespective of whether the edge has been operating for some time or has just been put to use. The edge is *as good as new*.

The probability that the edge will survive time $t+\Delta t$, given that it has survived time t, can be determined from the conditional probability formula:

$$\Pr(T \geq t + \Delta t | T > t) = \frac{R(t+\Delta t)}{R(t)} = \frac{\exp[-\lambda(t+\Delta t)]}{\exp(-\lambda t)} = \exp(-\lambda\, \Delta t) \tag{7.11}$$

where $R(t) = \exp(-\lambda t)$ is the probability that the edge will survive time t and $R(t+\Delta t) = \exp[-\lambda(t+\Delta t)]$ is the probability that the edge will survive time $t+\Delta t$. In other words, the probability that the edge will survive a time interval $t+\Delta t$ given that it has survived time t is always equal to the probability that the edge will simply survive the time interval $(0,\Delta t)$. This condition is approximately fulfilled for edges which practically do not degrade or wear out with time.

The negative exponential distribution can also be presented in the form

$$F(t) = 1 - \exp[-(t/\mathrm{MTTF})] \tag{7.12}$$

where $F(t)$ is the cumulative distribution of the time to failure and MTTF is the mean time to failure. The negative exponential distribution of the times to failure is characterised by a single parameter MTTF. The MTTF is *the average time to the first failure* and can be estimated from failure data by dividing the total accumulated operational time T (the sum of all operational times) by the number of failures k:

$$\mathrm{MTTF}' = \frac{T}{k} \tag{7.13}$$

where MTTF′ is the estimator of the unknown true MTTF. The negative exponential distribution is often specified as a function of the hazard rate $\lambda = 1/\mathrm{MTTF}$.

The rate λ has dimension 'time unit^{-1}' and is the proportion of items in service that fail per unit time interval.

7.1.3 Other Time-to-Failure Models

Because of its flexibility, the *Weibull distribution* is often assumed as a time-to-failure model:

$$F(t) = 1 - \exp[-(t/\eta)^{\beta}] \tag{7.14}$$

where $F(t)$ is the cumulative distribution of the time to failure; β (shape parameter) and η (characteristic life/scale parameter) are constants determined from experimental data. This model is commonly used in the case where the hazard rate depends on the age of the component.

The Weibull distribution is flexible and permits describing failure modes from any of the regions of the bathtub curve (Ebeling, 1997).

For example, specifying $\beta < 1$ corresponds to an early-life failure mode (decreasing hazard rate); $\beta = 1$ corresponds to a failure mode in the useful life region of the bathtub curve (constant hazard rate) and $\beta > 1$ corresponds to a failure mode in the wearout region of the bathtub curve (increasing hazard rate). Furthermore, by varying the values of the shape parameter β, the Weibull distribution can approximate well a number of other distributions – *negative exponential*, *normal* and *log-normal*.

The parameters in Eq. (7.14) characterise a single failure mode. A particular edge may have several failure modes characterised by a time-to-failure distribution given by the Weibull distribution. For example a power transmission line may fail due to overloading, lightning, fire, extreme weather conditions, malicious acts, etc. Different sets of values for the parameters β and η then need to be supplied in the input file, in order to characterise the time-to-failure distribution associated with each failure mode.

The negative exponential distribution can be obtained as a special case of the Weibull distribution, for $\beta = 1$. The negative exponential model also applies to components which fail due to overstress (the load exceeds the strength of the component). If the critical load applications exceeding the strength of the component follow a homogeneous poisson process, the time to failure of the component follows the negative exponential distribution.

Another possible time-to-failure distribution is the *log-normal distribution*:

$$f(t) = \frac{1}{t\sigma'\sqrt{2\pi}} \exp\left[\frac{-[\ln(t) - \bar{t}']^2}{2\sigma'^2}\right] \tag{7.15}$$

where $\bar{t}'$ and σ' are, respectively, the mean and the standard deviation of the natural logarithms of the times to failure characterising the particular failure mode. In Eq. (7.15), $f(t)$ is the probability density function. If the time to failure t is log-normally distributed, then $\ln(t)$ follows the normal distribution. The log-normal distribution is often appropriate in cases where the failure mechanism is fatigue. The log-normal time-to-failure distribution is also appropriate for describing the time to failure associated with degradation processes, where the degree of further damage is proportional to the extent of the existing damage.

Each edge (i,j) is also characterised by *a time-to-repair distribution*

$$G_{ij}(t) = \Pr(G \leq t) \tag{7.16}$$

which gives the probability that the time to repair G (the downtime) will be smaller than or equal to a specified time t. The most commonly specified repair time is the constant repair time. In the case of random repair times, a common model for describing the repair time distribution is the log-normal distribution (7.15). The log-normal distribution is skewed, with a long upper tail, which reflects correctly some problem repairs taking a long time.

Another common model for describing the distribution of the time to repair is the normal (Gaussian) distribution:

$$g(t) = \frac{1}{\sigma\sqrt{2\pi}} \exp\left(-\frac{(t-\mu)^2}{2\sigma^2}\right) \tag{7.17}$$

In Eq. (7.17), $g(t)$ is the probability density function of the repair times.

The normal distribution of the repair times is characterised by two parameters, the mean μ of the repair times and the variance σ^2. For the sake of simplicity, negative exponential times-to-failure distributions and constant repair times will normally be assumed for the edges of the repairable networks. On failure of edge (i,j), its flow capacity is reduced to zero:

$$c(i,j) = 0 \tag{7.18}$$

The variation of the throughput flow is the basis for determining important performance measures of repairable flow networks. The first performance measure will be referred to as *throughput flow reliability*. The throughput flow reliability R_{th} is measured by *the probability*

$$R_{\text{th}} = \Pr(q_r \geq q_{\text{th}}) \tag{7.19}$$

that upon demand, the maximum throughput flow q_r in the repairable network, in the presence of edge failures, will be at least equal to a specified threshold level q_{th}. The threshold level is non-negative and does not exceed the maximum throughput flow q_0, characterising the static network, where no edge failures exist $(0 \leq q_{\text{th}} \leq q_0)$.

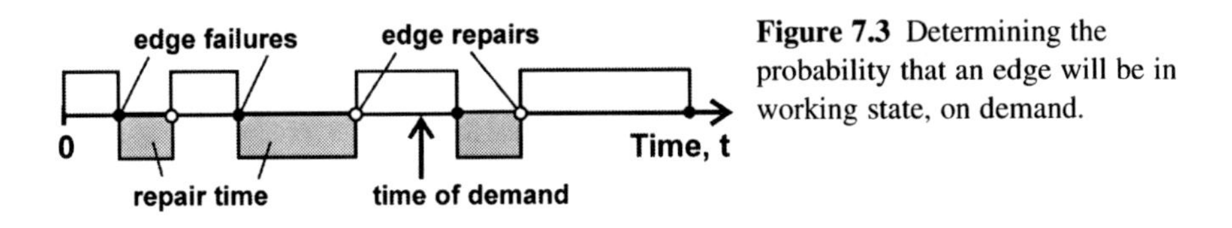

Figure 7.3 Determining the probability that an edge will be in working state, on demand.

It is assumed that the edges work independently from one another (the case where the time to failure of edges depend on a common cause will be considered in Chapter 14). This means that if a working state has been sampled for a particular edge i, the probability of sampling a working state for another edge j does not depend on the state of edge i. For the sake of simplicity, constant hazard rates (negative exponential time-to-failure distribution) and constant repair times will be assumed in presenting the algorithms. A particular edge 'i' is characterised by a constant hazard rate λ_i (negative exponential time-to-failure distribution) and a time-to-repair distribution with mean time to repair, $\text{MTTR}_i = d_i$. The mean time to failure of the edge is $\text{MTTF}_i = 1/\lambda_i$.

The probability that the demand at a specified point in time will sample a working state for the edge (Figure 7.3) is given by (Trivedi, 2002):

$$p_i = \alpha_i = \text{MTTF}_i/(\text{MTTF}_i + \text{MTTR}_i) \tag{7.20}$$

where α_i denotes the average availability of the edge. Substituting in Eq. (7.20) $\text{MTTF}_i = 1/\lambda_i$ and $\text{MTTR}_i = d_i$, and simplifying, yields

$$p_i = (1 + \lambda_i d_i)^{-1} \tag{7.21}$$

7.2 Probability of a Source-to-Sink Flow on Demand

7.2.1 Analytical Methods

7.2.1.1 Analytical Expressions for Series and Parallel Arrangement of the Edges

Determining the probability that a source-to-sink ($s{-}t$) flow will exist on demand, reduces to determining the probability that, on demand, there will be a directed path through working edges from the source to the sink. This probability is an important parameter characterising communication networks, where the existence of information flow between two communication nodes is of significance and not the actual amount of the transmitted data.

In this case, it is sufficient to check whether there exists a connection between the source and the sink by following forward edges only.

In determining the probability of existence of $s{-}t$ flow, the classical analytical methods developed for system reliability analysis (Billinton and Allan, 1992; Ebeling,

1997; Hoyland and Rausand; 1994; Ramakumar, 1993; Todinov, 2007) can be used. For example, *the network reduction method, the decomposition method*, and also methods based on *minimal paths sets* and *minimal cut sets* can be used.

A section of edges in series transmits flow if all edges in the section are in working state.

The probability p_i that on demand, the ith edge will be in working state is given by Eq. (7.20). Consequently, the probability R_{ser} that, on demand, the section of n edges will transmit flow is

$$R_{\text{ser}} = \prod_{i=1}^{n} p_i \tag{7.22}$$

The probability that on demand, a section of n edges in series will transmit flow is smaller than each of the availabilities p_i, characterising the separate edges. The probability of $s-t$ flow on demand is smaller than the worst edge availability $p_{\min}$.

$$R_{\text{ser}} = \prod_{i=1}^{n} p_i < p_{\min} \tag{7.23}$$

Therefore, improving the probability of a flow on demand should start with improving the worst edge availability. Increasing p_i in expression (7.20) can be done either by improving the reliability of the ith edge, which means increasing the MTTF_i, or by reducing the repair time of the edge, which means reducing the mean time to repair, MTTR_i.

A section of edges in parallel transmits flow on demand, if all at least a single edge in the section is in working state (Figure 7.4B). The complementary event of the event 'at least one edge is in working state' is the event 'none of the edges is in working state'. Hence, the probability that on demand, at least one edge will be in working state is equal to one minus the probability that none of the edges will be in working state. In other words, the probability R_{par} that, on demand, the section of n edges in parallel will transmit flow is

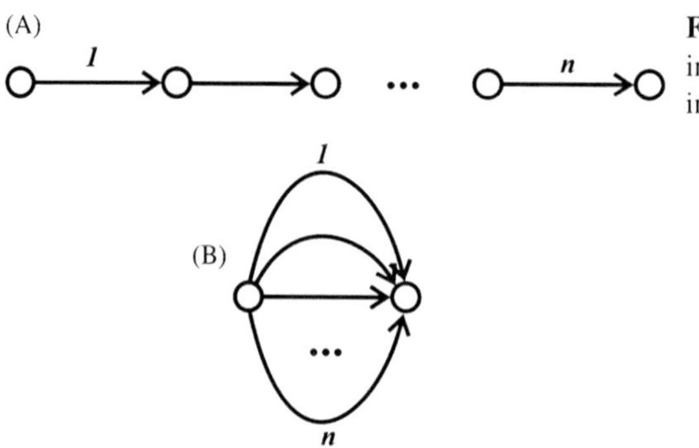

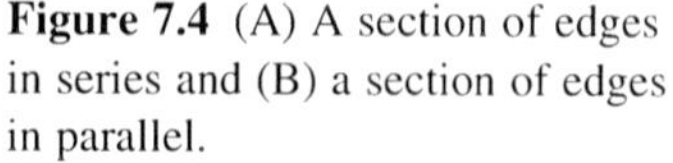

Figure 7.4 (A) A section of edges in series and (B) a section of edges in parallel.

$$R_{\text{par}} = 1 - \prod_{i=1}^{n}(1 - p_i) \tag{7.24}$$

These relationships can have useful applications for repairable flow networks whose topologies are compared and for which there are no data related to the availabilities of the separate edges.

Consider for example the two networks '1' and '2' built of type A, type B and type C edges/components (Figure 7.5). Type A edges have a higher availability than type B edges and type B edges have a higher availability than type C edges. It is important to know which network is characterised by a higher probability that on demand there will be flow from the source s to the sink t.

The answer to this question can be provided easily if the availabilities of edges A, B and C are denoted by a, b and c, respectively. The probability that on demand, there will be s–t flow in network '1' is $R_1 = 1 - (1 - ab)(1 - c) = c + ab - abc$. The probability that, on demand, there will be s–t flow in network 2 is $R_2 = 1 - (1 - ac)(1 - b) = b + ac - abc$. The difference $R_2 - R_1 = b + ac - c - ab = b - c - a(b - c) = (b - c)(1 - a)$. From $b > c$ and $1 - a > 0$ $(a < 1)$, it follows that $R_2 - R_1 > 0$. Therefore, the second network is characterised by a higher probability that on demand, there will be flow from the source to the sink.

7.2.1.2 Network Reduction Method

For networks with a series–parallel arrangement of the edges, the throughput flow reliability can be determined by *the network reduction method*. The system in Figure 7.6 consists of four types of components A, B, C and D, characterised by availabilities a, b, c and d, respectively. The reliabilities of the edges are shown as labels on the edges in Figure 7.6.

Following the network reduction method, the edges (3,5) and (5,6) are substituted by a single edge characterised by a probability c^2 of transmitting flow on demand (Figure 7.6B). Next, the section consisting of the two parallel edges from node 3 to node 6 is substituted by a single edge (3,6) characterised by a probability of transmitting flow on demand $R_{36} = 1 - (1 - c^2)(1 - b)$ (Figure 7.6C). Next, section (2,3,6) and section (2,4,6) are replaced with edges characterised by a probability of transmitting flow on demand $R_{236} = a \times R_{36}$ and $R_{246} = a \times b$. The result is the network in Figure 7.6D. Finally, the probability of s–t flow on demand, for the network in Figure 7.6D, is

$$R = d \times [1 - (1 - R_{236})(1 - R_{246})] \tag{7.25}$$

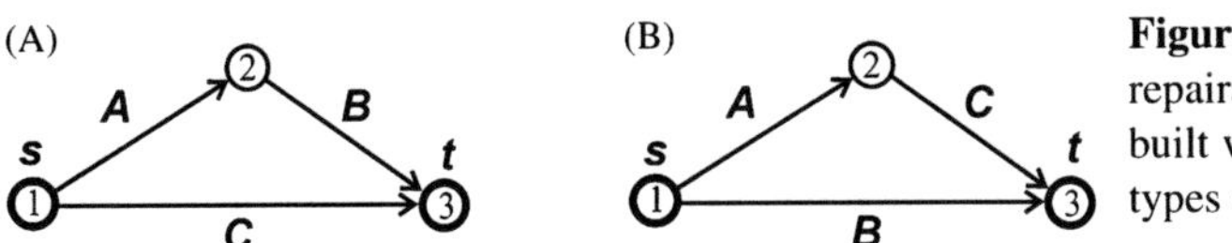

Figure 7.5 Two competing repairable flow networks built with three different types of edges.

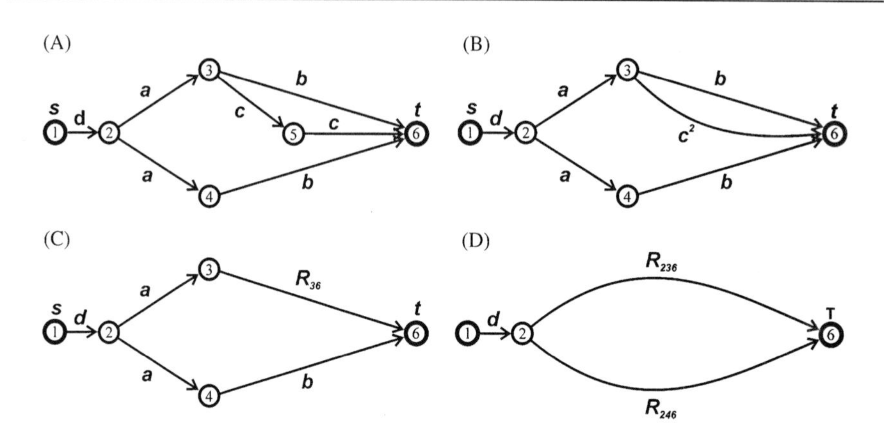

Figure 7.6 A repairable flow network whose reliability can be determined by the network reduction method.

7.2.1.3 Decomposition Method

The decomposition method is based on conditioning a complex network on the possible states of a key component K_1 or on the possible combinations of states of several key components. As can be verified from the Venn diagram in Figure 7.7, the event S denoting that '$s-t$ flow exists on demand' can be presented as a union of two mutually exclusive events: (i) $K_1 \cap S$ the key edge is in working state and there will be $s-t$ flow on demand and (ii) $\overline{K}_1 \cap S$ the key edge is in failed state and there will be $s-t$ flow on demand.

The idea of this method is to decompose the initial network into two networks $K_1 \cap S$ and $\overline{K}_1 \cap S$ with simpler topology. According to the total probability theorem (DeGroot, 1989; Ross 2002), the probability $\Pr(S)$ of an $s-t$ flow on demand can be presented as a sum of the probabilities of the two mutually exclusive events:

$$\Pr(S) = \Pr(S \cap K_1) + \Pr(S \cap \overline{K}_1) \tag{7.26}$$

Since $\Pr(S \cap K_1) = \Pr(S|K_1)\Pr(K_1)$ and $\Pr(S \cap \overline{K}_1) = \Pr(S|\overline{K}_1)\Pr(\overline{K}_1)$, it follows:

$$\Pr(S) = \Pr(S|K_1)\Pr(K_1) + \Pr(S|\overline{K}_1)\Pr(\overline{K}_1) \tag{7.27}$$

In Eq. (7.27), $\Pr(S|K_1)$ is the probability of $s-t$ flow given that the key component is in working state at the time of demand; $\Pr(S|\overline{K}_1)$ is the probability of $s-t$ flow, given that the key edge is in a failed state at the time of demand. $\Pr(K_1)$ and $\Pr(\overline{K}_1)$ are the probabilities that the key edge will be in working state and in failed state, respectively. Similarly, if two key edges K_1 and K_2 have been selected, the probability of $s-t$ flow on demand can be presented as a sum of the probabilities

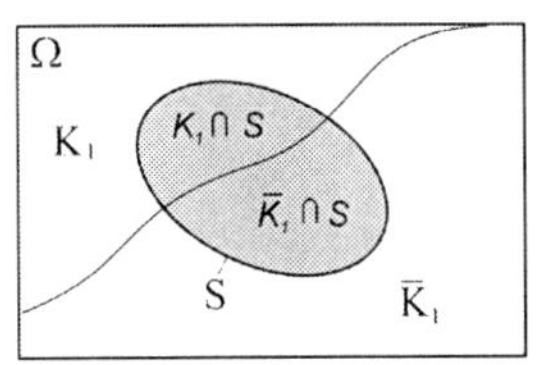

Figure 7.7 The event S ($s-t$ flow on demand exists) is the union of two mutually exclusive events: $K_1 \cap S$ and $\overline{K}_1 \cap S$.

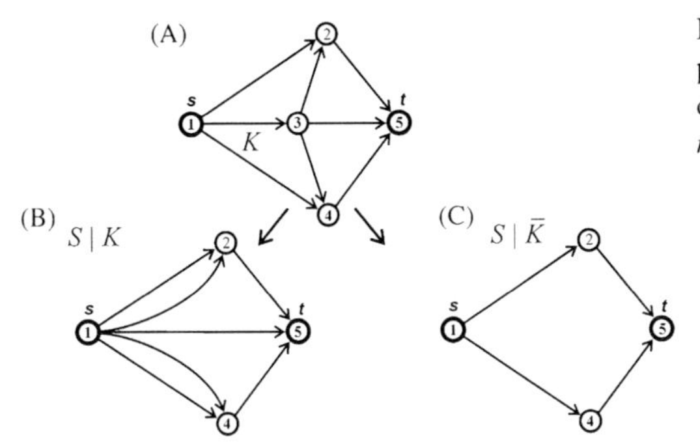

Figure 7.8 Evaluating the probability of source-to-sink flow on demand, by using the *decomposition method*.

of four mutually exclusive events, which correspond to the four distinct combinations of states, related to the key edges:

$$\begin{aligned}\Pr(S) &= \Pr(S|\mathrm{K}_1\mathrm{K}_2)\Pr(\mathrm{K}_1)\Pr(\mathrm{K}_2) + \Pr(S|\mathrm{K}_1\overline{\mathrm{K}}_2)\Pr(\mathrm{K}_1)\Pr(\overline{\mathrm{K}}_2)\\ &+ \Pr(S|\overline{\mathrm{K}}_1\mathrm{K}_2)\Pr(\overline{\mathrm{K}}_1)\Pr(\mathrm{K}_2) + \Pr(S|\overline{\mathrm{K}}_1\overline{\mathrm{K}}_2)\Pr(\overline{\mathrm{K}}_1)\Pr(\overline{\mathrm{K}}_2)\end{aligned}$$

The network in Figure 7.8, for example, cannot be presented with series and parallel arrangements of the edges. Its throughput flow reliability can however be evaluated analytically by the decomposition method. The network incorporates eight unreliable edges (with availabilities $0 \le p \le 1$). The throughput flow is greater than zero only if a directed $s-t$ path exists between the source s and the sink t.

If edge (1,3) is selected as a key edge K, the probability of system success $\Pr(S)$ (existence of $s-t$ flow) can be determined from the equation

$$\Pr(S) = \Pr(S|K)\Pr(K) + \Pr(S|\overline{K})\Pr(\overline{K}) \tag{7.28}$$

The probability $\Pr(S|K)$ of the system in Figure 7.8B can be estimated as a throughput flow reliability of a series–parallel system. Since the probability of a flow on demand through the parallel section (1,2) and (1,4) is $p_d = 1-(1-p)^2$, the probability of throughput flow on demand for the network in Figure 7.8B becomes

$$\Pr(S|K) = 1-(1-p)\times(1-p_d p)^2 \tag{7.29}$$

and the probability of throughput flow on demand for the network in Figure 7.8C becomes

$$\Pr(S|\overline{K}) = 1 - (1-p^2)^2 \tag{7.30}$$

Since $\Pr(K) = p$ and $\Pr(\overline{K}) = 1 - p$, the substitution in Eq. (7.28) yields the probability of $s-t$ flow on demand for the initial system in Figure 7.8A:

$$\Pr(S) = p \times [1 - (1 - p)(1 - p_d p)^2] + (1 - p) \times [1 - (1 - p^2)^2] \tag{7.31}$$

Substituting in Eq. (7.31) $p = 0.65$ yields $p_d = 0.877$ and $\Pr(S) = 0.84$.

If the probability of $s-t$ flow on demand for any of the simpler networks is difficult to calculate, another decomposition can be made by selecting another key edge K_2 and so on, until trivial networks are obtained, for which the probability of an $s-t$ flow can be evaluated easily. Consider the network in Figure 7.9A, which consists of seven identical edges with availabilities p. An $s-t$ flow is present if an $s-t$ path consisting of forward edges in working state exists between the source s and the sink t.

Although this network is not a trivial network, it can be simplified if the key edge K_1 is selected. The probability of an $s-t$ flow $\Pr(S)$ can be determined from Eq. (7.27). Because the probability $\Pr(K_1) = p$ and $\Pr(\overline{K}_1) = 1 - p$, the probability of existence of $s-t$ flow on demand becomes

$$\Pr(S) = p \times \Pr(S|K_1) + (1 - p) \times \Pr(S|\overline{K}_1) \tag{7.32}$$

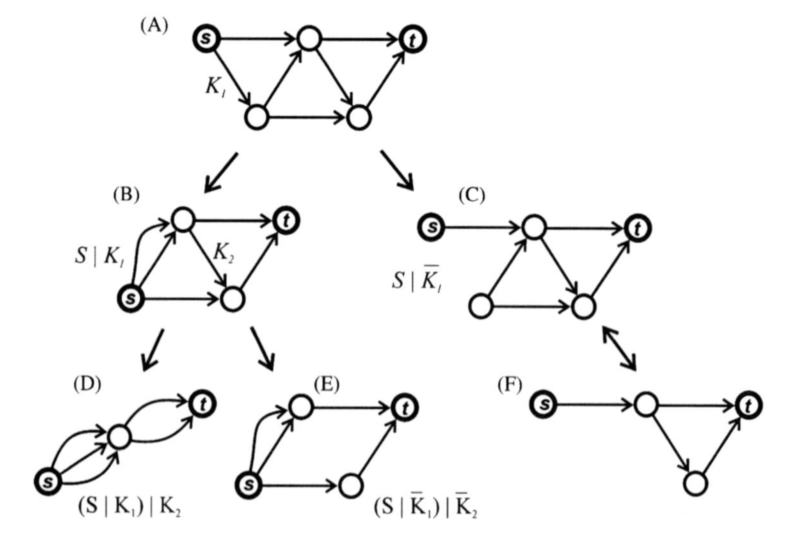

Figure 7.9 Determining the probability of $s-t$ flow on demand by using the decomposition method based on sequential selection of two key edges.

The probability $\Pr(S|\overline{K}_1)$ of an $s-t$ flow on demand for the network shown in Figure 7.9C can be determined easily because it is equivalent to the network shown in Figure 7.9F whose edges have a simple series–parallel arrangement. The probability of a directed $s-t$ path for the network shown in Figure 7.9C is therefore

$$\Pr(S|\overline{K}_1) = p \times [1-(1-p)(1-p^2)] \tag{7.33}$$

For $p=0.7$, Eq. (7.33) evaluates to $\Pr(S|\overline{K}_1)=0.59$.

The network shown in Figure 7.9B, which results from the decomposition of the network shown in Figure 7.9A, is not a trivial network. However, by selecting another key edge K_2, it can in turn be decomposed into two trivial networks (Figure 7.9D and E), whose throughput flow reliabilities can be determined easily. Consequently,

$$\Pr(S|K_1) = \Pr[(S|K_1)|K_2] \times \Pr(K_2) + \Pr[(S|K_1)|\overline{K}_2] \times \Pr(\overline{K}_2) \tag{7.34}$$

Because $\Pr[(S|K_1)|K_2] = [1-(1-p)^3] \times [1-(1-p)^2] = 0.885$, $\Pr[(S|K_1)|\overline{K}_2] = 1 - [1-(1-(1-p)^2)p](1-p^2) = 0.81$, $\Pr(K_2) = p$ and $\Pr(\overline{K}_2) = 1-p$.

Substituting these values in Eq. (7.34) yields

$$\Pr(S|K_1) = 0.885 \times p + 0.81 \times (1-p) = 0.864 \tag{7.35}$$

Finally, the substitution in Eq. (7.32) gives

$$\Pr(S) = p \times 0.862 + (1-p) \times 0.59 = 0.78$$

for the probability of $s-t$ flow on demand, in the initial network from Figure 7.9A.

This example shows that the decomposition method can be used for analysis of complex networks. However, the decomposition method also has significant limitations. It is not suitable for large networks. Each selection of a key edge splits a large network into two networks, each of which may in turn be split into two networks and so on. For a large number n of edges in the initial network, the number of product networks generated from the selection of key edges quickly increases exponentially and becomes unmanageable for large n.

Apart from the considered methods, system reliability tools based on *minimal cut sets* and *minimal paths* (Barlow and Proschan 1965, 1975; Hoyland and Rausand 1994; Ramakumar 1993) can also be used for determining the probability that on demand, there will be $s-t$ flow. Similarly, the main drawback of analyses based on minimal paths and minimal cut sets is that the number of minimal paths and cut sets increases exponentially with increasing the size of the network. Even for moderately large networks, the increase of the number of paths and cut sets leads to a combinatorial explosion.

In summary, the discussed analytical methods for analysis of repairable flow networks are not suitable for large and complex networks. The network reduction method is not suitable for topologically complex networks while the decomposition method is not suitable for large networks. Another severe limitation is that the discussed methods are only suitable for determining the probability of nonzero $s-t$ flow, not for determining the probability of $s-t$ flow of a certain minimal magnitude.

7.2.2 Monte Carlo Simulation Methods

In the cases where the analytical solution is impossible or difficult, because of the size and complexity of the network, a Monte Carlo simulation can be used to determine the probability of a source-to-sink flow on demand. This probability is equal to the probability of existence of a path from the source to the sink, which consists of forward edges only, each of which is in working state and has nonzero flow capacity.

Indeed, if no $s-t$ path containing forward edges only exists in a network with empty edges, no augmentable $s-t$ path including backward edges exists either. This is because, at the start of the path augmentation, all backward edges are empty and a path containing an empty backward edge cannot be augmented with flow.

The next algorithm, in pseudo-code, determines the probability of existence of source-to-sink flow.

Algorithm 7.1

```
function Find_dir_st_path_nonzero_cap();
function real_random();

function Probability_of_st_flow();
{
  success_cnt = 0;
  for i = 1 to num_trials do
  {
  for j = 1 to m do
     {
     tmp = real_random();
     if (tmp > p[j]) then reduce the flow capacity of edge j to zero;
     }
  path_exists = Find_dir_st_path_nonzero_cap();
  if (path_exists = 1) then
     success_cnt = success_cnt + 1;
  Restore the flow capacities of the failed edges to their original levels;
  }
  return success_cnt/num_trials;
} ■
```

The array $p[]$ contains the probabilities that the separate edges will be in working state on demand. In a nested loop controlled by the variable 'j', the state of the separate edges (working/failed) is determined at the time of demand. The state of the jth edge is tested by generating a uniformly distributed random number between 0 and 1 from the statement 'tmp = ***real_random()***', and comparing it with the probability p [j] that the jth edge will be in working state. The number of edges is m. A good algorithm for the standard function ***real_random()*** can be found in Park and Miller (1988).

The edges are characterised by constant hazard rates (negative exponential time-to-failure distribution). Consequently, the probability that the demand at a specified point in time will sample a working state for an edge is given by Eq. (7.20).

All probabilities p_j ($j = 1,2, \ldots ,m$) characterising the edges are pre-calculated and stored in the array $p[]$. If the generated random number 'tmp' uniformly distributed between 0 and 1 is greater than $p[j]$, then the jth edge is in a failed state and is essentially removed or deleted from the flow network by reducing its flow capacity to zero. If the converse is true, the edge remains in the network with its original flow capacity.

After the state of all edges has been determined, the function ***Find_dir_st_path_nonzero_cap()*** establishes whether there exists a directed $s-t$ path from the source to the sink, through forward edges with nonzero flow capacity. The ratio of the number of trials for which such a directed $s-t$ path exists and the total number of simulation trials is a measure of the probability of a source-to-sink flow. At the end of each simulation trial, the flow capacities of all failed edges are restored to their original level.

Consequently, central to the algorithm for determining the throughput flow reliability of the repairable flow network is a modification of the algorithm for determining a directed $s-t$ path, presented in Chapter 2.

In the case of very small unavailabilities characterising the edges, the precision of the presented Monte Carlo crude sampling can be increased by applying '*stratified sampling without replacement*'. This approach is discussed in more detail in the next section.

7.3 Probability of a Source-to-Sink Flow on Demand, of Specified Magnitude

Most often, the point of interest is the probability of a nonzero source-to-sink flow of specified magnitude. This performance characteristic has also been referred to as *threshold flow rate reliability* (Todinov, 2011a). The threshold flow rate reliability $R_{th} = \Pr(q_{max,r} \geq q_{th})$ is defined as the probability that upon demand, the maximum throughput flow $q_{max,r}$ in the repairable network, in the presence of component failures, will be at least equal to a specified threshold throughput flow q_{th} ($0 \leq q_{th} \leq q_{max,0}$). The threshold throughput flow is non-negative and does not exceed the maximum possible throughput flow $q_{max,0}$ characterising the network, in the absence of any edge failures.

The threshold throughput flow reliability in a repairable flow network with edges working independently from one another can be determined by a function whose algorithm in pseudo-code is given next.

Algorithm 7.2

```
function max_flow();
function real_random();

function Threshold_throughput_flow_reliability();
{
 success_cnt = 0;
 for i = 1 to num_trials do
  {
 for j = 1 to m do
     {
     tmp = real_random();
     if (tmp > p[j]) then reduce the flow capacity of edge j to zero;
     }
 maxf = max_flow();
 if (maxf > = threshold_flow) then success_cnt = success_cnt + 1;
 Restore the flow capacities of the failed edges to their original levels;
 }
 return success_cnt/num_trials;
}
```
■

Similar to the previous algorithm, all probabilities p_j ($j = 1,2, \ldots ,m$) are pre-calculated and stored in the array p[]. If the generated random number ‘tmp’, uniformly distributed between 0 and 1, is greater than $p[j]$, then the jth component is in a failed state and is essentially removed from the flow network by reducing its

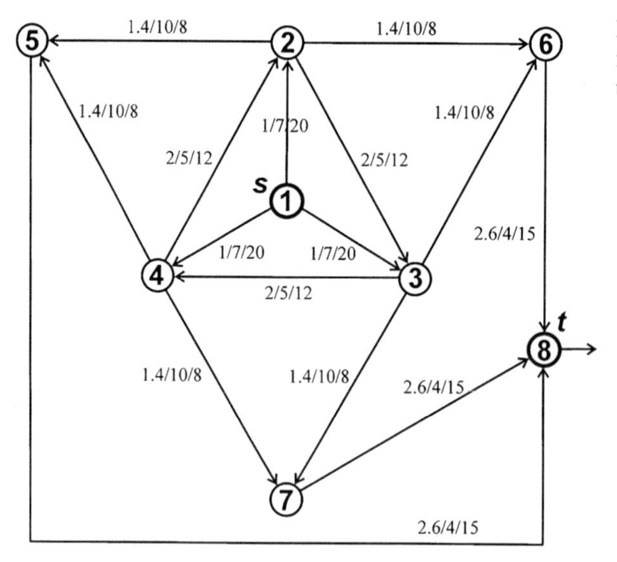

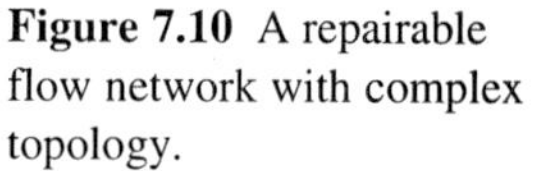
Figure 7.10 A repairable flow network with complex topology.

flow capacity to zero. If the converse is true, the component remains in the network with its full flow capacity.

After the state of all components has been determined, the flow maximisation function ***max_flow()*** establishes the maximum flow in the resultant static flow network after reducing the flow capacity of the failed components.

The ratio of the number of trials for which the maximum throughput flow 'maxf' is equal to or greater than the specified threshold flow rate, and the total number of simulation trials is a measure of the threshold throughput flow reliability for the repairable flow network – the probability that, on demand, the throughput flow will be equal to or will exceed the threshold value. At the end of each simulation trial, the capacities of all failed edges are restored to their original levels.

For the repairable flow network shown in Figure 7.10, the adopted format for the attributes of each edge is $\lambda/d/c$. The first number λ is the hazard rate of the edge in years^{-1}. The second number d is the downtime for repair in days and the third number c is the edge flow capacity. The capacities of the edges are in thousands of flow units per day. If no component failures occur, the flow maximisation algorithm yields a maximum flow of 45,000 flow units per day. For the network shown in Figure 7.10, the probability that on demand, the throughput flow will be equal to the maximum throughput flow of 45,000 units per day is 0.69.

The proposed method for determining the threshold throughput flow reliability is useful for flow networks with redundant capacities, for which no simple analytical solutions exist. This method is particularly useful for comparing network topologies and selecting the topology characterised by the largest threshold throughput flow reliability. The method is also useful in deciding whether the resources allocated for purchasing extra redundancy are justified.

The presented method for revealing the threshold throughput flow reliability is particularly relevant to telecommunication networks. Failures of hosts, routers and communication lines are frequent and are inevitably associated with downtime for repair. Consequently, by their nature, real telecommunication networks are repairable flow networks.

Failures of routers or communication lines lead to disappearance of throughput flow capacity and, as a result, the expected magnitude of the throughput data flow will not always be guaranteed. As a result, the quality of service received from the network, which is a key performance characteristic, is affected seriously. The problem is particularly acute for telecommunication networks oriented towards media applications, which require a high throughput data flow rate.

There are several important constraints posed by telecommunication networks, which need to be addressed during modelling. In telecommunication networks, both the nodes (representing routers or hosts) and communication lines are unreliable. This constraint can be accounted for by noting that failure of an unreliable node in a telecommunication network essentially eliminates all edges incident to the failed node. Next, the links in telecommunication networks are undirected links. In telecommunication networks, at a random point of demand, some of the bandwidth capacity of the communication lines may have been engaged by existing data transfers. This constraint can also be incorporated by simulating existing traffic along communication lines at the time of the random demand.

In the case of very small unavailabilities characterising the edges, the precision of the presented Monte Carlo crude sampling can be increased by applying '*stratified sampling without replacement*' (Vose, 2000). This method guarantees that sampling of a failed state will be made, irrespective of how small the unavailability of the edge is.

The essence of this approach is in splitting a distribution into intervals of equal probability, whose number n is equal to the number of simulation trials. Once an interval has been selected by generating a random number from 1 to n, it is never selected again until the end of the simulation. There are three important factors which permit a very efficient implementation of the method of stratified sampling without replacement: (i) unavailability is simulated by sampling from a uniform distribution in the interval (0,1); (ii) all sub-intervals corresponding to a working state of the edge are identical to one another and (iii) all sub-intervals corresponding to a failed state of the edge are also identical to one another. Suppose, for the sake of simplicity that the total number of sub-intervals in the interval (0,1) is n and each sub-interval is sufficiently small and homogeneous. In other words, sampling anywhere in a selected sub-interval corresponds to either a working state or to a failed state. Suppose that n_f is the number of sub-intervals corresponding to a failed state and n_w is the number of sub-intervals corresponding to a working state ($n_w + n_f = n$). Sampling without replacement then reduces to generating an integer random number x between 1 and n. If $x \leq n_w$, the edge is in a working state. To guarantee that the selected sub-interval will not be selected again, the number of sub-intervals n_w corresponding to a working state is then decremented by 1 ($n_w := n_w - 1$). If $x \geq n - n_w$, the edge is in a failed state. In this case, to guarantee that the sampled interval will not be sampled again, n_f is decremented by 1, ($n_f := n_f - 1$).

Because exactly one sub-interval has been blocked (eliminated), the next simulation step continues with $n - 1$ number of sub-intervals. The random number is now generated within the range $[1, n - 1]$, to reflect the circumstance that the sampling is without replacement – i.e., the sub-interval selected in the previous simulation is never selected again. The simulation continues until the last remaining sub-interval is selected.

In this way of sampling, for each edge, it is guaranteed that no matter how small the unavailability of the edge is, a failed state of the edge will certainly be sampled.

8 Reliability Networks

8.1 Series and Parallel Arrangement of the Components in a Reliability Network

The operation logic of engineering systems can be modelled by reliability networks, which in turn can be modelled conveniently by graphs. The nodes are notional (perfectly reliable), whereas the edges correspond to the components and are characterised by a failure rate.

The common system in Figure 8.1A consists of a power block PB, control module CM and a mechanical device MD.

Because the system fails whenever any of the components fails, the components are said to be logically arranged in series. The next system shown in Figure 8.1B is composed of two power generators *E1* and *E2* working simultaneously. Because the system fails whenever both generators fail, the generators are said to be logically arranged in parallel.

System failure is present if the system fails to deliver one or more specified functions, e.g. if at least one of the production units part of the system stops production. System failures usually require immediate corrective action (e.g. intervention for repair or replacement) in order to return the system to operating condition. Each system failure is associated with losses due to the cost of intervention, cost of repair and the cost of lost production.

For the simple production system shown in Figure 8.1C, an example of system failure is the failure of the power block PB, failure of the mechanical device or failure of both control modules (CM1 and CM2). In all these cases, the mechanical device MD stops functioning.

However, failure of the control module CM1 does not induce a system failure. The redundant control module CM2 will still maintain control over the mechanical device MD and the system will be operational.

For a reliability network with a single start (s) and end node (t), the system it models is operational if and only if a path through working components exists from the start node to the end node (Figure 8.1).

Reliability networks with a single start node (s) and a single end node (t) *can also be interpreted as single-source single-sink flow networks with components with integer capacity greater than zero*. The system is in operation if and only if, on demand, a path with unit flow can be augmented from the source s to the sink t (Figure 8.1). In this respect, reliability networks with a single start node and a

Flow Networks. DOI: http://dx.doi.org/10.1016/B978-0-12-398396-1.00008-8

single end node can be analysed by the algorithms developed for determining the reliability of the throughput flow of repairable flow networks.

8.2 Building Reliability Networks: Difference Between a Physical and Logical Arrangement

Unlike repairable flow networks, the reliability networks may not match the functional block diagram of the modelled system. This is the reason why alongside the analysis of reliability networks, an emphasis will also be put on building reliability networks.

The fact that the components in a particular system are logically arranged in parallel does not necessarily mean that they are physically connected in parallel. A typical example is a two-engine aircraft, capable of flying on one engine only. The converse is also true. Components arranged physically in parallel may not necessarily be logically arranged in parallel. Parallel pipelines transporting toxic chemicals are a good example. Accident/failure associated with a release of toxic substance occurs whenever at least one of the pipelines looses containment. As a result, a significant difference exists between a physical (functional) arrangement and a logical arrangement. This is illustrated in Figure 8.2 for a system of seals.

Although the physical arrangement of the seals is in series (Figure 8.2A and B), their logical arrangement with respect to the failure mode 'leakage in the environment' is in parallel (Figure 8.2C). Indeed, leakage in the environment is present only if both seals fail.

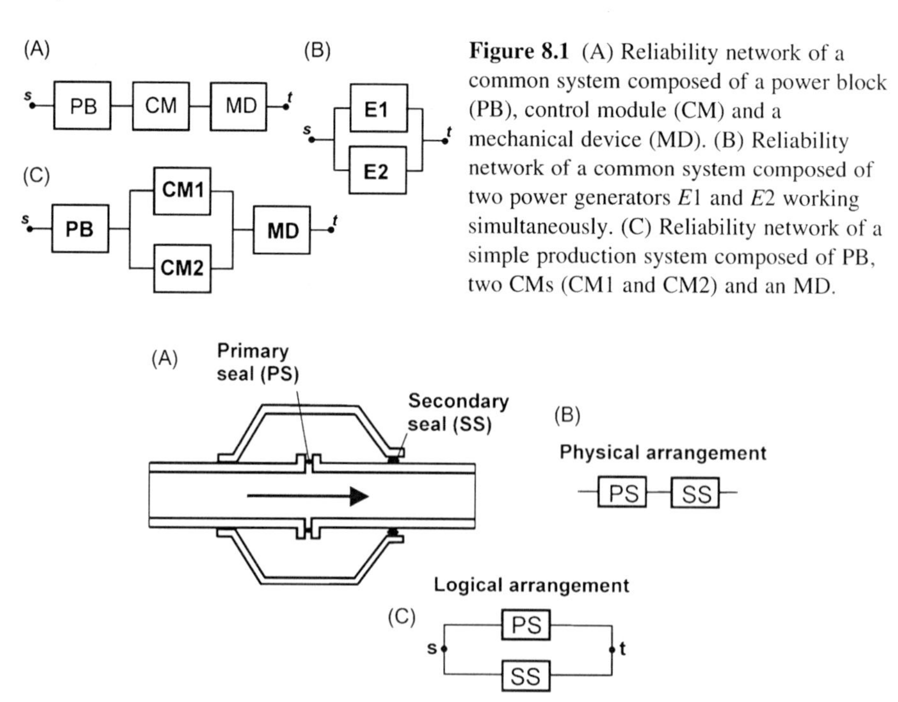

Figure 8.1 (A) Reliability network of a common system composed of a power block (PB), control module (CM) and a mechanical device (MD). (B) Reliability network of a common system composed of two power generators $E1$ and $E2$ working simultaneously. (C) Reliability network of a simple production system composed of PB, two CMs (CM1 and CM2) and an MD.

Figure 8.2 Seals that are physically arranged in series but logically arranged in parallel.

Conversely, components can be physically arranged in parallel, but their logical arrangement is in series. This is illustrated by the seals in Figure 8.3. Although the physical arrangement of the seals is in parallel, their logical arrangement with respect to the failure mode *leakage in the environment* is in series. Leakage in the environment is present if at least one seal stops working (sealing).

Reliability networks are built by using the top-down approach. The system is divided into several large blocks, logically arranged in a particular manner. Next, each block is further detailed into several smaller blocks. These blocks are in turn detailed and so on, until the desired level of indenture is achieved for all blocks.

This approach will be illustrated by the system shown in Figure 8.4, which represents toxic liquid travelling along two parallel pipe sections. The O-rings *O*1 and *O*2 seal the flanges, the pairs of seals (*A*1,*B*1) and (*A*2,*B*2) seal the sleeves.

The first step in building the reliability network of the system shown in Figure 8.4 is to note that despite the fact that, physically, the two groups of seals (*O*1,*A*1,*B*1) and (*O*2,*A*2,*B*2) are arranged in parallel, logically they are arranged in series with respect to the function 'preventing a leak to the environment' because both of the two groups of seals must prevent the toxic liquid from escaping in the environment. Failure to isolate the toxic liquid is considered at the highest indenture level: at the level of the two groups of seals.

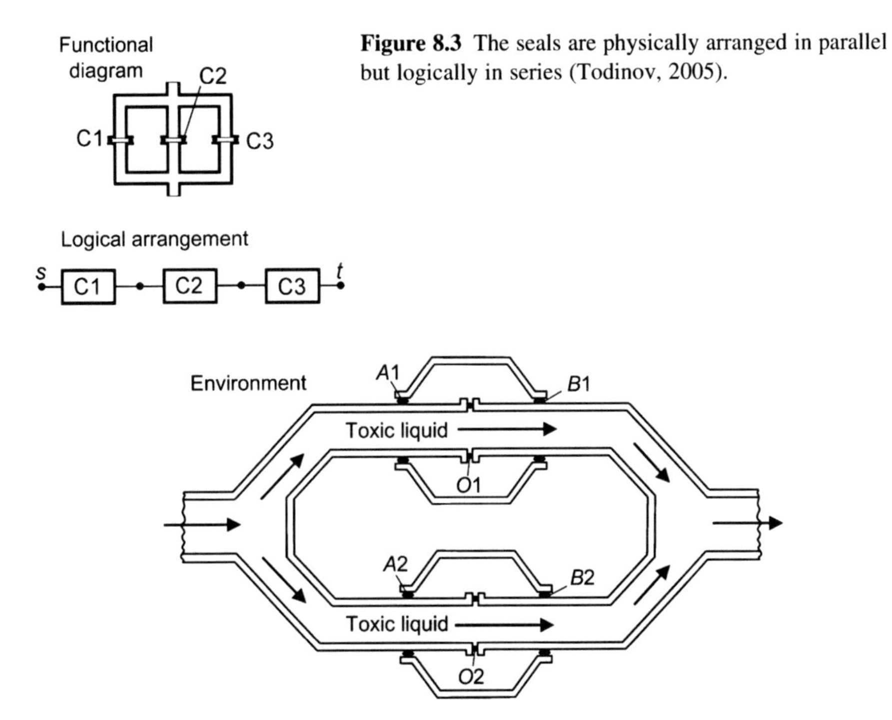

Figure 8.3 The seals are physically arranged in parallel but logically in series (Todinov, 2005).

Figure 8.4 A functional diagram of a system of seals isolating toxic liquid from the environment.

Within each of the two groups of seals, the O-ring seal works in parallel with the pair of seals (*A*,*B*) on the sleeves. For the first group of seals, for example, it is sufficient that the O-ring *O*1 works or the pair (*A*1,*B*1) works, in order to guarantee that the first group of seals (*O*1,*A*1,*B*1) works.

Consequently, the O-ring seal and the group of seals *A*,*B* are arranged in parallel in each of the two groups of seals (Figure 8.5B).

Finally, within the pair of seals (*A*1,*B*1), both seals *A*1 and *B*1 must work, in order to guarantee that the pair (*A*1,*B*1) works. The seals *A*1 and *B*1 are therefore logically arranged in series. This reasoning can be extended for the second group of seals and the reliability network of the system of seals is shown in Figure 8.6.

Suppose now that the O-rings, *O*1 and *O*2, sealing the flanges, are characterised by hazard rates $\lambda_{O1} = \lambda_{O2} = 0.15$ years^{-1} and the pairs of seals (*A*1,*B*1) and (*A*2,*B*2) on the sleeves are characterised by constant hazard rates $\lambda_{A1} = \lambda_{B1} = \lambda_{A2} = \lambda_{B2} = 0.3$ years^{-1}. A common problem is to determine the probability that within a specified period of continuous operation, for example 2 years, there will be no toxic liquid release in the environment.

For a constant hazard rate λ, the reliability $r(t)$ of a component associated with a particular operation time interval t is given by the negative exponential distribution (Ebeling, 1997; Elsayed, 1996):

$$r(t) = \exp(-\lambda t) \tag{8.1}$$

If the reliability of the O-ring for 2 years of operation is denoted by r_o and the reliability of each seal *A and B* by r_s, the reliability *R* of the seal assembly for a period of 2 years operation becomes

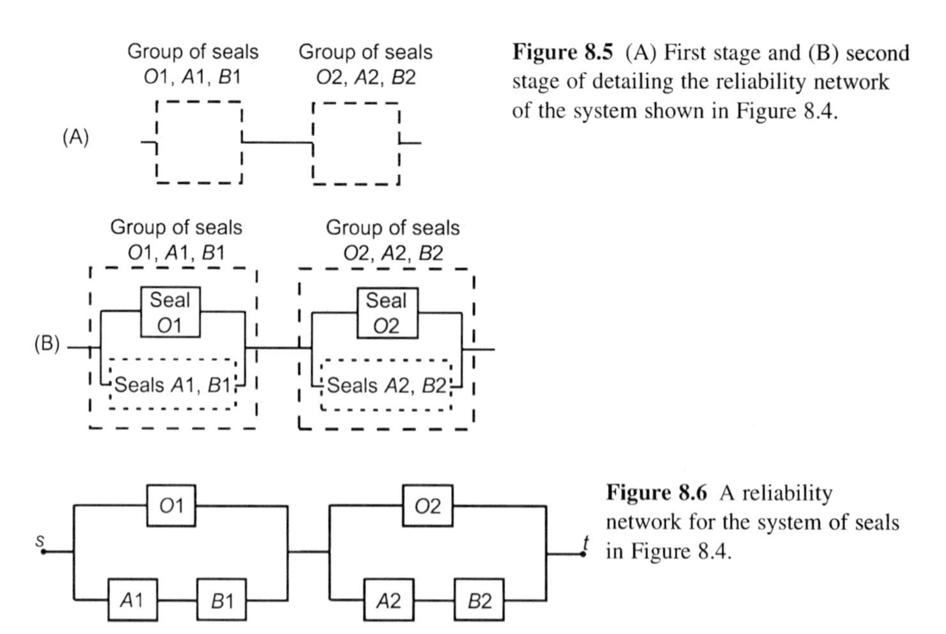

Figure 8.5 (A) First stage and (B) second stage of detailing the reliability network of the system shown in Figure 8.4.

Figure 8.6 A reliability network for the system of seals in Figure 8.4.

$$R = [1-(1-r_o)(1-r_s^2)]^2 \tag{8.2}$$

Considering that $r_o = \exp(-0.15 \times 2) = 0.74$ and $r_s = \exp(-0.3 \times 2) = 0.55$, substituting the numerical values in Eq. (8.2) gives

$$R = [1-(1-0.74)(1-0.55^2)]^2 = 0.67$$

for the reliability of the system of seals, associated with 2 years of continuous operation.

The next example features two valves on a pipeline, physically arranged in series (Figure 8.7A). With respect to isolating the production fluid, they are logically arranged in parallel (Figure 8.7B). With respect to enabling the flow through the pipeline, they are logically arranged in series (Figure 8.7C).

Indeed, to stop the flow through the pipeline, at least one of the valves must work on demand; therefore, the valves are logically arranged in parallel with respect to the function 'isolating the production fluid'. However, to enable the flow through the pipeline, both valves must open on demand. Therefore, in this case, the logical arrangement of the valves is in series (Figure 8.7C).

Figure 8.8 features the functional diagram of a system of pipes with six valves, working independently from one another, all of which are initially open. Each valve is characterised by a certain probability that if a command for closure is sent, the valve will close and stop the fluid passing through it.

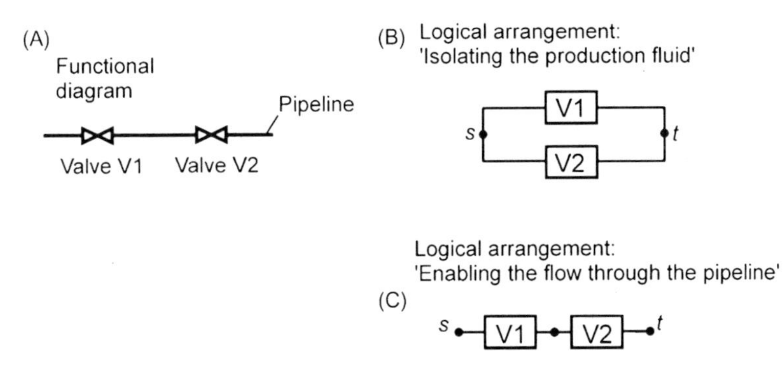

Figure 8.7 Physical and logical arrangement of (A) two valves on a pipeline with respect to two functions: (B) Isolating the production fluid and (C) Enabling the flow through the pipeline.

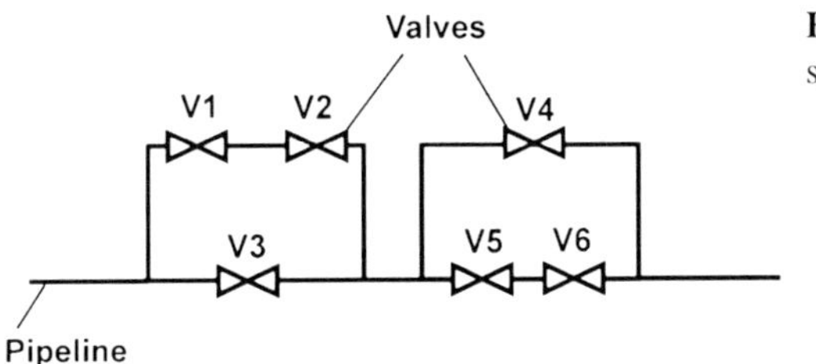

Figure 8.8 A functional diagram of a system of valves.

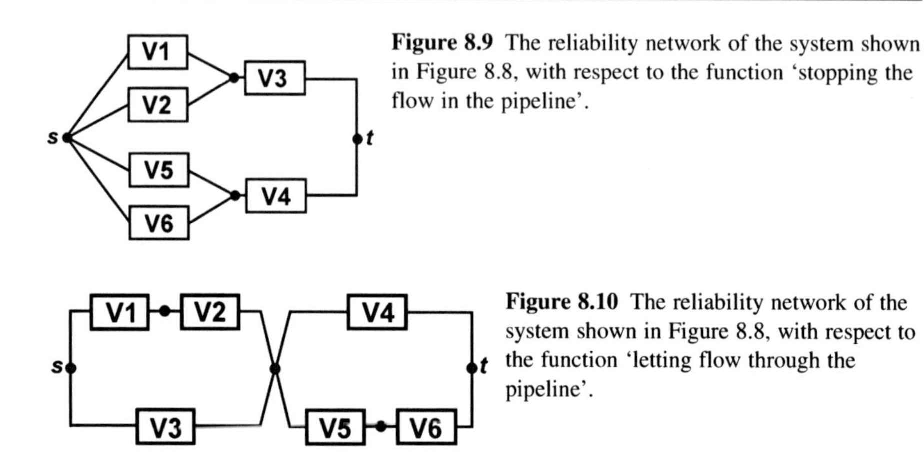

Figure 8.9 The reliability network of the system shown in Figure 8.8, with respect to the function 'stopping the flow in the pipeline'.

Figure 8.10 The reliability network of the system shown in Figure 8.8, with respect to the function 'letting flow through the pipeline'.

The reliability network related to the function stopping the flow in the pipeline is shown in Figure 8.9. The block of valves (V1,V2,V3) and the block of valves (V4,V5, V6) are arranged in parallel because the flow through the pipeline is stopped if either block stops the flow. The block of valves (V1,V2,V3) stops the flow if both blocks of valves (V3) and (V1,V2) stop the flow in their corresponding sections. Therefore, the blocks (V1,V2) and (V3) are logically arranged in series. The block of valves (V1,V2) stops the flow if either valve V1 or V2 stops the flow in their common section.

Similar reasoning applies to the block of valves (V4,V5,V6). The reliability network of the system in Figure 8.8 is shown in Figure 8.9.

Interestingly, regarding the function stopping the fluid in the pipeline, valves or blocks of valves arranged in series in the functional diagram are arranged in parallel in the reliability network. Accordingly, valves or blocks arranged in parallel in the functional diagram are arranged in series in the reliability network. It can be shown that this is valid for the function 'stopping the fluid in the pipeline' for *any* functional diagram involving valves arranged in series and parallel.

There are also cases where the physical arrangement coincides with the logical arrangement. Consider again the system of valves in Figure 8.8, with all valves initially closed. With respect to the function 'letting flow (any amount of flow) through the pipeline', the reliability network is shown in Figure 8.10. It matches the functional diagram in Figure 8.8.

8.3 Complex Reliability Networks Which Cannot Be Represented As a Combination of Series and Parallel Arrangements

There is a view among some reliability practitioners that a combination of series and parallel arrangements is sufficient for representing the reliability networks of engineering systems, particularly mechanical systems. Many simple engineering systems, however, have reliability networks that cannot be described in terms of combinations of series–parallel arrangements.

Consider the simple communication system shown in Figure 8.11A, where the same message is transmitted from two identical sources a_1 and a_2, to two identical receivers b_1 and b_2 (Figure 8.11A). The message can be sent directly from a_1 to b_1 or from a_2 to b_2, but cannot be sent directly from a_1 to b_2 or from a_2 to b_1. Instead, the message must be sent first to the transmitter c, as is shown in Figure 8.11A.

The logical arrangement of the components in this system can be represented by the *reliability network* shown in Figure 8.11B. The operation logic of the system can be modelled by a set of nodes (the filled circles shown in Figure 8.11B) and components ($a1$, $a2$, $b1$, $b2$ and c). The communication system works (the message is transmitted) only if there exists a path through working components between the start node s and the end node t. Alternatively, the communication system works if and only if, on demand, there exists nonzero throughput s–t flow in the flow network from Figure 8.11B, whose edges have unit capacities.

Now consider a communication system where a message is transmitted from three identical sources a_1, a_2 and a_3 to three identical receivers b_1, b_2 and b_3, with the help of two transmitters c_1 and c_2 (Figure 8.12A). The possible connections through which the message can be transmitted from the sources to the receivers and between the transmitters have been marked by dashed lines (Figure 8.12A).

The corresponding reliability network is shown in Figure 8.12B. This reliability network also cannot be presented with combinations of series and parallel arrangements.

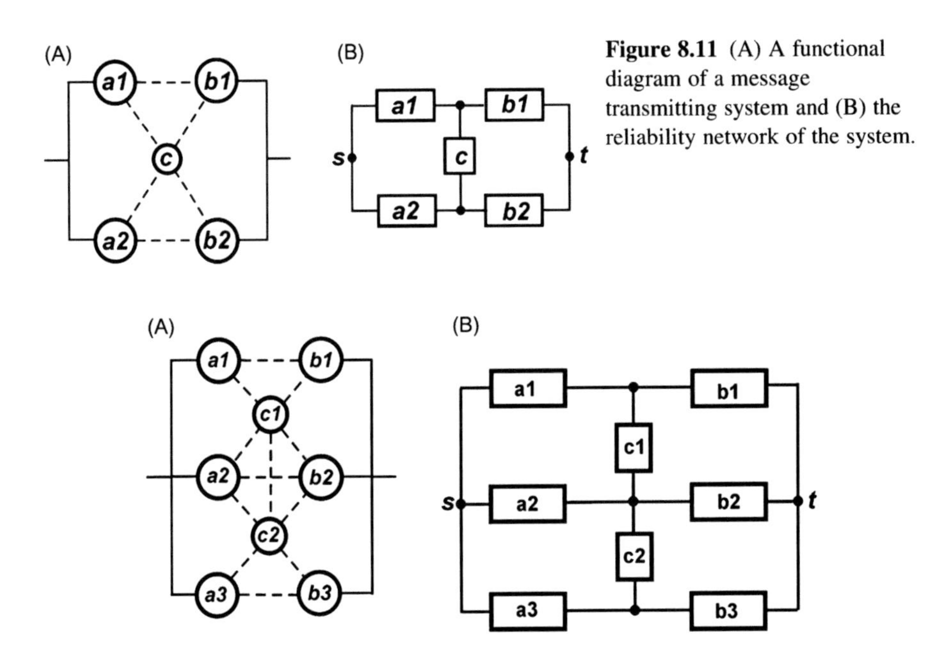

Figure 8.11 (A) A functional diagram of a message transmitting system and (B) the reliability network of the system.

Figure 8.12 (A) A functional diagram of a message transmitting system; (B) the reliability network of the system in (A).

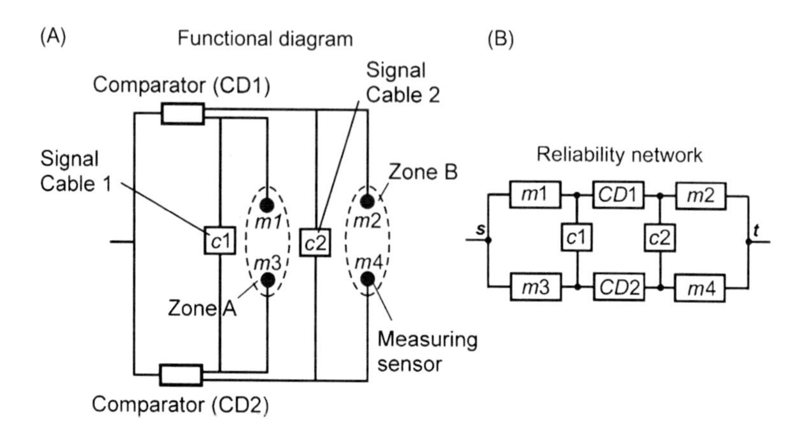

Figure 8.13 (A) A safety–critical system based on comparing measured quantities in two zones and (B) its reliability network.

The reliability network shown in Figure 8.12B can be modelled with a repairable flow network where all edges have greater than zero capacities. The system works if and only if, in the network from Figure 8.12B, there exists a path that can be augmented with unit flow from the source s to the sink t.

Systems with bridges similar to the ones discussed for communication networks are also typical for safety–critical systems based on comparing signals from sensors reading the value of a parameter (pressure, concentration, temperature, water level, etc.) in different zones. If the difference in the parameter levels characterising the two zones exceeds a particular critical value, a signal is issued by a special device (comparator). The reliability networks of some safety–critical systems of this type, cannot be presented as a series–parallel system.

Indeed, the complex safety-critical system shown in Figure 8.13A compares the temperature (pressure) of two different zones A and B, measured by the sensors ($m1$, $m2$, $m3$ and $m4$). If the temperature (pressure) difference is greater than a critical value, the difference is detected by one of the comparators CD1 or CD2, and a system shut-down signal is sent. The two comparators and the two pairs of sensors have been included to increase the robustness of the safety–critical system. For the same purpose, the signal cables $c1$ and $c2$ have been included, whose purpose is to transmit the signal from the sensors to the comparators. If, for example, sensors $m1$, $m2$ and comparator CD2 have failed, the system will still be operational. Because of the existence of signal cables, the measured temperature (pressure) by the remaining operational sensors $m3$ and $m4$ will be fed to comparator CD1 through the signal cables $c1$ and $c2$ (Figure 8.13). As a result, if excessive temperature (pressure) difference between the two zones exists, the comparator CD1 will still be capable of issuing a shut-down signal. If sensors $m1$ and $m4$ fail for example, and also comparator CD1 fails, the system is still operational because the excessive temperature (pressure) difference will be detected by sensors $m3$

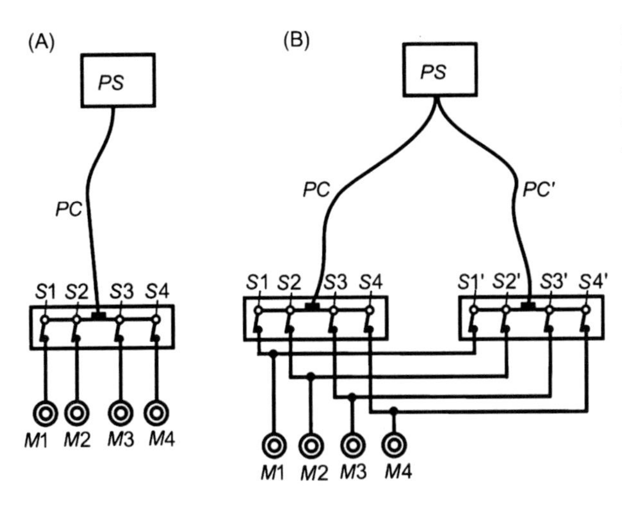

Figure 8.14 A functional diagram of a power supply to four electric motors: (A) without redundancy and (B) with redundancy.

and $m2$, and through the signal cable $c2$, the temperature (pressure) reading will be fed to comparator CD2.

The system will be operational whenever an $s-t$ path through working components exists in the reliability network shown in Figure 8.13B. The reliability network shown in Figure 8.13B cannot be reduced to combinations of series, parallel or series–parallel arrangements. Telecommunication systems and electronic control systems may also have very complex reliability networks, which cannot be represented with series–parallel arrangements.

Traditionally, reliability networks have been presented as networks with a single start node s and a single terminal node t (Andrews and Moss, 2002; Billinton and Allan, 1992; Blischke and Murthy, 2000; Ebeling, 1997; Hoyland and Rausand, 1994; Ramakumar, 1993). This traditional representation, however, is insufficient to model the failure logic of some systems. There are systems for which the correct representation of the logic of failure requires more than a single terminal node. Consider for example the simple system with redundancy in Figure 8.14A, that consists of a power supply PS, power cable (PC), block of four switches (S1, S2, S3 and S4) and four electric motors M1, M2, M3 and M4.

Because the reliability of the switches and the reliability of the power cable is often a problem (the power cable for example, can often be accidentally damaged), these components can be duplicated. In other words, redundancy can be introduced. In many safety–critical applications, all electric motors must be operational on demand. Typical examples are fans or pumps cooling chemical reactors, pumps dispensing water in the case of fire, life support systems, automatic shutdown systems and control systems. The reliability on demand of the system shown in Figure 8.14A can be improved significantly, by making cheap and low-reliability components redundant (the power cable and the switches) (Figure 8.14B). For the system shown in Figure 8.14B, the electric motor $M1$ for example will still operate if the power cable PC or the switch $S1$ fails because power supply will be maintained through the alternative power cable PC′ and the switch $S1'$. The same

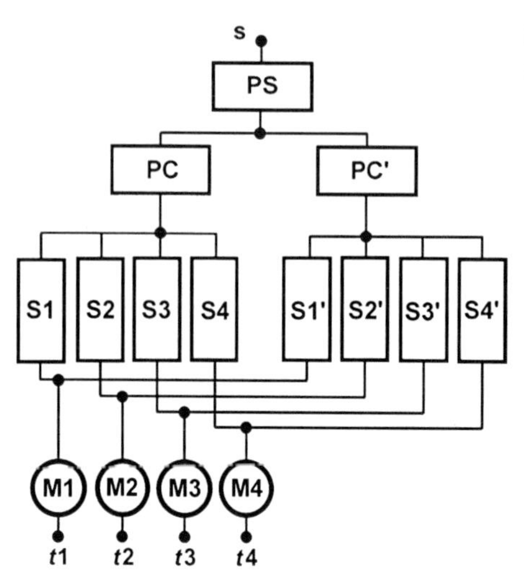

Figure 8.15 A reliability network of the system with redundancy from Figure 8.14B.

applies for the rest of the electric motors. The power supply to an electric motor will fail only if both power supply channels fail. The reliability network of the system in Figure 8.14B is shown in Figure 8.15. It has one start node s and four terminal nodes $t1$, $t2$, $t3$ and $t4$. The system is in working state if a path through working components exists between the start node s and each of the terminal nodes $t1$, $t2$, $t3$ and $t4$.

The reliability network shown in Figure 8.15 (which fails whenever at least one of the electric motors fails to operate on demand) cannot be presented as a series–parallel system. It is a system with complex topology.

Furthermore, in traditional reliability networks, only undirected edges are used (Andrews and Moss, 2002; Billinton and Allan, 1992; Blischke and Murthy, 2000; Ebeling, 1997; Hoyland and Rausand, 1994; Ramakumar, 1993). This traditional representation is insufficient to model correctly the logic of operation and failure. Often, introducing directed edges is necessary to emphasise that the edge can be traversed only in one direction but not in the opposite direction. Consider for example the computer network shown in Figure 8.16A, which consists of a router R, switches $K1-K4$ and four servers $S1-S4$.

Assume for the sake of simplicity that the connecting cables are perfectly reliable. As a result, the reliability of the system shown in Figure 8.16 is determined by the reliability of the router, the switches and the servers. Suppose that a message sent through the router must reach all four servers. The reliability of the system is then defined as '*the probability that the message will reach all servers*'. Because for the system in Figure 8.16A failure of any device (router, switch or server) means that the message will not be received by all servers, the system shown in Figure 8.16A, fails whenever any of the devices fails.

Similarly to the system in Figure 8.14A, the reliability of the system shown in Figure 8.16A can be improved significantly by making some of the devices redundant (e.g. the router and the switches) and by allowing dual communication channels to each server. As a result, from the system shown in Figure 8.16A, the system

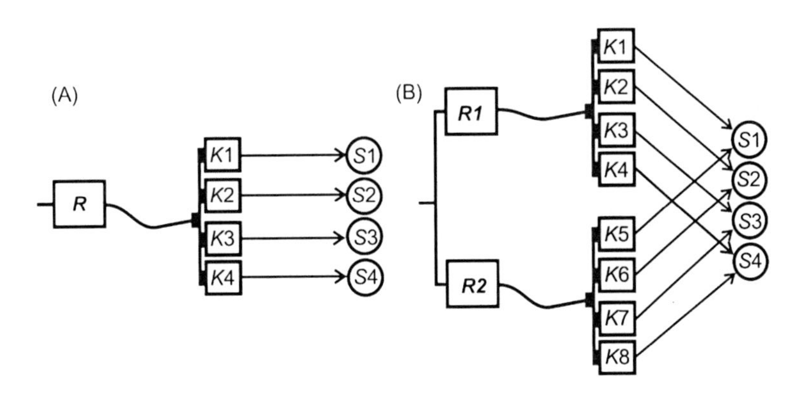

Figure 8.16 An example of a computer network including routers, switches and servers.

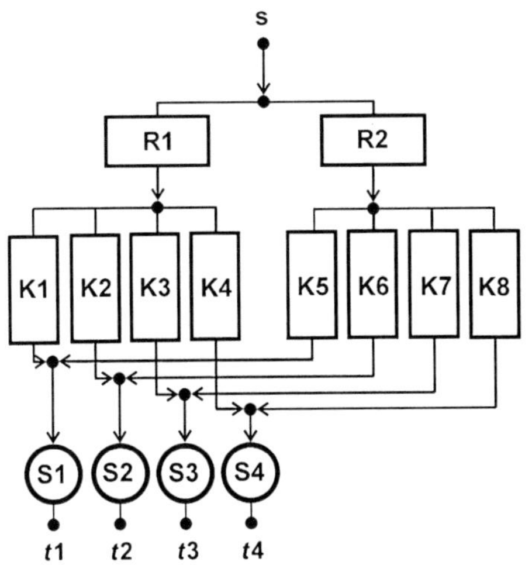

Figure 8.17 A reliability network of the computer network from Figure 8.16B.

shown in Figure 8.16B is obtained. For the system shown in Figure 8.16B, for example, the server $S1$ will still receive the message if the router $R1$ or the switch $K1$ fails. The message will be received through the alternative router $R2$ and the switch $K5$. The same applies to the rest of the servers. The message will not be received only if both communication channels fail.

Despite the seeming similarity between the reliability network of the computer system shown in Figure 8.17 and the reliability network of the system shown in Figure 8.15, there are essential differences. The system shown in Figure 8.15 for example will be fully operational after the failure of power cable PC' and switches $S2$, $S3$ and $S4$ (Figure 8.18A). In contrast, after the failure of router $R2$ and switches $K2$, $K3$ and $K4$, (Figure 8.17) only server $S1$ will receive the message (Figure 8.18B). This is because unlike the current in the power supply system, the message transmitted after the switch $K1$ cannot reach the other servers by travelling backwards through switch $K5$ (Figure 8.18B). This essential difference has been reflected *by introducing directed edges in the reliability network*.

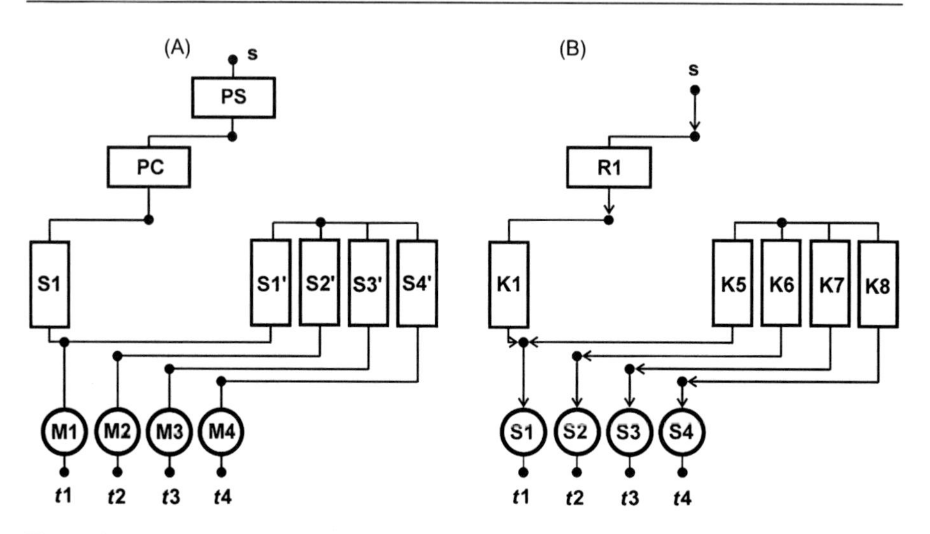

Figure 8.18 An illustration of the difference between the reliability networks shown in Figures 8.15 and 8.17.

Furthermore, it can also be shown that even mechanical systems, with no moving parts, may have reliability networks that cannot be represented as a combination of series–parallel arrangements. Such a system has been represented in Figure 8.19A. The mechanical system consists of a hub '1', on which the parallel plates 2 and 3 have been mounted. Each plate is connected to the static supports 8 and 9 by the links 4, 5, 6 and 7. The plates 2 and 3, and the links attached to each plate provide redundancy. Failure occurs if the hub 1 fails or both plates 2 and 3 fail or one of the supports fails or two links connected to the same support fail.

However, failure of a single plate does not cause a system failure and failure of any two links connected to different supports does not cause a system failure. The reliability network which satisfies each of these constraints is shown in Figure 8.19B. The probability that the system will be working at the end of a specified time interval is equal to the probability that at the end of the specified time interval, there will exist paths through working components from the start node *s* to each of the terminal nodes *t*1 and *t*2.

As can be verified, the reliability network of this simple mechanical system cannot be represented with series and parallel arrangements.

The next example features a system where both directed and undirected edges are necessary for describing correctly the logic of system operation. The safety–critical system shown in Figure 8.20 features two power generators *G*1 and *G*2, delivering current to two electric motors *M*1 and *M*2. The system is in operation if at least a single electric motor is in operation. The identical, independently working power generators *G*1 and *G*2 are controlled by four identical electronic control units *ECU*1, *ECU*2, *ECU*3 and *ECU*4, powered by two power units *PU*1 and *PU*2 (Figure 8.20). The redundant electronic control units guarantee that there will be a control signal to the generators even if some of them are in a failed state.

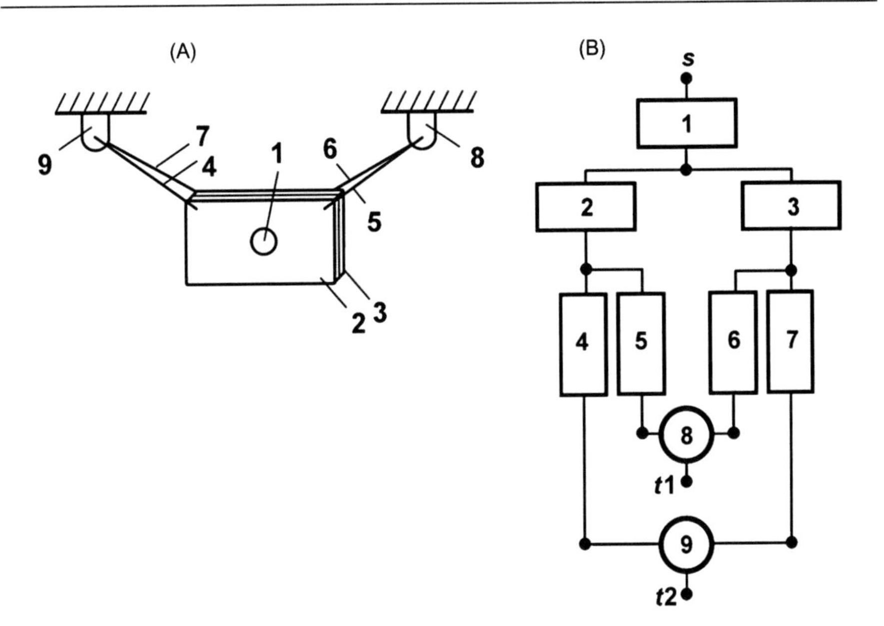

Figure 8.19 A simple mechanical system, whose reliability network cannot be represented with series and parallel arrangements.

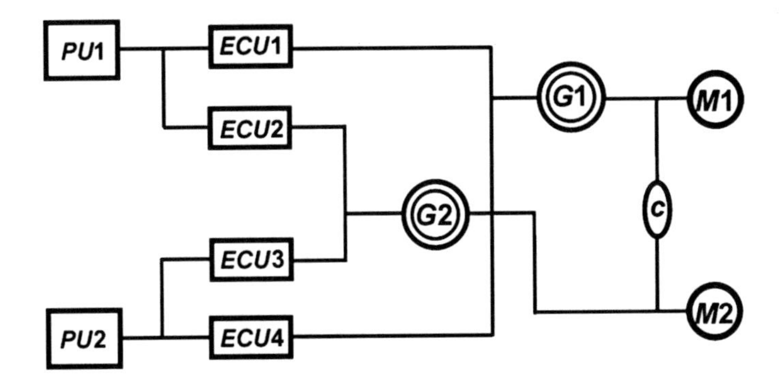

Figure 8.20 Two power generators $G1$ and $G2$ powering two electric motors $M1$ and $M2$. The power generators are controlled by four electronic control units $ECU1-ECU4$, powered by the units $PU1$ and $PU2$.

To further reduce the risk of system failure, a bridge c has also been included. The bridge guarantees the system's operation in the case where both the electric motor $M1$ and the power generator $G2$ are in a failed state at the end of a specified time interval or in the case where both the electric motor $M2$ and the power generator $G1$ are in failed state.

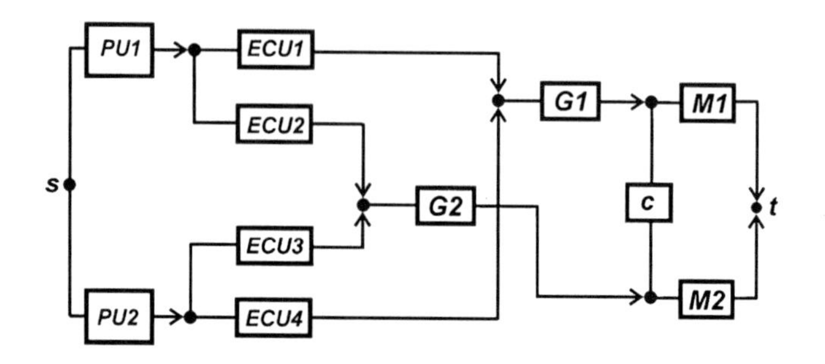

Figure 8.21 Reliability network of the system shown in Figure 8.20. Both directed and undirected edges are necessary to correctly represent the logic of the system's operation.

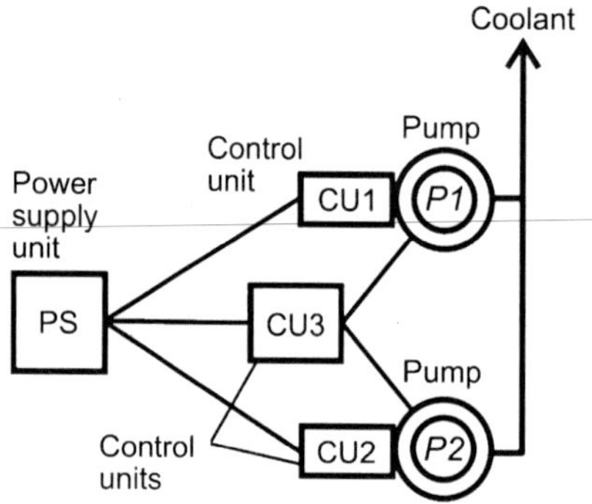

Figure 8.22 Two pumps $P1$ and $P2$ powered by a power supply unit PS and three independently working control units $CU1$, $CU2$ and $CU3$. The pumps deliver coolant to a chemical reactor.

The reliability network of the system from Figure 8.20 is shown in Figure 8.21. As can be verified, both directed and undirected edges are necessary to represent correctly the logic of operation. The electronic control units, for example, cannot be represented by undirected edges. Otherwise, this would mean that a control signal will exist for generator $G1$, if the power unit $PU1$ is in failed state and the power unit $PU2$ is in working state and the electronic control units $ECU1$, $ECU2$ and $ECU3$ are in working state. This is impossible because the control signal cannot travel from $ECU3$ to $G1$ through $ECU2$. The directed edges are necessary *to forbid* such redirection. However, the power cable shown in Figure 8.21 cannot be represented by a directed edge, because the power can travel in both directions, from $G1$ to $M2$ and from $G2$ to $M1$. The edge representing the bridge 'c' must be an undirected edge.

Finally, there exist even systems whose logic of operation requires reliability networks where more than two connecting nodes are necessary for some of the components. The system shown in Figure 8.22, for example, features two pumps $P1$ and $P2$, delivering coolant to a chemical reactor. The identical, independently working pumps $P1$ and $P2$ are controlled by two built-in identical control units $CU1$ and $CU2$. Because the reliability of the control units is insufficient, a separate,

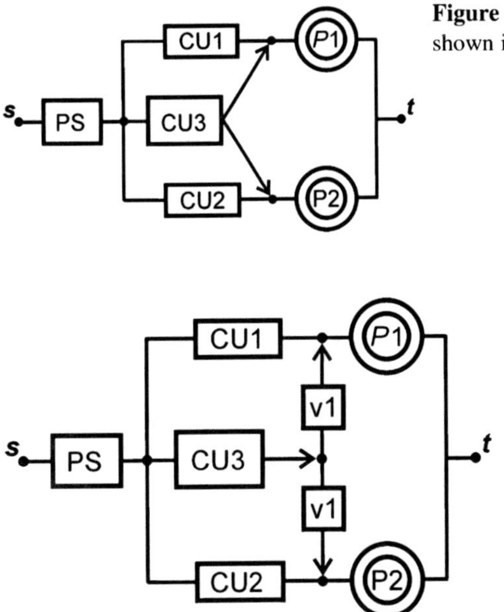

Figure 8.23 Reliability network for the system shown in Figure 8.22.

Figure 8.24 Reliability network for the system shown in Figure 8.23.

external control unit $CU3$ has also been included to provide extra control to both pumps, as shown in Figure 8.22.

The reliability network of the system from Figure 8.22 is shown in Figure 8.23.

As can be verified, failure of the external control unit $CU3$ eliminates control to both pumps $P1$ and $P2$. Consequently, the $CU3$ component has essentially *three connecting nodes* and cannot be modelled as an edge of a graph, where each edge has two connecting nodes. However, by introducing two virtual, perfectly reliable components $v1$ and $v2$, the reliability network from Figure 8.23 can be represented by a graph (Figure 8.24), where each component has exactly two connecting nodes. The virtual components $v1$ and $v2$ never fail. They have been introduced only to enable a graph representation of the network.

8.4 Evaluating the Reliability of Complex Systems

A central question posed for reliability networks is how to calculate the probability of surviving a particular time period of operation a, given the time-to-failure distribution $F_{ij}(t)$ of each component (edge) (i,j). The answer to this question is given by the methods for system reliability analysis discussed in the reliability literature. There exist a number of methods for system reliability analysis oriented mainly towards systems with simple topology. Such are, for example, the method of *network reduction* and the *event-tree method* (Billinton and Allan, 1992); the decomposition *method* (Hoyland and Rausand, 1994; Ramakumar, 1993); and methods based on minimal paths and cut sets (Hoyland and Rausand, 1994). These methods have significant limitations. They are not suitable for large and complex systems. For example, the network reduction method is not suitable for topologically complex systems while the decomposition method is not suitable for large systems.

Methods based on minimal path sets and minimal cut sets are very common (Billinton and Allan, 1992; Hoyland and Rausand, 1994; Kuo et al., 2001; Ramakumar, 1993).

A path is a set of components which, when working, connect the start node with the end node through working components, thereby guaranteeing that the system is in a working state.

A *minimal path* is a path from which no component can be removed without disconnecting the link it creates between the start node and the end node. Consequently, minimal paths are free of loops. In other words, in each minimal path, a particular node may appear only once.

The system reliability can be determined as *the probability of the union of all paths* through an expression known as '*inclusion–exclusion expansion*' (Ebeling, 1997).

A *cut set* is a set of components whose failures cause the system to fail. A *minimal cut set* is a cut set for which no component can be returned in working state without returning the system into working state. The probability of system failure is determined as the probability of the union of all minimal cut sets in the system, through the inclusion–exclusion expansion expression (Ebeling, 1997).

Both the minimal paths method and the minimal cut set method require all paths or cut sets in the system to be known in advance. In Chapter 1, we demonstrated that *the number of minimal paths and cut sets increases exponentially with increasing the size of the network.*

Even for moderately large systems, the total number of paths or cut sets is impossible to determine. This constitutes the main drawback of methods based on minimal paths and minimal cut sets. Although they work well for small-size systems, for moderately large and large systems they are not feasible.

System reliability, however, can be defined as a probability of existence of a path through working components from the start node to each of the end nodes in the reliability network, at the end of the specified time interval of operation. *A reliability network can be interpreted as a repairable flow network whose edges have integer capacities equal to the number of edges in the reliability network and availabilities equal to the reliabilities of the edges in the reliability network.* A unit flow can be sent from the source to each of the end nodes of the flow network, whenever a path through working components exists, from the start node to each of the end nodes in the reliability network.

The requirement for each edge to have integer capacity equal to the number of edges in the network, is necessary to guarantee that there can be no bottleneck in the flow network caused by augmenting paths with unit flow from the source s to each of the terminal nodes.

Indeed, the maximum throughput flow in the repairable flow network modelling the reliability network is not larger than the capacity of its minimal cut. Because the integer capacity of a single edge in the repairable flow network is equal to the number of edges in the network, the capacity of the minimal cut cannot be smaller than the number of edges in the network. However, the number of end nodes cannot be larger than the number of edges in the network.

Consequently, the minimal cut in the repairable flow network can never be saturated by augmenting paths with unit flow from the source to each of the end nodes. This means that there can be no bottleneck caused by augmenting paths with unit flow from the source s to each of the terminal nodes.

Considering this, any reliability network can be modelled as a repairable flow network with edges with integer capacities equal to the number of edges in the network. The availability of each edge is assumed to be equal to the probability that the edge will survive without failure the specified time interval a. Consequently, the system reliability analysis on complex networks can be performed by the tools developed for repairable flow networks. The task of determining the system reliability is now replaced with the task of determining *the probability that, on demand, unit flow paths can be augmented from the source to each of the terminal nodes*.

The analysis of reliability networks, however, can be further simplified by noticing that reliability can simply be determined as *probability of existence on demand of directed paths with nonzero flow capacity from the source node to each of the terminal nodes*. The probability of existence of a directed path with nonzero flow capacity is estimated on the basis of a Monte Carlo simulation. The next algorithm, in pseudo-code, determines the reliability of a complex system with k terminal nodes.

Algorithm 8.1

```
function paths_to_all_sinks();
function real_random();

function system_reliability();
{
  success_cnt = 0;
  for i = 1 to num_trials do
  {
    for j = 1 to m do
      {
        tmp = real_random();
        if (tmp > p[j]) then reduce the flow capacity of edge j to zero;
      }
    paths_exists = paths_to_all_sinks();
    if (paths_exists = 1) then success_cnt = success_cnt + 1;
    Restore the flow capacities of all failed edges to their original levels;
  }
    return success_cnt/num_trials;
}  ■
```

The array $p[]$ contains the availabilities of the separate edges, which are equal to the probabilities that the separate edges will be in working state at the end of a specified time interval with length a. If the time-to-failure distribution of the ith edge is given by $F_i(t)$, then $p[i] = 1 - F_i(a)$.

In a nested loop controlled by the variable 'j', the state of the separate edges (working/failed) is determined. The state of the jth edge is tested by generating a uniformly distributed random number between 0 and 1 from the statement 'tmp = ***real_random()***', and comparing it with the probability p [j] that the jth edge will be in working state. The number of edges is m.

All availabilities p_j ($j = 1,2, \ldots ,m$) characterising the edges are pre-calculated and stored in the array $p[]$. If the generated random number 'tmp', uniformly distributed between 0 and 1, is greater than p [j] then the jth edge is in a failed state and is removed from the network by reducing its flow capacity to zero. If the converse is true, the edge remains in the network.

After the state of all edges has been determined, the function ***paths_to_all_sinks()*** establishes whether there exist directed paths consisting of edges with nonzero flow capacity, from the source to all terminal nodes, after essentially removing the failed edges by reducing their flow capacities to zero. The ratio of the number of trials for which directed paths to the terminal nodes exist and the total number of simulation trials, is a measure of the reliability of the system. At the end of each simulation trial, the capacities of all failed edges are restored to their original level.

In the case of very small availabilities (reliabilities) characterising the edges, the precision of the presented Monte Carlo crude sampling can be increased by applying the '*stratified sampling without replacement*' discussed in Chapter 7.

The algorithm of the function determining the existence of paths from the source to each of the terminal nodes is similar to the algorithm from Chapter 2, introduced for determining the existence of a path from the source to a single terminal node. A connection between two neighbouring nodes i and j exists only if edge (i,j) has not been removed because of failure. Removing an edge upon failure is done by setting its flow capacity to zero. If edge (i,j) exists (has not been removed), a connection from node i to node j is present if edge (i,j) is an undirected edge or if it is directed from node i to node j. If edge (i,j) is directed from node j to node i, no connection from node i to node j exists. The possible connections among edges are specified in the adjacency arrays nb[].

The variable 'sinks_visited' counts the number of terminal nodes that have been visited by the breadth-first search algorithm.

Because the largest node indices are always reserved for the terminal node, if the number of terminal nodes is 'num_sinks' the statement '**if** (node >= n − num_sinks + 1) **then**' checks whether the visited node is a terminal node. Every time a terminal node (sink) is visited, the variable 'sinks_visited' is incremented. On visiting a terminal node, it is also marked in the marked[] array as visited by the statement 'marked[node] = 1;' Nodes marked by '1' in the marked[] arrays are never visited again. Therefore, there is no possibility that the same terminal node can be visited twice. The statement '**if** (sinks_visited = num_sinks) **then** **return** 1;' checks whether the number of visited terminal nodes is equal to the actual number 'num_sinks' of all terminal nodes. If this is the case, then the system is operational and the function returns '1'.

Algorithm 8.2

```
function paths_to_all_sinks()
  {
     marked[1] = 1; sinks_visited = 0;
     qhead = 0; qend = 1; queue[1] = 1;

     for i = 2 to n do marked[i] = 0;
   while (qhead < qend) do
   {
     qhead = qhead + 1;
    r_node = queue[qhead];
   for i = 1 to nb[r_node][0] do
     {
       node = nb[r_node][i];
       if (marked[node] = 0 and capacity of edge (r_node, node) > 0) then
                 {
                 marked[node] = 1;
                 if (node >= n - num_sinks + 1) then {
                                   sinks_visited = sinks_visited + 1;
                                   if (sinks_visited = num_sinks) then return 1;
                                                }
                 else {
                                 qend = qend + 1;
                                 queue[qend] = node;
                                       }
                 }
     }
   }
   return 0;
 }
```

Consider now the system shown in Figure 8.22. For such systems, it is important to know the probability that within a specified time interval t_a: (a) there will be supply of coolant to the reactor, which is present if at least one pump delivers coolant and (b) there will be a full supply of coolant, which is present if both pumps are working.

Suppose, for the sake of simplicity, that the connecting cables are perfectly reliable and the separate components are characterised by a negative exponential time-to-failure distribution $F(t) = 1 - \exp(-\lambda t)$, where λ is the hazard rate and t is the time. Let us denote by λ_{CU} the hazard rates of the control units $CU1$, $CU2$ and $CU3$. Accordingly, λ_P denotes the hazard rate of the pumps and λ_{PS} is the hazard rate of the power supply. Then, the reliabilities R_i of the separate components are:

$R_{CU1} = R_{CU2} = R_{CU3} = R_{CU} = \exp(-\lambda_{\mathrm{CU}} t_a)$, $R_{P1} = R_{P2} = R_P = \exp(-\lambda_P t_a)$ and $R_{\mathrm{PS}} = \exp(-\lambda_{PS} t_a)$, correspondingly.

The reliability network of the system is shown in Figure 8.25A.

If S denotes the event 'the system works', the probability $P(S)$ that the system works (the reliability of the system) can be obtained by the *decomposition method,* by selecting the external control unit $CU3$ as a key component and by conditioning the probability that the system will work on the state of the key component. According to the total probability theorem:

$$P(S) = P(S|CU3) \times P(CU3) + P(S|\overline{CU3}) \times P(\overline{CU3}) \tag{8.3}$$

where $P(S|CU3)$ is the probability that the system is operational, given that component $CU3$ works; $P(S|\overline{CU3})$ is the probability that the system is operational given that component $CU3$ does not work; $P(CU3)$ is the probability that component $CU3$ works and $P(\overline{CU3})$ is the probability that component $CU3$ does not work. The probability $P(S|CU3)$ that the system in Figure 8.25B works, given that component $CU3$ works, is $P(S|CU3) = R_{\mathrm{PS}} \times [1 - (1 - R_P)^2]$. The probability $P(S|\overline{CU3})$ that the system in Figure 8.25C works, given that component $CU3$ does not work is $P(S|\overline{CU3}) = R_{PS} \times [1 - (1 - R_{CU} R_P)^2]$.

Finally, substituting the expressions in Eq. (8.3) yields

$$P(S) = R_{CU} \times R_{PS}[1 - (1 - R_P)^2] + (1 - R_{CU}) \times R_{PS}[1 - (1 - R_{CU} R_P)^2] \tag{8.4}$$

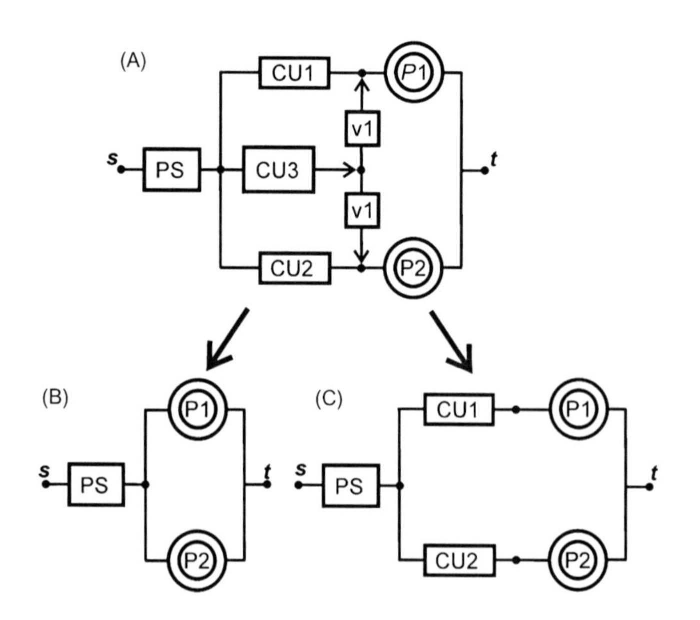

Figure 8.25 Reliability network decomposition of the system shown in Figure 8.23.

For the sake of simplicity, let us assume that all components are characterised by the same hazard rate of 0.05 day^{-1}, and $t_a = 3$ days. The reliability of each component, associated with the time interval $t_a = 3$ days, is then $R = \exp(-\lambda t_a) = 0.86$.

After substituting the numerical values for the component reliabilities in Eq. (8.4), $P(S) = 0.837$ is obtained for the system reliability

The first step in determining the reliability of the system by using Algorithm 8.1 is to calculate the availabilities of the separate components, equal to their reliabilities associated with the specified time interval t_a.

The reliability of the system shown in Figure 8.24 can then be modelled by the repairable flow network shown in Figure 8.26, where edge (1,2) represents the power supply *PS*; edge (2,3) represents the control unit *CU*1; edge (2,5) is the control unit *CU*2; edge (2,4) is the external control unit *CU*3; edge (3,6) is the pump *P*1 and edge (5,6) is the pump *P*2. Edges (4,3) and (4,5) are perfectly reliable and have been introduced to represent correctly the system logic.

The availabilities of the separate edges are $p[1] = \ldots = p[6] = 0.86$.

A supply of coolant to the reactor is present if and only if, on demand, there exists a directed path from the source *s* to the sink *t*. Executing Algorithm 8.1 gives 0.836 for this reliability which confirms the analytical result obtained from the system reliability analysis.

In addition, the described approach also delivers the probability of a full fluid supply (both pumps working. Directed paths must now exist to both sinks t1 and t2, for a full fluid supply to be present. Running Algorithm 8.1 yields 0.613 for this probability. Again this result can be verified by the decomposition method.

If edge (2,4) is selected as a key edge, the probability that there will be directed paths to each of the sinks t1 and t2 can be obtained by the *decomposition method*, which essentially conditions this probability on the state of edge (2,4). Thus, if *S* denotes the event 'there is a directed path to each of the sinks', according to the total probability theorem:

$$P(S) = P(S|(2,4)) \times P((2,4)) + P(S|\overline{(2,4)}) \times P(\overline{(2,4)}) \tag{8.5}$$

where $P(S|(2,4))$ is the probability that there is a directed path to each of the sinks, given that edge (2,4) is working; $P(S|\overline{(2,4)})$ is the probability that there is a

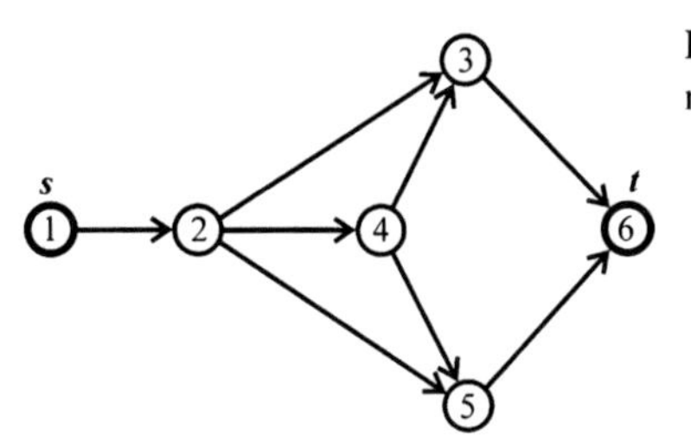

Figure 8.26 The repairable flow network which models the system shown in Figure 8.24.

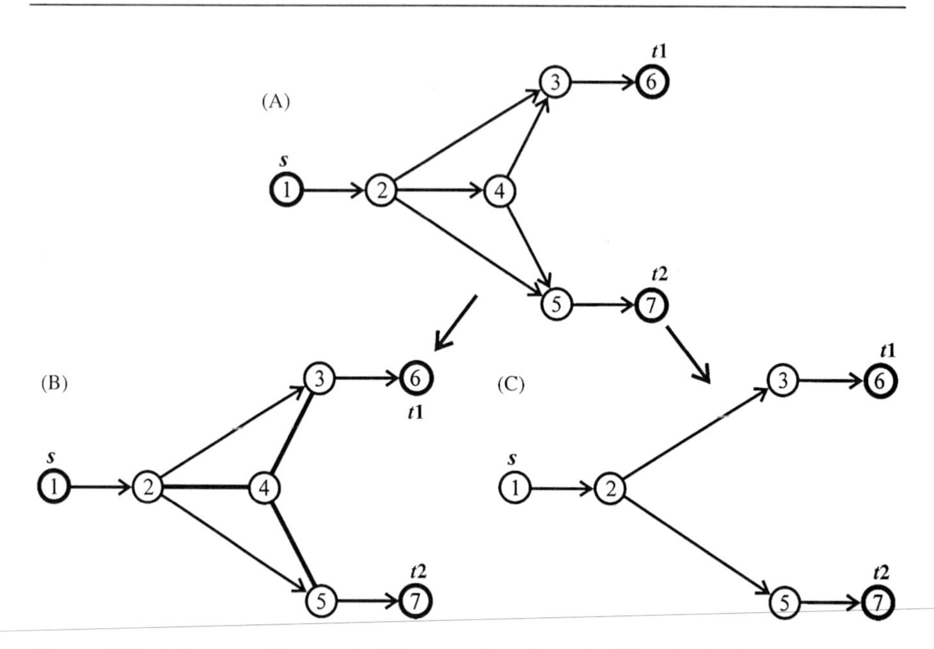

Figure 8.27 A decomposition method for revealing the probability that both pumps in the system from Figure 8.24 will be working.

directed path to each of the sinks, given that edge (2,4) is not working; $P((2,4))$ is the probability that edge (2,4) is in working state and $P(\overline{(2,4)})$ is the probability that edge (2,4) is in failed state. $P(S|(2,4)) = R^3$, because the edges (1,2), (3,6) and (5,7) must be in working state in order to guarantee a directed path to each of the sinks. Edges (4,3) and (4,5) are perfectly reliable. They correspond to the virtual, perfectly reliable components $v1$ and $v2$ in Figure 8.25A. Edge (2,4) is also working, according to the conditioning assumption. For the network in Figure 8.27C, $P(S|\overline{(2,4)})$ is the probability that there will be a directed path to each of the sinks. This probability is $P(S|\overline{(2,4)}) = R^5$, because all five edges must be in working state so that directed paths to each of the sinks exist.

Finally, the substitution of the expressions in Eq. (8.5) yields

$$P(S) = R \times R^3 + (1 - R) \times R^5 \quad (8.6)$$

Substituting the numerical values of the component reliabilities gives $P(S) = 0.613$. This result confirms the simulation result.

The system shown in Figure 8.16A can be modelled by the repairable flow network shown in Figure 8.28A. Suppose, for the sake of simplicity, that the times to failures of all components follow the negative exponential distribution, with a parameter MTTF = 300 h. It is required to calculate the reliability of the system associated with 24 h of operation.

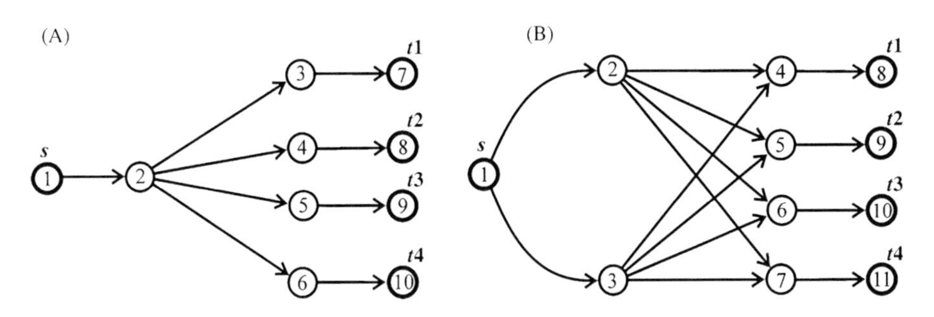

Figure 8.28 The reliabilities of the systems shown in Figure 8.16 can be analysed successfully by repairable flow networks.

Edge (1,2) corresponds to the router R, edges (2,3), (2,4), (2,5) and (2,6) correspond to the switches $K1-K4$ and edges (3,7), (4,8), (5,9) and (6,10) correspond to servers $S1-S4$. The system is in operation whenever, on demand, a directed path exists from the source s to each of the four sinks, $t1$, $t2$, $t3$ and $t4$.

The system shown in Figure 8.16B is a complex system and cannot be modelled by a series-parallel network. Its reliability, however, can also be revealed easily by applying Algorithm 8.2 to the network shown in Figure 8.28B, which is the reliability network of the system in Figure 8.16B.

The execution of Algorithm 8.1 yields 0.488 for the reliability of the system shown in Figure 8.16A, and 0.678 for the reliability of the system shown in Figure 8.16B.

The result related to the system in Figure 8.16A can be verified easily. The reliability of a single component is $p = \exp(-24/300) = 0.9231$.

In order for the system to be operational on demand, all of the devices must be working. This means that the reliability of the system in Figure 8.16A is simply given by $p^9 = 0.487$, which confirms the simulation result.

The result related to the system in Figure 8.16B can also be verified. This can be done by conditioning the corresponding reliability network in Figure 8.28B, on the state of edges (1,2) and (1,3). There are four different, mutually exclusive outcomes: (i) edge (1,2) working, edge (1,3) working; (ii) edge (1,2) working, edge (1,3) failed; (iii) edge (1,2) failed, edge (1,3) working and (iv) edge (1,2) failed, edge (1,3) failed. Depending on these outcomes, the resultant networks are shown in Figure 8.29.

The network in Figure 8.29A corresponds to the outcome where both edges (1,2) and (1,3) are working. As a series−parallel network, the probability of existence of directed paths to each sink is $p^4[1-(1-p)^2]^4$. The network shown in Figure 8.29B corresponds to each of the two outcomes where exactly one of the edges (edge (1,2) or edge (1,3)) is working. The network shown in Figure 8.29B is a simple network, for which the probability of existence of paths to each sink is

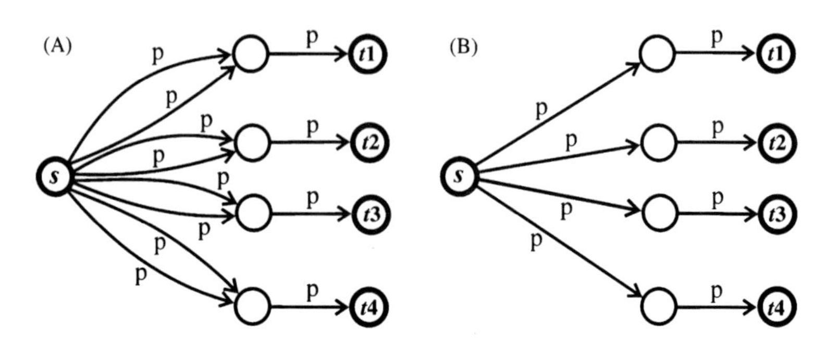

Figure 8.29 The product networks after conditioning the network from Figure 8.28B on the state of edges (1,2) and (1,3).

p^8. The conditioning on the four distinct outcomes, formed by the possible states of the two edges, yields:

$$P(S) = p^2 \times p^4[1-(1-p)^2]^4 + p(1-p) \times p^8 + (1-p)p \times p^8 + (1-p)(1-p) \times 0 \tag{8.7}$$

for the probability that on demand, there will be directed paths from the source s to each of the four sinks. Substituting the numerical value $p = 0.9231$ in Eq. (8.7), yields $P(S) = 0.679$, which confirms the result obtained from running Algorithm 8.1.

9 Production Availability of Repairable Flow Networks

9.1 Discrete-Event Solver for Determining the Production Availability of Repairable Flow Networks

An important performance measure of repairable flow networks is the *production availability*. This is *the expected fraction of throughput flow transmitted during a specified time interval in the presence of component failures*. It is defined as the ratio

$$\psi = \overline{Q}_r / Q_0 \tag{9.1}$$

of the maximum expected throughput flow $\overline{Q}_r$ of the repairable flow network during a specified time interval $[0,a]$ and the maximum throughput flow Q_0 that could be transferred from source to sink, during this time interval in the absence of failures (Figure 9.1).

Another related performance characteristic is the cumulative distribution of the total throughput flow Q_r during a specified time interval:

$$\Phi(x) \equiv \Pr(Q_r \leq x) \tag{9.2}$$

The cumulative distribution $\Phi(x)$ of the total throughput flow reveals the variation caused by component failures and *gives the probability* $\Pr(Q_r \leq x)$ *that the total throughput flow* Q_r *in the presence of component/edge failures, during a specified time interval, will fall below a specified level* x.

The variation of the total throughput flow Q_r, transferred from source to sink of a repairable network, is a function of the times-to-failure and times-to-repair distributions of the components. To reveal the variation of the total throughput flow, a large number of failure–repair histories during the period of operation of the network must be simulated (Figure 9.2A).

On each edge failure, the flow capacity of the failed edge decreases to zero, which causes a decrease in the throughput flow of the network (Figure 9.2B). A maximisation of the throughput flow upon a component failure can be performed to make the best of the remaining residual capacity of the working edges. This strategy increases significantly the total throughput flow in the network.

Flow Networks. DOI: http://dx.doi.org/10.1016/B978-0-12-398396-1.00009-X

With increasing the number of components/edges in a flow network, the likelihood of component failures, whose repair times overlap, increases significantly. Tracking the variations of the total throughput flow caused by components whose repair times overlap is necessary for determining the production availability correctly. The problem has been illustrated in Figure 9.2B, which depicts a variation of the throughput flow rate in a repairable network, caused by multiple overlapping failures and repairs. If a component failure chokes the throughput flow, it causes a 'dent' or a decrease in the throughput flow rate (Figure 9.2B). Conversely, a return from repair of a failed component, if it results in an increase of the throughput flow, causes a rise in the throughput flow rate. Because overlapping failures and repairs can be nested in a very complex fashion, tracking the changes of the throughput flow rate is not an easy task.

This computational difficulty can be avoided if the value of the throughput flow rate is calculated on each failure of a component and return from repair. Furthermore, different failure–repair histories (Figure 9.2A) yield a different level of the throughput flow rate (Figure 9.2B) and, as a result, a different level of the total throughput flow is obtained, during the time of operation t_{op} of the network.

Another big advantage from building a simulation software tool for determining the expected amount of throughput flow in a repairable flow network is the possibility for establishing important links between topology and properties of repairable flow networks. Such a tool, for example, can be used for determining what type of network topology makes the throughput flow rate relatively insensitive to variations caused by component failures.

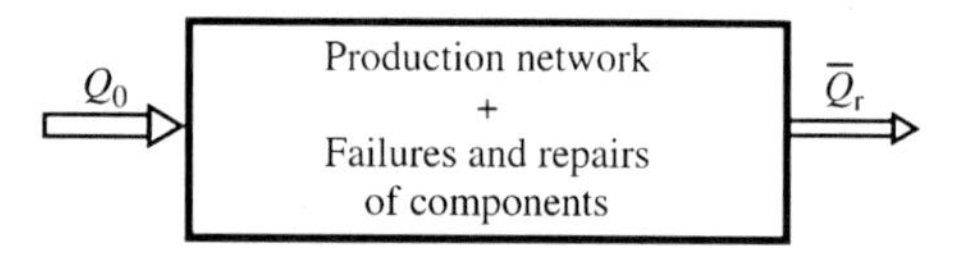

Figure 9.1 For a production network, the expected throughput flow $\overline{Q}_r$ in the presence of failures is always smaller than the throughput flow Q_0 in the absence of failures.

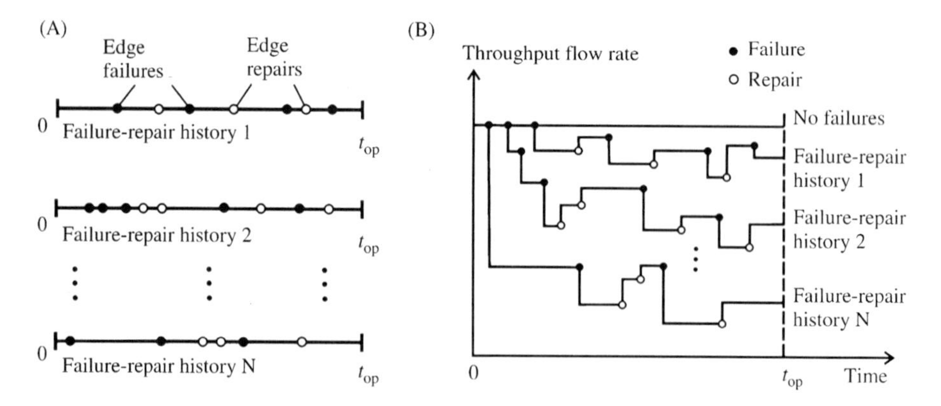

Figure 9.2 The variation of the throughput flow rate with time, as a result of the specific failure–repair history of the flow network.

The expected operating periods for some repairable flow networks can be many years, the number of sources of flow may reach hundreds, and the components may number thousands. To optimise production performance, a large number of alternative design layouts must be analysed in a short period of time. Furthermore, repairable flow networks (especially telecommunication networks) are characterised by a complex topology.

As a result, an important requirement from the discrete-event solvers determining the variation of the total throughput flow is a high computational speed and the capability of handling large repairable flow networks with complex topology.

Maximising the throughput flow in a repairable flow network, although important, does not reveal the variation of the total throughput flow caused by component failures. This variation, which is a key to determining the risk that the total throughput flow will fall below a required level, can be revealed by building a discrete-event solver, whose algorithm in pseudo-code is presented next (Todinov, 2011a).

Algorithm 9.1

function max_flow(); *// Returns the maximum throughput flow in the network*

for i = 1 **to** num_trials **do**
{
cur_flow = 0; cur_time = 0; prev_time = 0;
Generate new lives for all components and place the times to failure of the components in a linked list, in ascending order;

repeat
{
cur_time = *The smallest time to an event, taken from the head of the linked list;*
maxf = ***max_flow()***;
if (cur_time > t_{op}) **then**
{
cur_flow = cur_flow + (t_{op} − prev_time) × maxf;
break;
}
cur_flow = cur_flow + (cur_time − prev_time) × maxf;
if (*the event from the linked list is a component failure*) **then**
{
Reduce the flow capacity of the failed component to zero;
Place the time of return from repair of the component in the ordered linked list;
After a delay determined by the downtime for repair of the failed component, generate a new time to failure for the component and place it in the ordered linked list;
}
else if (*the event is a component repair*) **then** ***Update the capacity of the repaired component, to its initial level;***

```
        prev_time = cur_time;
    }
  until (a break-statement is encountered in the loop);
  cumul_array[i] = cur_flow;
  }
```

Sort the total throughput flows in cumul_array[], ***from all simulation trials***;
Determine the expected amount of transmitted flow, in the presence of component failures. ■

The algorithm includes a simulation loop with control variable i, which executes the block of statements in the braces 'num_trials' number of times. The variable 'prev_time' keeps track of the time of the previous event. The variable t_{op} is the specified total operation time of the network.

At the beginning of the simulation loop, times to failure are generated for all components in the network and are placed in a queue of events (linked list), in ascending order. A second **repeat-until** loop is subsequently entered, which is terminated when the end of the operating period of the network is reached. The smallest time to event is obtained at the head of the list of events. If a statement **break** is encountered in the body of the repeat-until loop, the algorithm execution continues with the next statement ('cumul_array[i] = cur_flow;') immediately after the repeat-until loop, skipping all statements between the statement **break** and the end of the loop. The statement **break** is activated when the current event occurrence is larger than the operating period t_{op} of the network.

Next, the length of the time interval from the previous event (within which no failure or return from repair events exist) is calculated and used to determine the amount of maximum throughput flow since the occurrence of the previous event.

The maximal throughput flow is calculated by calling the function **max_flow()**. The total throughput flow in the current simulation trial is accumulated in the variable 'cur_flow'.

If the current event is a 'component failure', the flow capacity of the component is reduced to zero. After a delay determined by the downtime for repair of the component, a new time to failure is generated for the component and placed in the ordered list of events. In the case where the event is a 'return from repair', the flow capacity of the repaired component is restored to its original level.

At the end of each simulation trial, the statement 'cumul_array[i] = cur_flow' stores the accumulated flow in the array 'cumul_array[]'. At the end of all simulation trials, this array is sorted in ascending order and used to plot the cumulative distribution of the total throughput flow. The expected amount of the total throughput flow during the entire operating period t_{op} of the network is also obtained.

The proposed algorithm *does not impose any restrictions on the times-to-failure distributions of the edges and on the times-to-repair distributions*. For example, the times to failure of the edges may follow a negative exponential, Weibull, log-normal distribution, etc. Similarly, the times-to-repair distributions characterising the edges could be log-normal, Gaussian, negative exponential, Weibull or constant.

The developed discrete-event solver has been applied for calculating the variation of the total throughput flow in a section of a repairable gas production network (Figure 9.3). The considered gas production section consists of three sources (initial injection stations) *s*1, *s*2 and *s*3. Each injection station (source) has a production capacity of 70×10^3 m^3/day. From the initial injection stations, through a system of pipelines, compressors and valves, the gas is delivered to the sink *t*. Each edge models a pipeline section, with an unreliable compressor transporting the gas through it. Each pipeline section is therefore associated with a particular failure rate, as well as flow capacity. It is also assumed that failure of a pipeline section (edge) causes the flow through the corresponding edge to stop.

For the sake of simplicity, constant hazard rates have been assumed for the pipeline sections (edges) and the initial injection stations. On each edge, the hazard rates (in years^{-1}) and the capacities of the edges (in thousands of cubic metres per day) have been given as two consecutive numbers. For example, 4/60 stands for a hazard rate 4 year^{-1} and flow capacity of 60,000 m^3 a day. The initial injection stations have hazard rates equal to 2 year^{-1}.

For the sake of simplicity, the downtime for repair of each edge has been assumed to be the same: $d = 11$ days. If no failures occur, the maximum throughput flow in the network is 150×10^3 m^3/day.

The proposed discrete-event solver also handles repairable networks with complex topology, including cycles. For example, the production network shown in Figure 9.3A contains the cycle (4,6,5,4).

In Figure 9.3B, the gas production network with three sources (*s*1, *s*2 and *s*3) from Figure 9.3A has been transformed into the single-source network from Figure 9.3B.

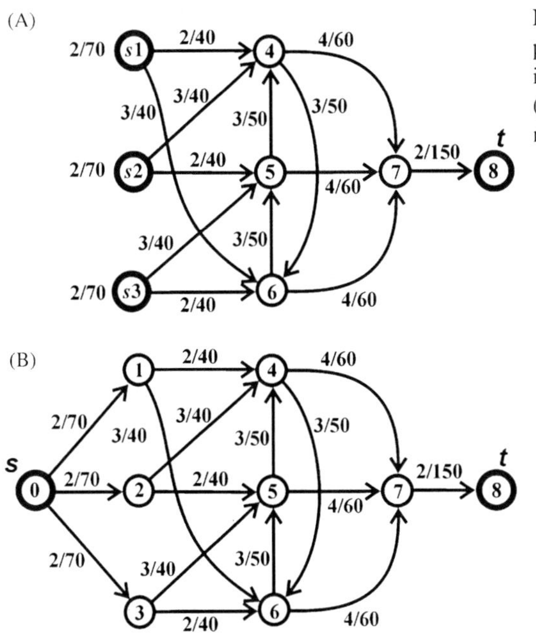

Figure 9.3 (A) A repairable gas production network based on three initial injection stations *s*1, *s*2 and *s*3. (B) Transforming the gas production network into a single-source network.

The edges (0,1), (0,2) and (0,3), connecting the sources $s1$, $s2$ and $s3$ to the super-source s, are characterised by the failure rates and the flow capacities of the sources $s1$, $s2$ and $s3$ they replace.

Two important characteristics for the network shown in Figure 9.3 were determined. The first is the ratio defined by Eq. (9.1), of the expected amount of total throughput flow for 1 year, in the presence of edge failures and the total amount of throughput flow for 1 year, in the absence of failures. For the network shown in Figure 9.3, $\psi = \overline{Q}/Q_{\max} = 0.854$ has been determined by the discrete-event solver. The second characteristic $\Phi(x)$ is the distribution of the total throughput flow during 1 year of operation, in the presence of failures. For the network shown in Figure 9.3, the distribution of the total throughput flow, during 1 year of operation, is shown in Figure 9.4.

The lost flow caused by component failures can be accounted for easily. This makes the solver particularly suitable for tracking the lost flow in case of heavily overlapping repair times. The cost of intervention for replacing failed components, as well as the cost of lost production, can also be tracked. This has been demonstrated in Todinov (2009), where the total distribution of the potential losses from failures for multi-commodity production networks with tree topology, has been obtained. From the obtained distribution, the probability that the potential losses from failures will exceed a particular critical value can be determined. This probability is an important risk measure related to the production economics, particularly to the production economics of oil and gas production fields.

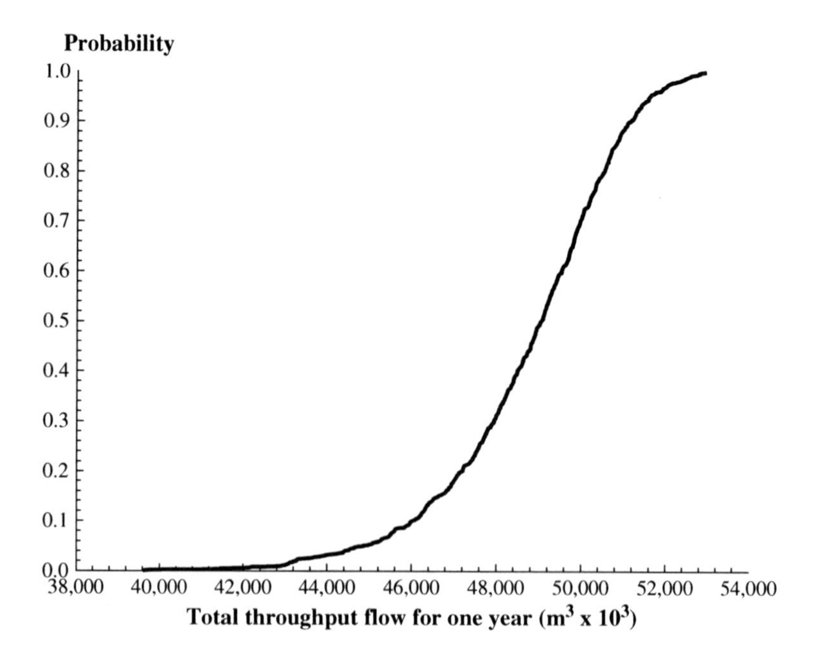

Figure 9.4 Distribution of the total throughput flow during 1 year of operation, of the network in Figure 9.3.

The discrete-event simulator can be applied for determining the variation of the throughput flow in production systems, communication networks, power networks transportation networks, etc. The discrete-event solver can even be modified to control communication networks by optimal redirection of the information flow upon component failures, thereby maintaining the required level of service.

9.2 A Fast Algorithm for Determining the Production Availability of Repairable Flow Networks

In the common case of times to failures following the negative exponential distribution, to calculate the production availability ψ in Eq. (9.1), a novel and very fast algorithm has been proposed in Todinov (2012b), whose computational speed *does not depend on the length of the operating interval, on the failure frequencies or on the lengths of the repair times.*

Suppose that the network is composed of independently working edges, each characterised by a mean time to failure MTTF_i and mean time to repair MTTR_i. The unavailability of the ith edge is given by $q_i = \text{MTTR}_i/(\text{MTTF}_i + \text{MTTR}_i)$ (Trivedi, 2002).

Define probabilities $q_i = \text{MTTR}_i/(\text{MTTF}_i + \text{MTTR}_i)$ with which the individual edges will be removed/deleted from the network, at a given (fixed) time of demand $x = \theta$. The maximum throughput flow in the network, at time $x = \theta$, after removing the edges, is denoted by $Q^{\text{T}}(\theta)$. The average of the maximum throughput flow rates $Q_i^{\text{T}}(\theta)$ on demand, calculated after removing with probabilities q_i the separate edges of the network, will be denoted by $\overline{Q}^{\text{T}}(\theta)$. This quantity is defined by

$$\overline{Q}^{\text{T}}(\theta) = \lim_{n\to\infty}\left(\frac{1}{n}\sum_{i=1}^{n} Q_i^{\text{T}}(\theta)\right)$$

Suppose that Q_0 is the maximum throughput flow rate in the network, in the absence of failures. Then, the following theorem holds.

Theorem 9.1 *If the edge times to failure follow a negative exponential distribution on a specified time interval, the average production availability ψ of the repairable flow network is given by the ratio $\psi = \overline{Q}^{\text{T}}(\theta)/Q_0$.*

Proof For a large operational cycle with length a, the average production availability ψ of a repairable flow network is given by

$$\psi = \frac{1}{Q_0 \times a}\int_0^a Q^{\text{T}}(x)\text{d}x \tag{9.3}$$

This is the ratio of the integral of the maximum throughput flow rate $Q^{\mathrm{T}}(x)$ in the presence of failures, at time x, divided by the product of the maximum throughput flow rate Q_0 in the absence of failures and the length a of the specified time interval. The integral in Eq. (9.3), however, can be determined with any specified precision by a Monte Carlo simulation. This involves a sufficiently large number (n) of random samples of the time interval with length a, each of which determines the maximum throughput flow rate Q_i^T. According to the theory of Monte Carlo integration (Glasserman, 2003; Ross, 1997; Rubinstein, 1981),

$$\int_0^a Q^{\mathrm{T}}(x)\mathrm{d}x = \lim_{n\to\infty}\left(\frac{a}{n}\sum_{i=1}^{n} Q_i^{\mathrm{T}}\right) \tag{9.4}$$

Consequently, Eq. (9.3) can be presented as

$$\psi = \lim_{n\to\infty}\frac{(1/n)\sum_{i=1}^{n} Q_i^{\mathrm{T}}}{Q_0} \tag{9.5}$$

At a random point in time however (Figure 9.5), corresponding to the ith sampling, the maximum throughput flow Q_i^T in the network is determined by considering the state of the components in the network at that particular time. Some of the components (edges) will be in working state while some will be in failed state. The probability that a particular edge/component will be in failed state is $q_j = \mathrm{MTTR}_j/(\mathrm{MTTF}_j + \mathrm{MTTR}_j)$ (Trivedi, 2002). A failed edge has a zero capacity. Consequently, if the network consists of M edges, the maximum throughput flow Q_i^T in the network corresponding to the ith sampling is determined after removing the separate edges from the network with probabilities $q_j = \mathrm{MTTR}_j/(\mathrm{MTTF}_j + \mathrm{MTTR}_j)$, $j = 1,\ldots,M$. Because the failure times follow a negative exponential distribution, the probability that an edge will be in a working or a failed state does not depend on the actual sampling time along the time interval $(0,a)$. The pattern of failures and repairs will be the same, steady-state pattern, along the entire time interval $(0,a)$, and

$$\sum_{i=1}^{n} Q_i^{\mathrm{T}} = \sum_{i=1}^{n} Q_i^{\mathrm{T}}(\theta)$$

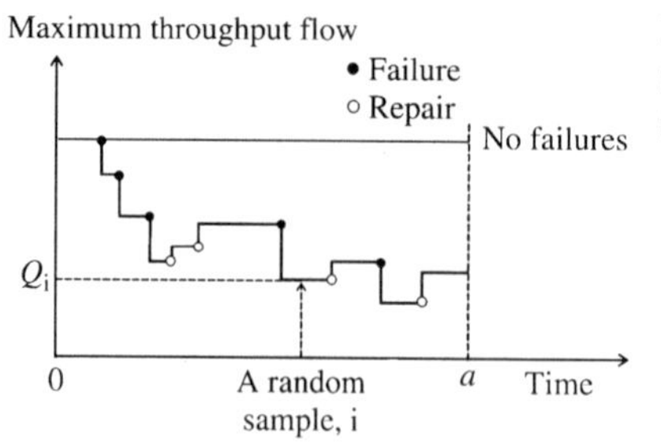

Figure 9.5 Variation of the maximum throughput flow as a function of the failure/repair history.

where $\sum_{i=1}^{n} Q_i^{\mathrm{T}}(\theta)$ is calculated at a fixed time of demand $x = \theta$, and $\sum_{i=1}^{n} Q_i^{\mathrm{T}}$ is calculated at random times of demand x_i in the interval (0,a). After the substitution in Figure 9.5,

$$\psi = \frac{\lim_{n\to\infty}(1/n)\sum_{i=1}^{n} Q_i^{\mathrm{T}}(\theta)}{Q_0} \tag{9.6}$$

is obtained, which proves the theorem. □

Theorem 9.1 creates the basis of extremely fast solvers for the production availability of complex repairable networks. The running time of these solvers is independent of the length of the operational interval, the failure frequencies of the edges and the lengths of their repair times. The efficiency of these solvers creates the possibility of embedding them in simulation loops performing topology optimisation of large and complex flow networks.

The production availability of a complex network, which includes independently working edges, characterised by constant hazard rates, can be determined very efficiently by Algorithm 9.2.

The array q[] contains the probabilities that the separate edges will be in a failed state on demand. In a nested loop controlled by the variable j, the state of the separate edges (working/failed) is determined at the time of demand. (M denotes the number of edges/components in the network.) The state of the jth edge is tested by generating a uniformly distributed random number between 0 and 1 by the statement 'tmp = ***real_random()***', and comparing it with the probability q[j] that the jth edge will be in a failed state. The edges are assumed to be with constant failure rates (negative exponential time-to-failure distribution). Consequently, the probability that a demand at a specified point in time will sample a failed state for the jth edge, can be approximated by $q_j = \mathrm{MTTR}_j/(\mathrm{MTTF}_j + \mathrm{MTTR}_j)$, where q_j is the average unavailability of the jth edge.

Algorithm 9.2

```
function max_flow();
function real_random();

function production_availability();
{
 total_flow = 0;
 for i = 1 to num_trials do
 {
   for j = 1 to m do //m is the number of edges in the network
     {
       tmp = real_random();
       if (tmp < q[j]) then reduce the flow capacity of edge j to zero;
     }
   maxf = max_flow();
   total_flow = total_flow + maxf;
```

```
        Restore the flow capacities of all failed components to their original values;
    }
        return total_flow/num_trials;
    } □
```

All probabilities q_j ($j = 1,2,\ldots,m$) are pre-calculated and stored in the array $q[]$. If the generated random number 'tmp', uniformly distributed between 0 and 1, is smaller than $q[j]$, the jth edge is in a failed state and is essentially removed from the network by setting its flow capacity to zero. If the converse is true, the edge remains in the network.

After the state of all edges has been determined, the function ***max_flow()*** calculates the maximum throughput flow in the resultant network (after setting to zero the flow capacities of the failed edges).

The expected value of the total throughput flow accumulated in the variable 'total_flow' is obtained by dividing the total throughput flow to the number of simulation trials. At the end of each simulation trial, the flow capacities of all failed edges are restored to their original levels.

The performance of the proposed fast algorithm can be demonstrated on the network in Figure 9.6. The labels on each edge are in the format λ/C and denote the hazard rate λ year^{-1} and the flow capacity C of the edge. The repair time of each edge has been assumed to be 10 days. The specified operation period is 25 years. One million simulation trials on a computer with processor *Intel(R) Core(TM) 2 Duo CPU T9900 @ 3.06 GHz* took only 0.671 s!

The calculated production availability for the network shown in Figure 9.6 was $\psi = 0.967$.

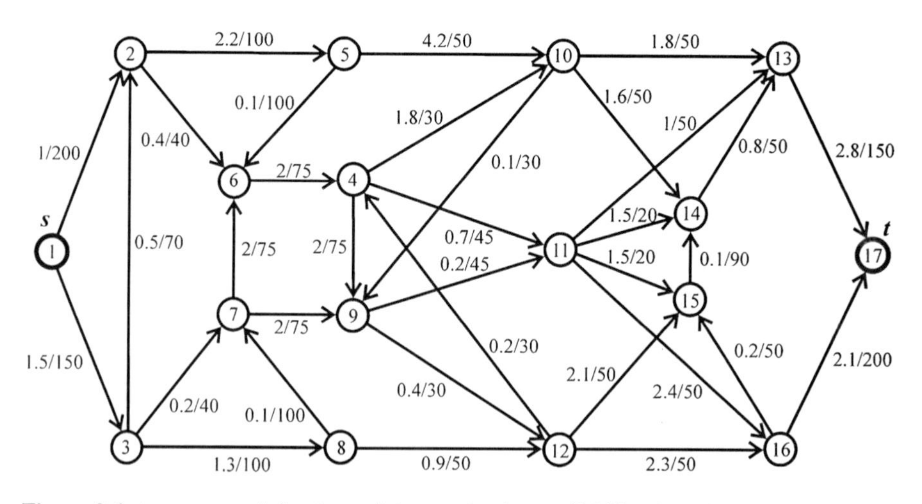

Figure 9.6 A test network for determining production availability, by using the fast algorithm.

9.3 Comparing the Performance of Competing Network Topologies

An important application of the fast availability algorithm proposed in this chapter is in comparing quickly the performance of competing network topologies and selecting the topology with the best performance.

Consider, for example, the competing production networks shown in Figure 9.7A and B, where, for the sake of simplicity, all edges have the same flow rate capacity of 40 flow units per day, hazard rate of four expected failures per year and downtime for repair 10 days. Edges (3,8) and (4,9) from network a and edges (2,8) and (4,10) from network b are redundant. Without a supporting software tool, it is difficult to infer which network topology is superior. Applying the proposed algorithm yields production availability of $\psi_a = 70.5\%$ for the network in Figure 9.7A and $\psi_b = 75.2\%$ for the network in Figure 9.7B. Despite the seemingly insignificant differences in the competing topologies, the impact on the production availability is significant!

Consider now the network shown in Figure 9.8 where again, all edges have the same flow capacity of 40 flow units per day, hazard rate of four expected failures per year and downtime for repair 10 days. If the redundancies are placed such as it has been shown in Figure 9.8, the availability of the network can be increased to 80.5%.

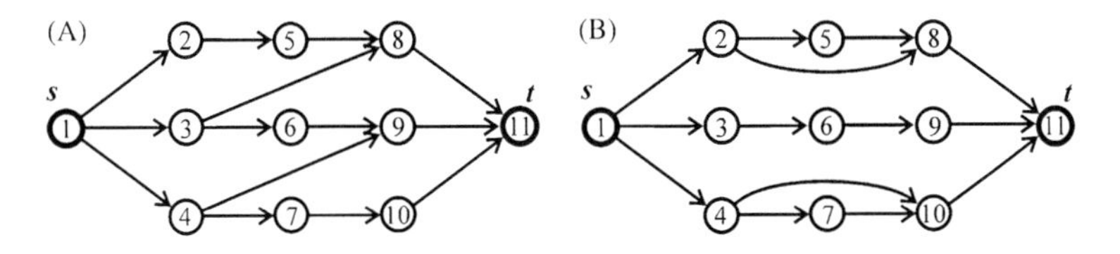

Figure 9.7 Two competing networks with different types of redundancy topology.

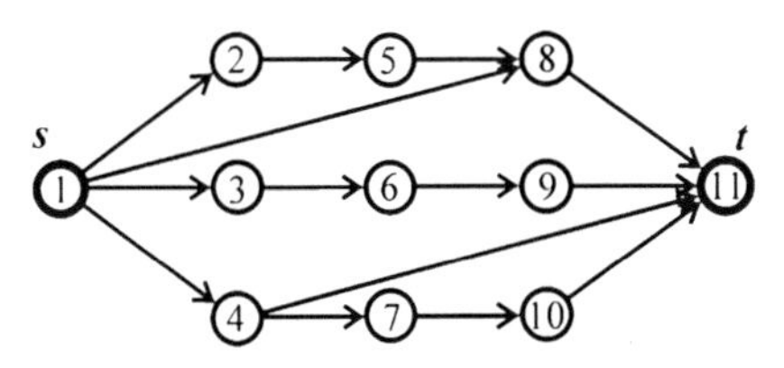

Figure 9.8 A network with the same number of redundant links and superior availability.

10 Link Between Topology, Size and Performance of Repairable Flow Networks

10.1 A Software Tool for Analysis and Optimisation of Repairable Flow Networks

In the general case, obtaining a closed-form solution for the throughput flow reliability or for the average availability of a complex flow network is very difficult or impossible. In this case, the throughput flow reliability and the production availability can be determined by simulation methods, whose algorithms have been presented earlier.

The high-speed solvers for revealing the performance of repairable flow networks have been embedded in a software tool with graphics user interface (Figure 10.1) by which the network topology is drawn on screen and the parameters characterising the edges and nodes are easily specified.

The software tool includes a menu for specifying the desired flow network and a menu from which an embedded service function is called to perform a desired service. There are embedded service functions for (i) determining the production availability of a repairable flow network; (ii) determining the threshold throughput flow reliability; (iii) determining the maximum throughput flow; (iv) topology optimisation; (v) discovering and removing directed loops of flow; (vi) determining system reliability; (vii) determining the minimum-cost throughput flow and (viii) determining the throughput flow with the smallest losses from failures, in a non-reconfigurable network.

The parameters characterising each edge and node can be specified individually, simply by double-clicking them (Figure 10.1). For each edge or node, a large number of failure modes can be specified and for each failure mode, the time-to-failure distribution and the time-to-repair distribution can be specified as well as their parameters.

A copy/paste function has been provided, for quickly transferring the parameters from one edge/node to another edge/node. Functions for zooming, panning, auto-arranging, saving and loading flow networks have also been provided. Attributes of edges and nodes can be viewed and listed on the screen. Nodes and edges can be easily deleted and added to the network, in order to achieve the desired topology and size. Deleting and inserting nodes and edges also permits easy modification of existing networks and generating multiple derivative networks.

Flow Networks. DOI: http://dx.doi.org/10.1016/B978-0-12-398396-1.00010-6

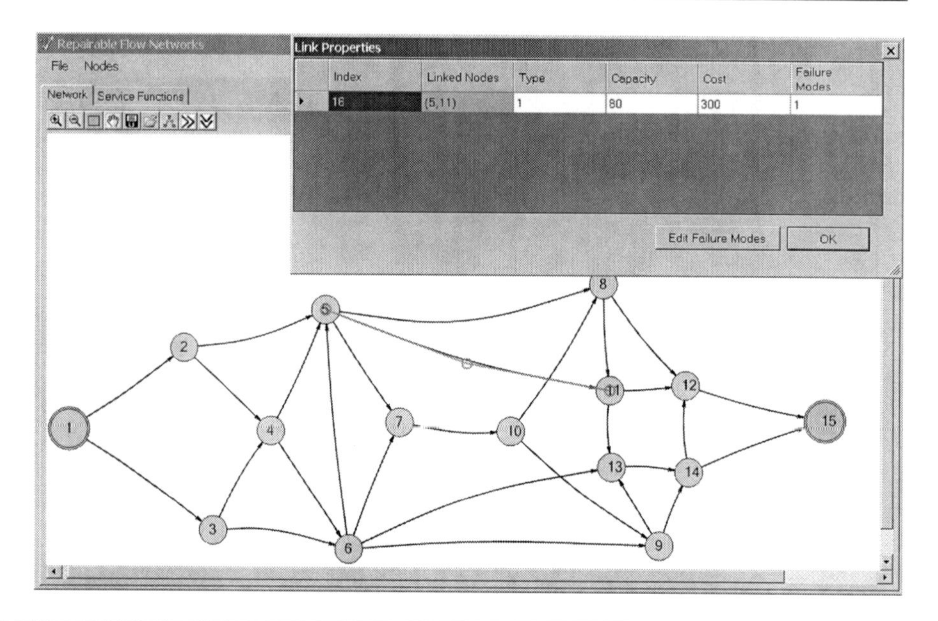

Figure 10.1 Specifying the parameters of an edge by using the developed software tool.

In short, using a graphics editor for specifying the network topology is vital in increasing the speed and quality of the analysis. This is due to (i) increased speed of building new networks; (ii) increased speed of modifying existing networks and producing multiple variants of the same network and (iii) eliminating input errors while specifying large and complex flow networks.

10.2 A Comparative Method for Improving the Performance of Repairable Flow Networks

Often, the absolute reliability of a flow network cannot be revealed. This is because of:

- The complexity of the physical processes and physical mechanisms underlying the failure modes of the separate edges, most of which remain unknown or are associated with large uncertainties.
- The complex influence and uncertainty associated with the environment, the operational cycles and common causes for accelerated deterioration.
- The variability associated with reliability-critical design parameters. Such are, for example, the times to failure, the state of deterioration of components, the impact of the environment, the specific duty cycles and the common cause factors.
- The non-robustness of the reliability prediction models, for which small variations in the input data may cause large variations in the predicted values.

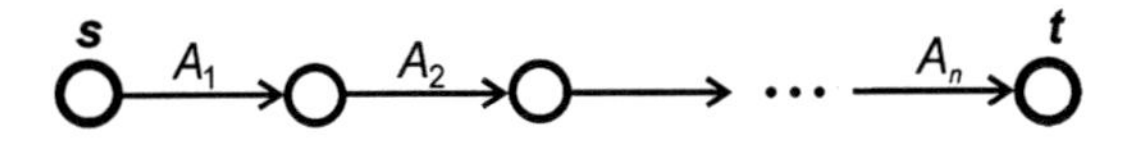

Figure 10.2 The throughput flow reliability of this network is equal to the product of the availabilities of the separate edges.

Key reliability-critical parameters are associated with uncertainty which does not permit revealing the absolute reliability level. Major sources of this type of uncertainty are associated with the times-to-failure and times-to-repair models and their parameters. Furthermore, even if this information were available for common, widely used analytical models, even a relatively small amount of uncertainty associated with the reliability parameters leads to large errors in the model predictions, which renders such predictions of questionable value. Here is an example illustrating the problem.

According to Eq. (7.22), the throughput flow reliability of a network consisting of n edges arranged in series (Figure 10.2) and working independently from one another is given by

$$R_{\text{th}} = \prod_{i=1}^{n} A_i \tag{10.1}$$

where A_i is the average availability of the ith edge (the probability that on demand, the edge will be in working state). For the special case, where the flow network is built with n identical edges, each characterised by availability A, the throughput flow reliability of the network becomes

$$R_{\text{th}} = A^n \tag{10.2}$$

Estimating the availability A of an edge, however, is always associated with uncertainty. Indeed, errors associated with determining the MTTF of the edge or the MTTR will cause errors in the estimated availability value A. An error ΔA in estimating the availability of a single edge will lead to a very large error $\Delta R_{\text{th}}/R_{\text{th}} = n(\Delta A/A)$ in the predicted throughput flow reliability for the network. Predicting the throughput flow reliability of large networks is also associated with uncertainty related to the operation cycle and the environment, the interdependencies caused by unknown common causes, etc. Capturing this uncertainty, necessary for a correct prediction of the throughput flow rate reliability, is a complex task. In the common case where the focus is on improving the reliability of designs, an efficient way of resolving this predicament is *not to attempt prediction of the absolute throughput flow reliability or production availability*. Competing designs are simply compared on the basis of their performance, which is calculated with the same predefined set of MTTFs and MTTRs for the separate edges. Despite the fact that the exact throughput flow reliability or average availability may remain unknown, *the relative performance level of the networks can be determined easily*. The relative performance level can then be used as a basis for evaluating the performance of the competing network topologies.

10.3 Investigating the Impact of the Network Topology on the Network Performance

A very important issue related to the design of repairable flow networks is the link between network topology and performance. An important question here is what features in the network topology and structure (e.g. connectivity, reliability of the components, spare capacity) make the throughput flow least sensitive to component failures.

Consider for example the two competing production network designs in Figure 10.3A and B, where the capacities of the edges are according to the figure and all edges have the same hazard rate of four expected failures per year and downtime for repair 10 days. By using the developed software tool for determining production availability, it is easy to establish which competing topology is superior and by how much. The software tool calculated production availability (expected fraction of the transmitted throughput flow in presence of failures) of $\psi_a = 79.2\%$ for the network in Figure 10.3A and only $\psi_b = 73\%$ for the network in Figure 10.3B. Despite the seemingly insignificant differences between the competing topologies, the impact on the production availability is significant.

One of the reasons for the superior performance of network topology 'a' is that redundant edges (2,7) and (3,7) bypass more unreliable edges compared to the corresponding redundant edges (4,7) and (5,7) from network topology 'b'. Indeed, failure of unreliable edge (2,4) in topology 'a' does not cause a loss of flow because the flow through the failed edge (2,4) can be redirected through edge (2,7). Failure of edge (2,4) in topology 'b', however, causes the flow through edge (2,4) to be lost, because there is no possibility of redirecting the flow along alternative paths. Consequently, in improving the availability of production networks, redundancies need to be placed in such a way that they bypass as many unreliable edges as possible.

Consider now the alternative network topologies in Figure 10.4. Edge (0,1) is perfectly reliable (never fails) and has a flow capacity 200 units. The rest of the edges are characterised by the same flow capacity of 100, the same failure rate of four failures per year and the same time for repair of 7 days. The network topologies

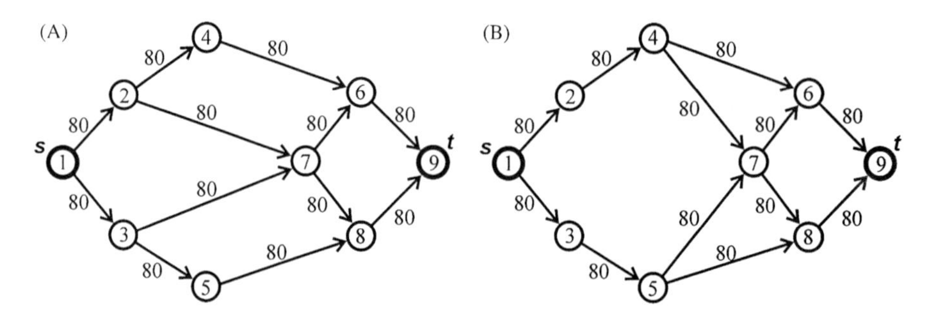

Figure 10.3 Two competing networks with different types of redundancy.

have different types of redundancy which consist of eight redundant edges incorporated in different ways. For the network in Figure 10.4A, the eight redundant edges form an extra flow path (1,3,6,9,12,15,18,21,23). For the network in Figure 10.4B, the eight redundant edges form the flow path (3,2,5,6,9,10,13,14,15) alternating between the two parallel branches. In the network from Figure 10.4C, the redundant edges form the cross-bridges (3,4), (2,5), (7,8), (6,9), (9,10), (8,11), (13,14) and (12,15).

The maximum throughput flow reliabilities characterising the networks, calculated by the developed software tool, are $R_A = 0.579$, $R_B = 0.62$ and $R_C = 0.521$, respectively. These are the probabilities that, on demand, the network will deliver the maximum possible throughput flow of 200 units. The average availabilities characterising the networks are $\psi_A = 74.5\%$, $\psi_B = 80\%$ and $\psi_C = 74.7\%$, correspondingly. These are the expected fractions of the transmitted throughput flow in the presence of failures. As can be verified, network topology 'B' is superior to

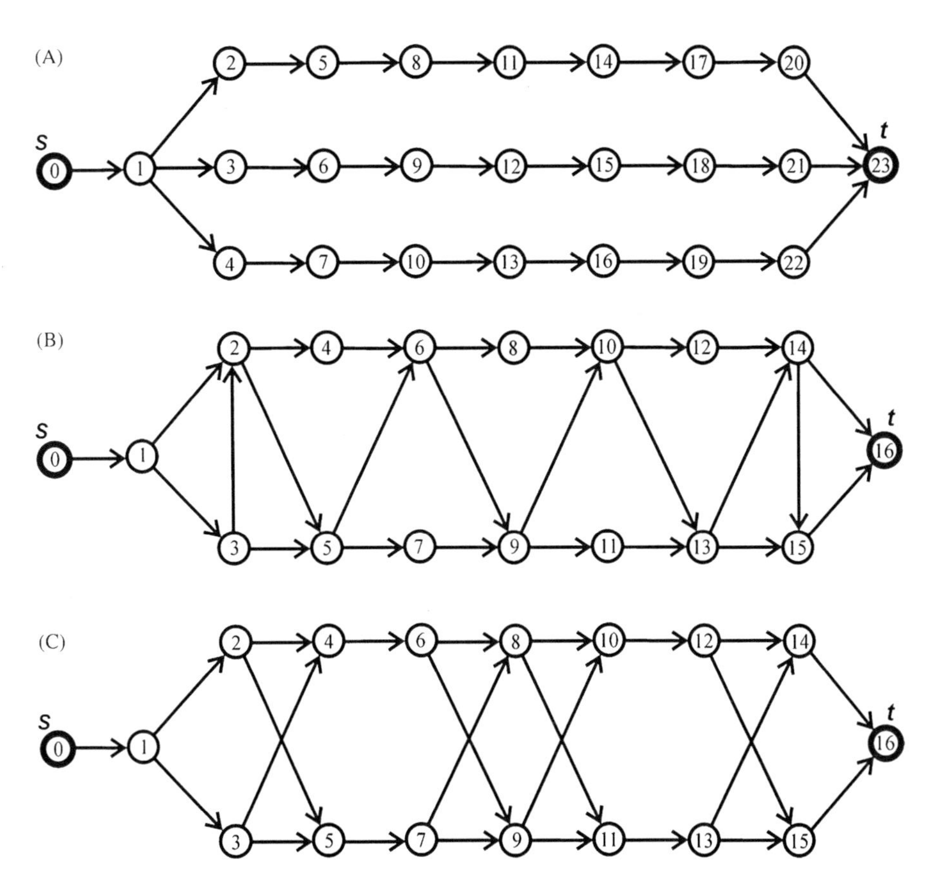

Figure 10.4 Three repairable flow networks with the same number of redundant edges arranged as: (A) an extra flow path; (B) an alternating flow path and (C) cross-bridges.

network topology 'A'. A seemingly insignificant alteration of network topology 'B', however, leads to network topology 'C', characterised by inferior performance.

With increasing the size of the networks, the performance figures decrease significantly. Consequently, the throughput flow reliability of repairable flow networks depends significantly not only on the network topology but also on the size of the networks. Selecting a particular design solution must therefore be based on revealing the performance parameters of the networks, by using the software tool.

The proposed software tool is particularly useful for comparing network topologies and selecting the topology characterised by the best performance. Such a comparison is also useful in deciding whether allocating additional resources for purchasing extra redundancy is justified.

Next, consider the repairable flow network from Figure 10.5A. Each edge is characterised by a throughput capacity of 100 units, hazard rate of four expected

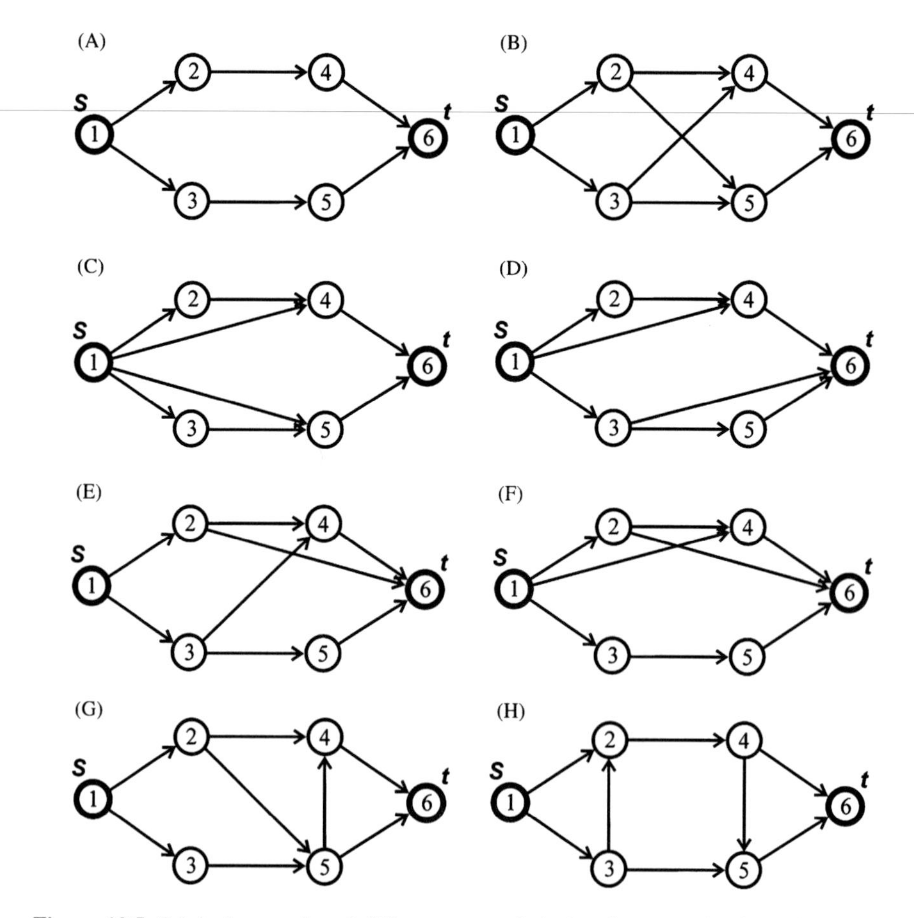

Figure 10.5 Original network and different ways of placing the two redundant edges.

number of failures per year and downtime for repair of 14 days. Suppose that a reliability improvement budget is sufficient for purchasing two redundant edges only, which can be placed as shown in Figure 10.5B–H. The performance measure used for assessing the networks is the maximum throughput flow reliability.

Interestingly, the network with the highest probability of 200 units throughput flow on demand is the network in Figure 10.5F, $R_f = 81.5\%$. For comparison, the network in Figure 10.5H is characterised by a probability of 200 units throughput flow on demand $R_f = 42.7\%$. Smaller than 81.5% probabilities of the required throughput flow, correspond to the networks in Figure 10.5B (53.1%), Figure 10.5C (70.4%), Figure 10.5D (70.4%) and Figure 10.5E (66.7%).

The reason for the superior throughput flow reliability of the network shown in Figure 10.5F becomes clear if a connection is made with the topic related to the number of disjoint $s-t$ paths in a network, discussed in Chapter 3. (Two paths are edge-disjoint if they do not share common edges.) The number of disjoint paths for the network shown in Figure 10.5F is three. These are the paths (s,4,*t*), (s,2,*t*) and (s,3,5,*t*). In contrast, the rest of the networks have only two disjoint paths. The extra disjoint path provides extra resilience of the network shown in Figure 10.5F against simultaneous edge failures and this explains its superior performance.

Repairable flow networks from different application areas impose particular constraints that need to be addressed during the design of discrete-event solvers. In production networks for example, oil and gas production networks, only the edges/components are unreliable. The nodes are notional (perfectly reliable) and are used only to define the topology of the network. Furthermore, the links in production networks are directed links because, as a rule, no reversal of flows or back flows are permitted. In contrast, in computer networks both, the nodes (representing routers) and the edges (representing communication lines) are unreliable.

Furthermore, parametric studies showed that flow networks with meshed topology have a superior throughput flow reliability on demand, compared to networks with tree topology. The reason is the alternative paths provided by the mesh topology. For networks with tree topology, such alternative paths are missing and failure of any edge results in the loss of the entire flow through the edge.

10.4 Investigating the Link Between Network Topology and Network Performance by Using Conventional Reliability Analysis

In some cases, inferences about the link between flow network topology and flow network performance can be made by using a standard system reliability analysis.

Let us consider the repairable flow networks with redundancy shown in Figure 10.6, which have different topologies but contain the same number of components. The same number of components selected for the competing topologies

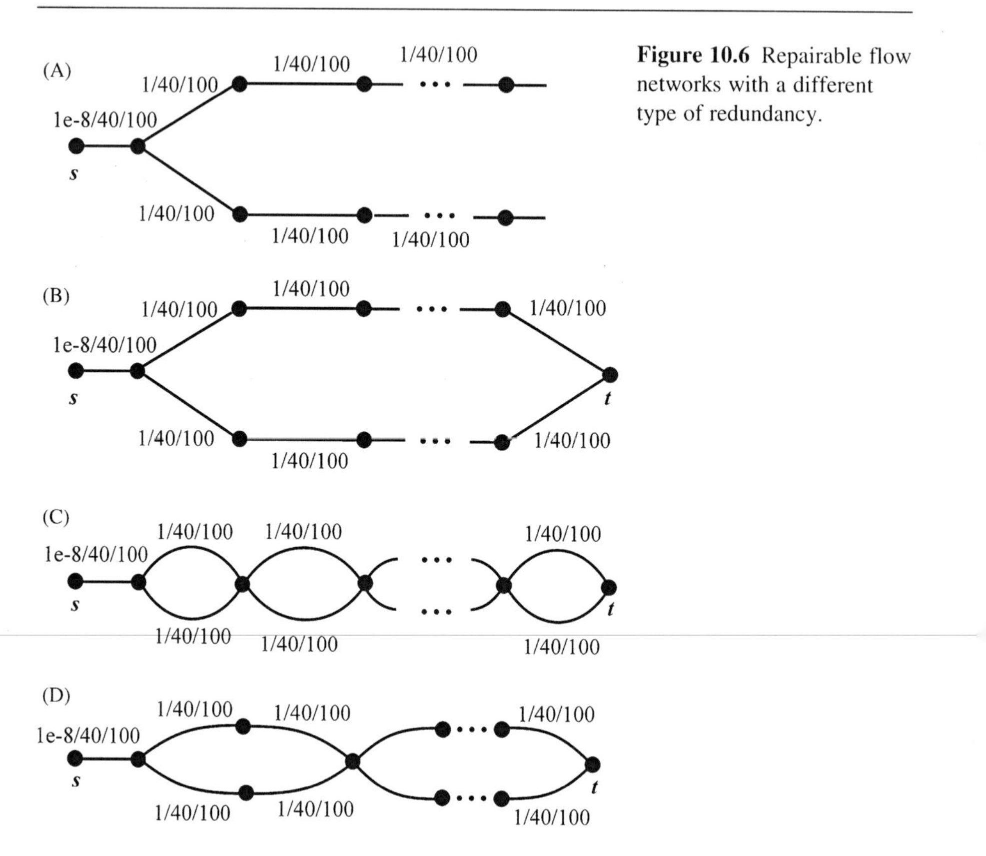

Figure 10.6 Repairable flow networks with a different type of redundancy.

provides an opportunity to reveal solely the effect of the network topology on the reliability of the throughput flow.

The edges in the networks from Figure 10.6 represent components. Again, these components are characterised by constant hazard rates (a negative exponential time-to-failure distribution). The adopted format related to the attributes of each edge includes three numbers separated by '/'. Thus, '*f/d/c*' on an edge means hazard rate '*f*', downtime '*d*' and flow rate capacity '*c*'.

The first component, connecting the source *s* with the rest of the network, is characterised by a very small failure rate of 1e-8, chosen deliberately to practically exclude failures associated with this component. The probability that the throughput flow on demand will be equal to the maximum possible flow of 100 units is equal to the probability that a path through working edges will exist, from the source *s* to the sink *t*. The probability of existence of such a path is, in fact, the reliability on demand of the network. The probability that a demand at a particular point in time will sample a working state for edge (i,j) is approximately equal to the average availability of the edge:

$$p_{ij} = p = \frac{\text{MTTF}_{ij}}{\text{MTTF}_{ij} + \text{MTTR}_{ij}} \tag{10.3}$$

where $\text{MTTF}_{ij} = 1/\lambda_{ij}$ is the mean time to failure of edge (i,j), λ_{ij} is the hazard rate and $\text{MTTR}_{ij} = d_{ij}$ is the mean time to repair.

Consider the repairable flow network consisting of two parallel branches in Figure 10.6A, which share a common node after every k edges. The number of edges in a single branch is n. The quantity n is selected to be a multiple of k. In other words, for a specified k, the ratio n/k is always an integer number. The throughput flow reliability R, will then be equal to the reliability on demand of the network, which is

$$R = [1-(1-p^k)^2]^{n/k} \tag{10.4}$$

because all edges are in a series–parallel arrangement and have the same availability $p_{ij} = p$. If a common node is introduced (after all 'n' edges, Figure 10.6B), then $k = n$, and Eq. (10.4) becomes

$$R_\text{B} = 1 - (1-p^n)^2 \tag{10.5}$$

which gives the throughput flow reliability of arrangement 'B'. If a node is shared after each edge, then $k = 1$, and Eq. (10.4) becomes

$$R_\text{C} = [1-(1-p)^2]^n \tag{10.6}$$

which is the throughput flow reliability of arrangement 'C'.

If a node is shared after every other edge, then $k = 2$ and Eq. (10.4) becomes

$$R_\text{D} = [1-(1-p^2)^2]^{n/2} \tag{10.7}$$

which gives the throughput flow reliability of arrangement 'D'. Plotting the throughput flow reliability as a function of the size of the network (the number of edges in a single parallel branch) is given in Figure 10.7.

Suppose that for a single edge, with a negative-exponential time-to-failure distribution, MTTF = 365 days and MTTR = 40 days. According to Eq. (10.3), the availability characterising a single edge is

$$p = \frac{\text{MTTF}}{\text{MTTF} + \text{MTTR}} = \frac{365}{365 + 40} \approx 0.9$$

As can be expected, arrangement 'C' is characterised by the largest maximum throughput flow reliability (Figure 10.7). Arrangement 'C' is less sensitive to failures in the parallel branches compared to arrangements 'B' and 'D'.

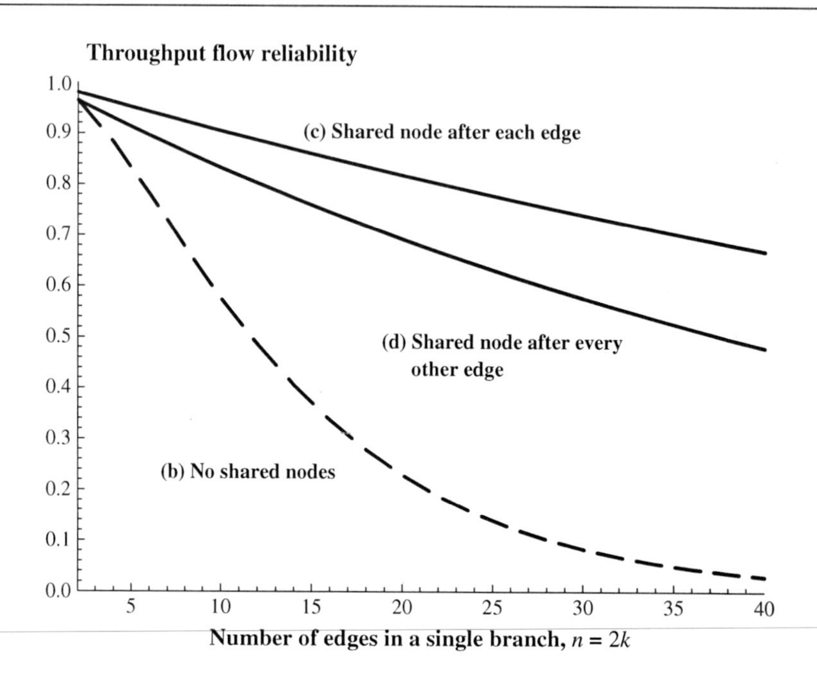

Figure 10.7 A comparison between the throughput flow reliabilities of the flow networks shown in Figure 10.6.

Indeed, while failures of two edges, each in a separate branch, always stops the throughput flow for the arrangement shown in Figure 10.6B, the throughput flow in the network from Figure 10.6C stops only if the two failed edges are from the same section composed of two parallel branches. There are n^2 different ways of having a single failure in each parallel branch. While any of these combinations invariably stop the throughput flow for the network shown in Figure 10.6B, only n of them actually stop the throughput flow in the network from Figure 10.6C. Furthermore, with increasing the size of the network, the throughput flow reliability decreases significantly for arrangement 'B', while the throughput flow reliability still remains at a high level for arrangement 'C'. In short, with increasing the size of the network, a network with redundancy where nodes are shared after each edge, maintains its throughput flow reliability at a high level.

Network 'D', for which a node is shared after every two edges, is characterised by a throughput flow reliability dependence located between the dependencies characterising networks 'B' and 'C'. The throughput flow reliability dependence for a network in which a node is shared after every three edges will be between dependences 'B' and 'D' (Figure 10.7) and so on.

From the analysis presented so far, it seems that increasing the connectivity of a network will always yield a significant improvement in the throughput flow rate reliability. To show that this is not necessarily true, consider the network shown in

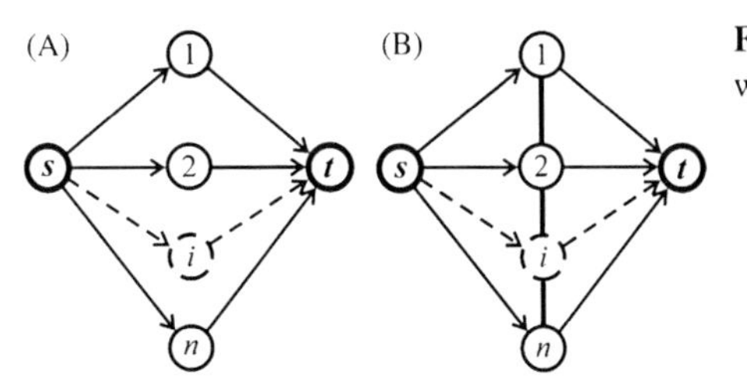

Figure 10.8 Two repairable flow networks with different connectivity.

Figure 10.8A. Assuming for simplicity that all edges are characterised by the same reliability on demand p given by Eq. (10.3), the probability of a source-to-sink flow for a network consisting of n pairs of edges (Figure 10.8A) is

$$R = 1 - (1 - p^2)^n \tag{10.8}$$

If perfectly reliable bridges are introduced so that each edge is connected to every other edge, the well-connected network shown in Figure 10.8B is obtained. The probability of a source-to-sink flow for the network shown in Figure 10.8B is

$$R = [1 - (1 - p)^n]^2 \tag{10.9}$$

The graphs of Eqs. (10.8) and (10.9) corresponding to availability $p = 0.1$ of a single edge are presented in Figure 10.9. Clearly, for edges characterised by a small availability, the benefit from increasing the connectivity of the network is significant.

However, if the average availability of the edges is increased to $p = 0.6$, the benefit from having a network with larger connectivity is reduced significantly (Figure 10.10). For edges characterised by a high availability (e.g. $p = 0.9$), the difference between the two curves is negligible and the effect from the extra connectivity (bridges) is insignificant.

These examples indicate that for some networks, an investment towards increasing connectivity is most efficient in cases where the availability of the edges/components is small.

For each particular repairable flow network, an assessment must be made regarding the benefits from reliability improvement through increased connectivity. If the impact is insignificant, the cost towards increasing the connectivity may not be justified by the small improvement of the throughput flow reliability.

10.5 Degree of Throughput Flow Constraint

The effect of edge failures on the maximum throughput flow is an important problem and has been studied before (Wollmer, 1968). Theorem 5.1 provides the basis for an importance measure characterising the edges in a repairable flow

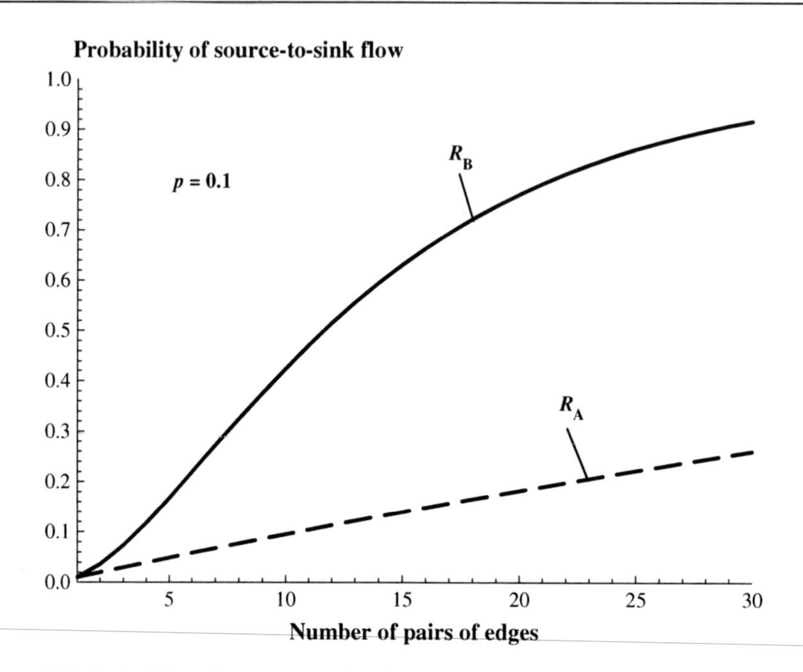

Figure 10.9 Probability of a source-to-sink flow on demand for the networks shown in Figure 10.8A and B, in the case of a small availability of a single edge ($p = 0.1$).

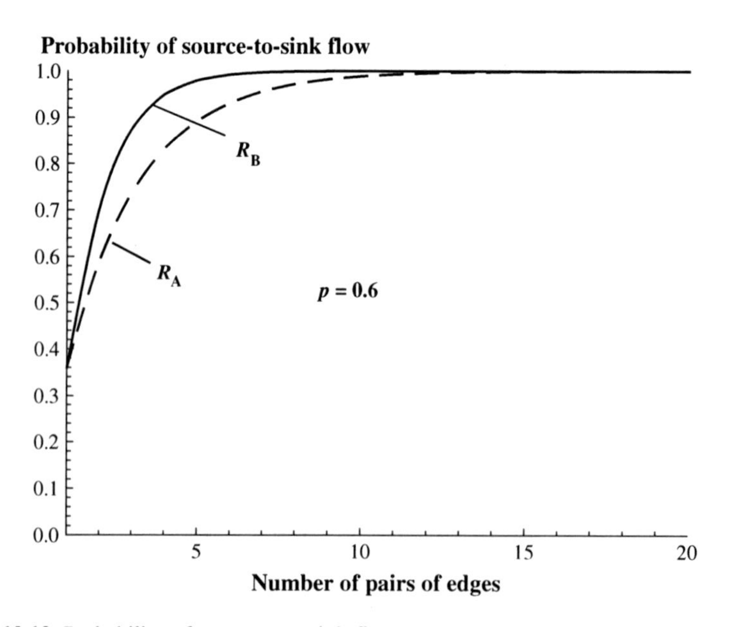

Figure 10.10 Probability of a source-to-sink flow on demand for the networks shown in Figure 10.8A and B, in the case of a moderate availability of a single edge ($p = 0.6$).

network – the degree of constraint which an edge failure exerts on the throughput flow. During the augmentation of the dual network after the failure of edge (e,d), part of the excess flow $f(e,d)$ is redistributed along alternative flow paths but part of it, $f'(e,d)$, cannot be redistributed. Consider the remaining excess flow $f'(e,d)$ which cannot be eliminated by augmenting paths in the dual network. The ratio of $f'(e,d)$ to the maximum throughput flow Q_{max} in the absence of failures will be called the '*degree of constraint on the throughput flow*'. It is measured by

$$G_{(e,d)} = \frac{|f'(e,d)|}{Q_{max}} \tag{10.10}$$

The degree of constraint on the throughput flow takes values from the interval [0,1]. Failures of edges characterised by a large degree of constraint, exert a strong impact on the throughput flow. Failures of edges with a very small degree of constraint exert a negligible impact on the throughput flow. Such is the case with sections including two parallel redundant edges with the same capacity, one of which is empty while the other carries all of the flow through the section. The degree of constraint on the throughput flow for any of the parallel edges is zero. If the entire throughput flow in a particular network passes through a single edge (e,d), failure of this edge will choke the entire throughput flow. Consequently, the degree of constraint on the throughput flow exerted by this edge is one ($G_{(e,d)} = 1$). For example, for a network with merging flows (tree topology), the edge collecting all of the streams from the sources has a degree of constraint one because its failure fully chokes the flows from all sources. Failures of edges which are low in the tree hierarchy constrain fewer sources of flow and their degree of constraint on the total throughput flow will be smaller. An important design principle for repairable flow networks can now be formulated: *the larger the degree of constraint on the throughput flow exerted by a component failure, the larger is the reliability level to which the component should be designed.*

11 Topology Optimisation of Repairable Flow Networks and Reliability Networks

11.1 Theoretical Basis of the Proposed Method for Topology Optimisation

The purpose of the topology optimisation of complex repairable networks is to determine a network topology which is characterised by a maximum throughput flow or a maximum production availability, attained within a specified budget for building the network. Given are a set of nodes, a set of possibilities for building edges between the nodes and the costs for building the edges. It is required to maximise the throughput flow or the production availability associated with a specified time interval, so that the cost of the flow network remains within the specified budget.

The methods for topology optimisation described in this chapter have been proposed by Todinov (2011a,d, 2012b) and are based on exploring the space of all alternatives 'locked' in the full-complexity network, by *pruning the full-complexity network* and using the *branch and bound* method.

The algorithm starts with the full-complexity flow network, which includes all possible connections between the nodes. Let M be the number of edges in the full-complexity network. All possible networks, locked in the full-complexity network, can then be presented as a union of M mutually exclusive and exhaustive sets $A_1, A_2, \ldots, A_M$ (Figure 11.1). For these, $A_i \cap A_j = \varnothing$ if $i \neq j$ and $A_1 \cup A_2 \cup \ldots \cup A_M = \Omega$ hold, where Ω is the set of all possible networks locked in the full-complexity network. The set A_i includes all networks locked in the full-complexity network, for which edge i is missing (has been deleted). In turn, each set A_i can in turn be presented as a union of mutually exclusive subsets $A_{i,i+1}, A_{i,i+2}, \ldots, A_{iM}$. The subset A_{ij} ($j > i$) includes all networks, locked in the full-complexity network, for which edges i and j ($j > i$) are missing and the index j is greater than the index i. For these subsets, $A_{i,i+1} \cup A_{i,i+2} \cup \ldots \cup A_{i,M} = A_i$ and $A_{ij} \cap A_{ik} = \varnothing$ if $j \neq k$ holds (Figure 11.1).

The subsets A_{ij} are in turn partitioned into smaller, nested, mutually exclusive subsets $A_{ij,j+1}, A_{ij,j+1}, \ldots, A_{ij,M}$ and so on, until no further partitioning can be done. Entering a large set A_i means removing edge i from the network. Entering a subset A_{ij} nested in the set A_i means removing edges i and j from the network; entering a subset A_{ijk} nested in the subset A_{ij} means removing edges i,j and k, and so on. The

Flow Networks. DOI: http://dx.doi.org/10.1016/B978-0-12-398396-1.00011-8

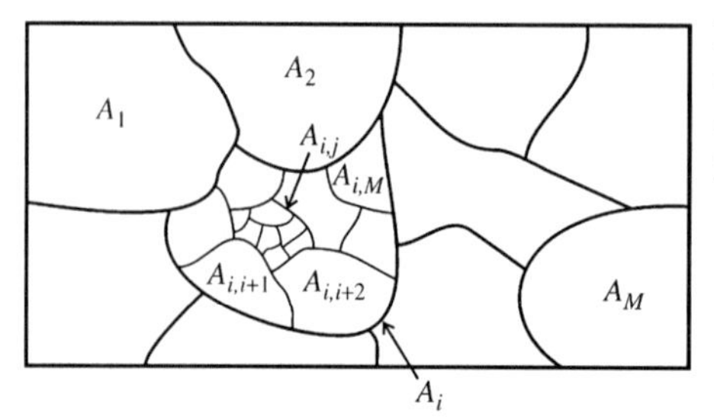

Figure 11.1 The space of all alternatives can be presented as a union of mutually exclusive and exhaustive subsets (Todinov, 2012b).

non-intersecting subsets contain all distinct combinations of edges to be removed. They are built in such a way that the indices k of the subsets nested in the current subset A_{ij} are always greater than the indices i and j of the current subset ($k>j>i$).

After entering a nested subset, the maximum throughput flow or the production availability of the resultant network (after removing the corresponding edges by setting their flow capacities to zero) is calculated. Suppose that the global maximum of the maximum throughput flow or production availability, for a network built within the specified budget, is $A_{\max}$. Suppose that the current global maximum for the maximum throughput flow or production availability (achieved within the specified budget for building the network) is $A_{c\max}$. (Initially, the current global maximum $A_{c\max}$ is set to zero.)

Note that, during the entire search in the nested subsets, the relationship $A_{c\max} \leq A_{\max}$ is fulfilled. Because the maximum throughput flow or production availability cannot possibly increase with edges being removed from the network, *if, after entering a nested subset, the calculated throughput flow or production availability A is smaller than the current achieved global maximum $A_{c\max}$, then there is no need to enter subsets nested in the current subset.* Entering nested subsets will only result in *a smaller throughput flow or production availability*, $A \leq A_{c\max}$. Therefore, the global maximum $A_{\max}$ of the maximum throughput flow or production availability cannot possibly be found in subsets nested in the current subset. This allows us to abort searching in subsets and cut down large subsets from the space of alternatives.

If, after entering a nested subset, the resultant network satisfies the budget constraint, and the calculated maximum throughput flow or production availability 'A' is larger than the current maximum $A_{c\max}$ ($A>A_{c\max}$), the current maximum is replaced by the throughput flow or production availability characterising the current subset ($A_{c\max} := A$). Further search into subsets nested in the current subset is aborted again, because no value greater than $A_{c\max}$ can possibly be obtained by removing edges from the network. Once a search into a subset has been aborted, it is never resumed again.

Following the described search method, it is guaranteed that the global maximum $A_{\max}$ will be found. Indeed, the bound and branch method is essentially an exhaustive search method in which large subsets are excluded from the search. It is guaranteed that the absolute maximum will be found because, during the exhaustive search, only subsets in which the absolute maximum A_{max} certainly cannot be found are excluded from the search.

11.2 Topology Optimisation Algorithm

Exploring the full-complexity network by pruning edges is essentially done by building recursively the tree of edges to be removed. The tree contains all distinct combinations of edges to be removed. It is built in such a way that the index i of any edge to be removed is always smaller than the indices of all edges in the subtree originating from edge 'i' (Figure 11.2).

The implementation of the described optimisation algorithm in C is given in Appendix A. To simplify the presentation, this implementation is relevant to the case when all edges have the same cost. The minimum number of edges that should be removed in order to comply with the allocated budget for building the network is given by the constant 'min_num_rmvd_edges'.

11.3 Solved Examples

11.3.1 *Maximising the Throughput Flow Within a Specified Cost for Building the Network*

For the network shown in Figure 11.3A for example, the labels stand for the flow capacity of the edges. For example, a label 80 means a flow capacity of 80,000 flow units a day. For the sake of simplicity, the same amount $500 has been assumed for the cost of each edge. The cost of the entire network is the sum of the costs of its edges.

The problem is to comply with a fixed budget of $5500 for components/edges (six edges are to be removed from the full-complexity network), and achieve at the same time a maximum throughput flow in the network. (The total cost of the edges should be within the specified budget.)

Applying the described optimisation algorithm to the full network shown in Figure 11.3A, resulted in the network from Figure 11.3B, whose cost does not exceed the specified $5500, and which has a maximum flow of 63,000 units per day. This network has been obtained by pruning six edges (components) from the full-complexity network shown in Figure 11.3A, characterised by a maximum flow of 68,000 units per day.

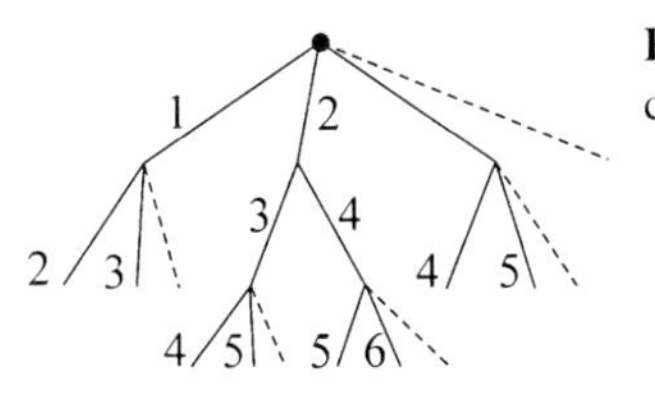

Figure 11.2 Tree of edges to be removed from the full-complexity network (Todinov, 2011a).

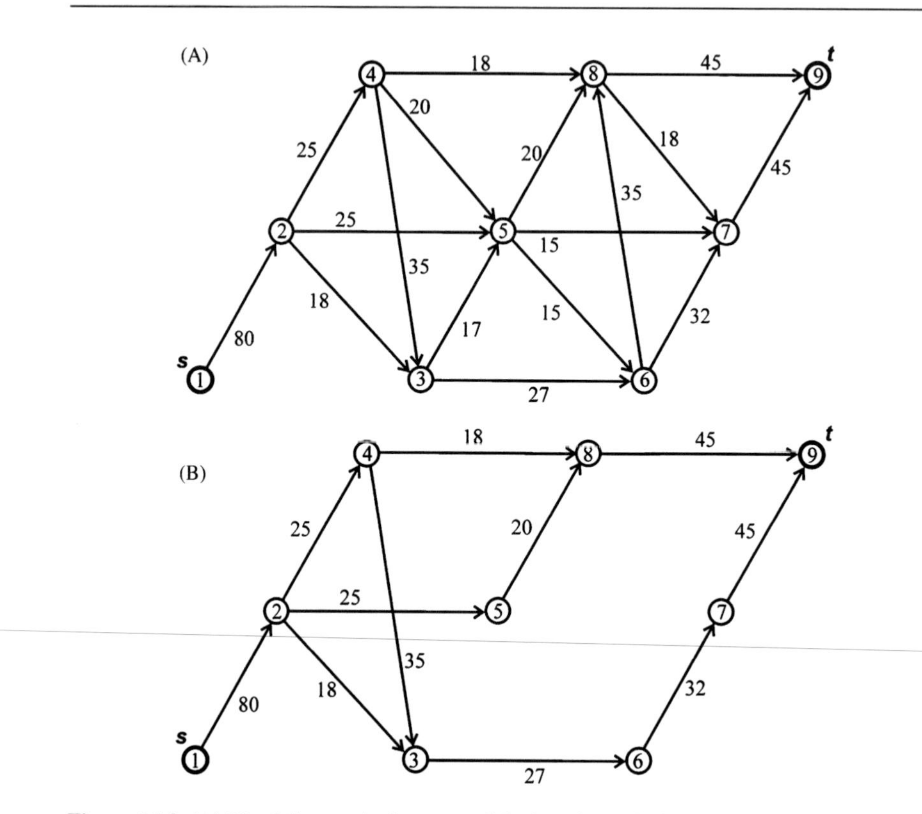

Figure 11.3 (A) The full-complexity network before the optimisation. (B) The optimal network combining cost not exceeding $5500, and a maximum flow of 63,000 units per day (Todinov, 2011a).

11.3.2 Maximising the Production Availability Within a Specified Cost for Building the Network

The algorithm proposed in Chapter 9 for which calculates the production availability is significantly faster than a conventional algorithm, which consists of sequentially generating thousands of failure–repair histories and which calculates the total amount of throughput flow delivered at the sink for each failure history. The advantage of the proposed algorithm to conventional algorithms is due to the several optimisation features boosting its performance. The most important optimisation feature is the method for calculating the production availability without actually generating the failure–repair history. The algorithm simply calculates the expected throughput flow associated with a stochastic removal of edges. Calculating the expected throughput flow associated with a stochastic removal of edges is a task considerably simpler than generating thousands of failure–repair histories and determining the production availability associated with each failure–repair history. Indeed, the time for calculating the expected throughput flow associated with a stochastic removal of edges is constant and independent of

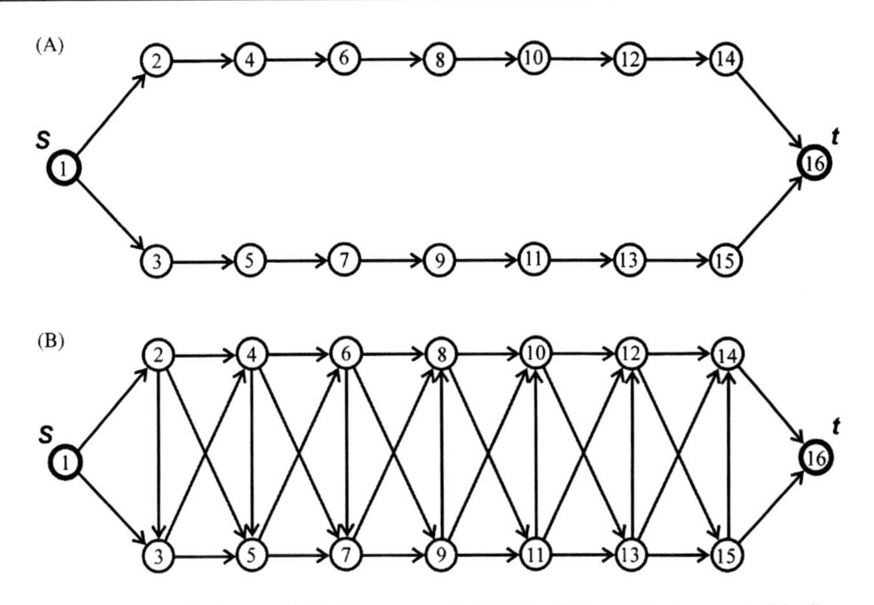

Figure 11.4 (A) A basic repairable flow network (B) The full-complexity repairable flow network.

the length of the specified time interval and the hazard rates of the edges. In contrast, for a conventional algorithm, the time for generating thousands of failure–repair histories and determining the production availability associated with each one of them is proportional to the length of the specified time interval and the hazard rates of the edges. The larger the specified time interval and the hazard rates of the edges, the larger is the calculation time.

The next important feature of the algorithm proposed for calculating production availability is the new algorithm for maximising the throughput flow after the stochastic removal of certain edges. First, the high speed of this algorithm is due to the throughput flow maximisation, which starts with an already optimised network, not with a network whose edges are empty. Second, the high speed of the algorithm is also due to the application of Theorem 5.3, which determines the maximum throughput flow, without actually calculating the feasible edge flows. This is a very important short cut which significantly speeds up the computation.

The advantages of the algorithm for calculating production availability result in a high computational speed of the algorithm for optimising the network topology to achieve a maximum production availability within a specified budget. Furthermore, unlike genetic algorithms or other heuristic algorithms, the *bound and branch method* always guarantees that the optimal solution will be found. At the same time, the proposed optimisation method is significantly faster compared to a full exhaustive search, in the space of the available alternatives.

Consider now the network shown in Figure 11.4A. For the sake of simplicity, all edges are characterised by the same flow capacity of 100, the same constant failure rate of four failures per year and the same time for repair of 7 days. Each

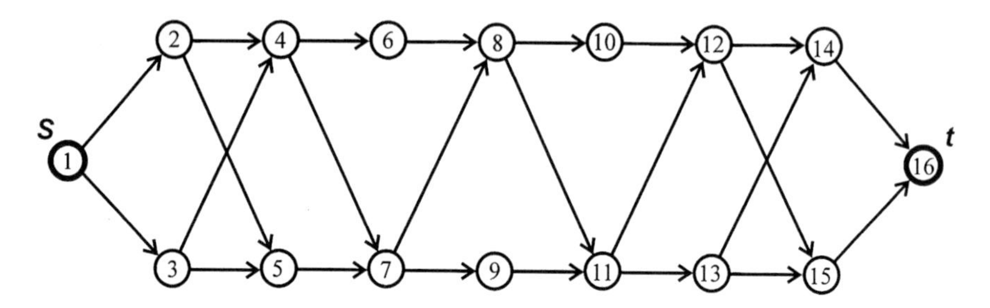

Figure 11.5 The optimal network, combining a maximum production availability of 0.81 and cost for purchasing redundant components not exceeding $24,000.

component/edge costs the same amount of $3000. The total budget for upgrading the network shown in Figure 11.4A to a more reliable network with redundancies is $24,000. This budget is sufficient for purchasing exactly eight redundant components/edges. All possibilities for implementing redundancies in the network from Figure 11.4A are given in Figure 11.4B, which is the full-complexity network, incorporating all possible connections for the redundant edges.

The problem is to comply with the fixed budget of $24,000 for purchasing eight edges (11 edges are to be removed from the set of all possible redundant edges in the full-complexity network), and achieve a maximum production availability of the repairable network. (The total cost of the remaining redundant edges should be within the specified budget of $24,000.)

Applying an optimisation based on the described stochastic optimisation algorithm, to the redundant edges only, in the full-complexity network shown in Figure 11.4B, resulted in the network shown in Figure 11.5, for which the cost of the eight redundant edges ((3,4), (2,5), (4,7), (7,8), (8,11), (11,12), (12,15) and (13,14)) is exactly $24,000, which has a maximum production availability of $\psi = 0.81$. This network has been obtained by pruning 11 edges from the possible redundant edges in the full-complexity network shown in Figure 11.4B, characterised by a production availability $\psi = 0.857$.

11.4 Topology Optimisation of Reliability Networks

Topology optimisation of reliability networks is particularly important for safety–critical systems which are widely used in industry, hospitals, construction engineering and public buildings. The loss from failure of a safety–critical system can be very high. This circumstance significantly affects the design of safety–critical systems which necessarily has to be a risk-based design (Todinov, 2007). Safety–critical systems must be highly reliable to guarantee a small risk of failure yet they must be designed at affordable cost, often within a limited budget. Existing design tools, however, *operate on a fixed system*

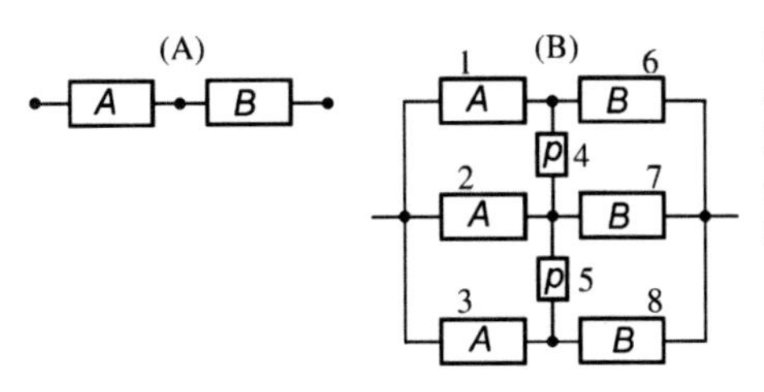

Figure 11.6 (A) An example of the reliability network of a basic safety–critical block and (B) the full-complexity reliability network, incorporating all possibilities (Todinov, 2011d).

architecture (Kuo et al., 2001; Tillman et al., 1985). They do not perform a repeated modification of the system topology and *do not* search in a large space of alternatives, in order to determine the system design, which delivers a maximum possible reliability attained within a specified budget for building the system. This is the reason why the existing computational tools *are not capable of supporting the optimal design of safety–critical systems.*

In the complex, safety–critical systems used today, there is a very large number of possibilities for selecting components with different reliabilities and costs, design configurations, cross-bridges and redundancies (e.g. active, standby, *k*-out-of-*n*).

In this huge space of alternatives, identifying the optimum set of alternatives for the components, the optimum system architecture and the necessary redundancies is not a trivial task. Without the right models and tools, design alternatives far from optimal will be selected – associated with either a significant risk of failure or a significant cost for building the system.

Consequently, an optimisation method for minimising the risk of failure of a safety–critical system within a specified budget for building the system is critically important.

The optimisation algorithm, developed for repairable flow networks, can also be used for optimising safety–critical systems, in order to obtain maximum reliability within a specified budget for building the system.

Consider a basic safety–critical block (Figure 11.6A) consisting of two components 'A' and 'B', logically arranged in series (e.g. component A could be an electronic circuit and component B could be an electro-mechanical device). Suppose that all possible ways of increasing the reliability of this safety–critical block are introducing up to three redundant safety–critical blocks working in parallel, introducing bridges '*p*' and including active redundancy (components working in parallel) for each component (Figure 11.6B). The full-complexity reliability network is shown in Figure 11.6B, where for each of the slots marked by numbers 1,2, ... ,8, three alternatives exist: alternative 1 – a single component in the slot; alternative 2 – a component and an active redundant component in the slot and alternative 3 – no component in the slot.

By varying the values characterising the slots, all possible reliability networks, locked in the full-complexity reliability network, can be obtained. Thus, the space of all possible alternatives for the full-complexity reliability network shown in

Figure 11.6B is 3^8. For larger networks, the size of the space of alternatives increases exponentially.

For the sake of simplicity, suppose that the separate components in the slots of the full-complexity reliability network are characterised by negative exponential time-to-failure distributions $F_i(t) = 1 - \exp(-t/\eta_i)$, where the mean times to failure are $\eta_A = 3$ years, $\eta_B = 7$ years and $\eta_p = 2$ years. The specified time of operation for the safety–critical block is $t = 2$ years. The costs of the single components are as follows: $c_A = \$130$, $c_B = \$390$ and $c_p = \$5$.

If, for a particular slot, an active redundancy is included, the cost of components for this particular slot is doubled. If no components are present in a particular slot, the cost of the components in the slot is zero. The total budget for purchasing components is \$1000.

Again, in order to optimise the topology of the safety–critical system, the idea is to start with the reliability network with full complexity, including all possible bridges and redundancies. Different combinations of components from the full-complexity reliability network are then pruned by the bound and branch algorithm. This operation permits prospective reliability networks embedded in the full-complexity reliability network to be explored, but not necessarily all of them. After pruning an edge, from the tree of the edges to be pruned, the reliability of the network for $t = 2$ years is calculated by using the system reliability algorithm described in Chapter 8. The topology optimisation of the reliability network can also be done by an exact, full exhaustive search recursive algorithm. This solution however is suitable only for networks with a relatively small size. The topology optimisation based on the described branch and bound method, does not require exploration of all possible networks locked in the full-complexity network.

If the calculated reliability of the pruned network is smaller than the achieved-so-far maximum reliability corresponding to a reliability network which satisfies the budget constraint, further pruning of components is not performed. Here again, in order to stop further branching and unnecessary exploration of the space of alternative reliability networks, we use the following fact (whose proof has been omitted). *By removing a component, the reliability of the resultant network can never increase*. This is true because *the reliability function is a monotonically increasing function in the reliabilities of the separate components*. This fact is used for constructing an algorithm, whose running time is superior to the running time of an algorithm based on a full-exhaustive search.

The important point is that if, for several pruned components, the reliability of the resultant reliability network is smaller than the achieved-so-far reliability maximum, there is no need for pruning other components from the reliability network in order to satisfy the budget constraint. Further pruning will only result in even smaller network reliability. This comparison permits continuing the search without having to descend on the sub-trees of pruned edges, where a larger system reliability cannot possibly be found. The tree of alternatives that follows the removal of a component leads to reliability networks with a smaller reliability than the achieved-so-far reliability maximum.

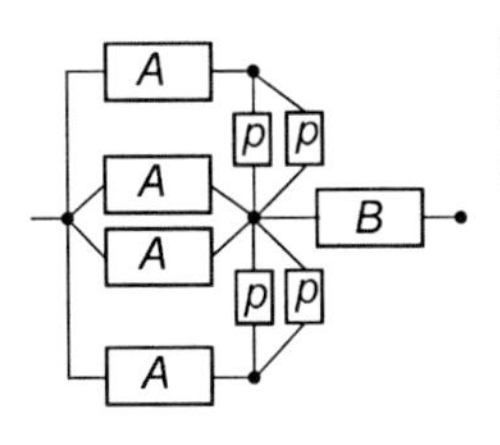

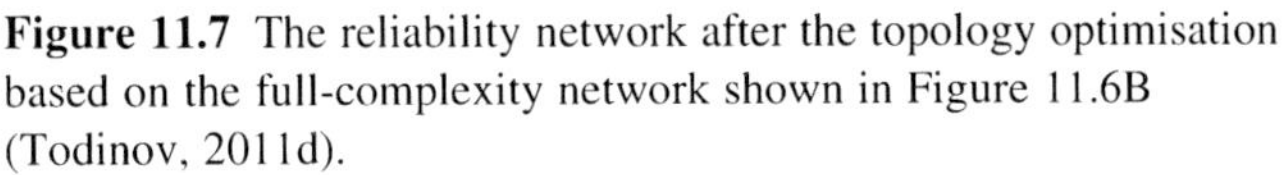
Figure 11.7 The reliability network after the topology optimisation based on the full-complexity network shown in Figure 11.6B (Todinov, 2011d).

If, as a result of the pruning, the resultant reliability network satisfies the budget constraint and the reliability of the resultant network is larger than the achieved-so-far global maximum (record), the current record is replaced by the calculated reliability characterising the resultant network. The implementation of the described optimisation algorithm in C is given in Appendix B (Todinov, 2011d).

For the network shown in Figure 11.6B, applying the topology optimisation procedure yields an optimum vector (1,2,1,2,2,0,1,0) for the components.

This means that in the optimal network topology, there are single components in slots 1, 3 and 7, two components logically arranged in parallel in slots 2, 4 and 5 and no components in slots 6 and 8. The optimal reliability network is shown in Figure 11.7. Its reliability for $t = 2$ years is 0.66, which is the maximum possible system reliability achieved within the specified budget of \$1000 and under the constraint that 'no more than two components can be arranged in parallel in any given slot'. From the optimal reliability network, the optimal system architecture can be reconstructed easily.

The presented method creates the attractive opportunity of increasing the safety margin of common safety–critical systems, without increasing the cost of current designs.

Alternatively, the method can also be used for exploring whether a higher level of reliability can be achieved at the current cost or even at a reduced cost for building the system. This creates the attractive possibility of maintaining the competitiveness of industrial companies and their position on the market.

Appendix A

In the algorithm, n denotes the number of all edges in the full-complexity network. The array 'redges[]' contains the indices of removed edges.

The recursive procedure ***Branch&Bound_prune***, with parameters 'depth' and 'next', organises the tree of unique combinations of edges to be removed. The function ***max_flow()*** calculates the maximum throughput flow while the function ***production_availability()*** calculates the production availability in a network from which a number of edges have been removed. These functions have been described in previous chapters. If the maximum throughput flow or production availability is smaller than the global maximum 'cur_value' achieved so far, the statement '*continue*' skips (prunes) the current combination of edges and no descending down the tree of removed edges is done. This is because a further descending (pruning) will only result in a reduction of the maximum throughput flow or production availability and the absolute maximum of the production availability cannot possibly be found on this sub-tree.

Descending on the tree of removed edges is initiated if the calculated flow from '*s = **max_flow()***' or the calculated production availability from '*s* = ***production_availability()***' is larger than the global maximum achieved so far and the budget constraint has not been satisfied. Then, after replacing the current record with the calculated throughput flow or production availability and saving the indices of the removed edges, a recursive call is initiated by '***Branch&Bound_prune***(depth + 1,j)', which increases the depth of search on the tree of edges by one.

The bottom of the recursion is reached when the number of removed edges becomes larger than the minimum number of edges 'min_num_rmvd_edges' to be removed.

Algorithm 11.2

(Algorithm 11.1 implemented in C)

```
void remove_edge(int k)
  { // Removes the edge with index k from the network by setting its flow capacity to zero }

void save_indices()
  { // Saves the indices of the removed edges }

void restore_removed_edges()
  { // Restores the removed edges in the network by restoring their original capacities}

double max_flow()
  { // Determines the maximum throughput flow in the network }

double production_availability()
  { // Determines the production availability of the network }

int const min_num_rmvd_edges;

void Branch&Bound_prune (int depth, int next)
  {
  int j,t;
  double s;

  if(depth > min_num_rmvd_edges) return;
  for(j = next + 1; j< =n; j++)
  {
    redges[depth] = j;
  for(t = 1; t< =depth; t++) remove_edge(redges[t]);
  s = max_flow()// or s = production_availability(); if production availability is
                                                             determined
 restore_removed_edges();
```

```
        if (s > cur_value) {
                if (depth = = min_num_rmvd_edges)
                        {
                         cur_value = s;
                         save_indices();
                        }
                             }
        else continue;
    Branch&Bound_prune(depth + 1, j);
    }
}

void main()
 {
   cur_value = 0;
   Branch&Bound_prune(1,0);
}
```

Appendix B

Again, to simplify the presentation of the algorithm, this implementation is relevant to the case where all edges have the same cost and the minimum number of edges that should be removed in order to comply with the allocated budget for building the network is given by the constant 'min_num_rmvd_edges'.

```
void remove_edge(int k)
   { // Removes the edge with index k from the network }

void save_indices()
   { // Saves the indices of the removed edges }

void restore_removed_edges()
   { // Restores the removed edges in the network }

double system_reliability()
   { // Returns the reliability of the network }

int const min_num_rmvd_edges;

void Branch&Bound_prune(int depth, int next)
   {
   int j,t;
   double s;

   if(depth > min_num_rmvd_edges) {
              return;
              }
   for(j = next + 1; j< = N; j+ +)
```

```
{
    redges[depth] = j;
for(t = 1; t< = depth; t+ +) remove_edge (redges[t]);
s = system_reliability();
restore_removed_edges ();
            if (s > cur_max_reliability) {
                    if (depth = = min_num_rmvd_edges)
                                {cur_max_reliability = s;
                                save_indices ();
                        }
                                            }
            else continue;
Branch&Bound_prune(depth + 1, j);
    }
}
    void main()
    {
    cur_max_reliability = 0;
    Branch&Bound_prune(1,0);
}
```

Again, in the algorithm, '*n*' is the number of edges in the full-complexity reliability network and redges[] is an array containing the indices of removed edges.

Similar to the case of a repairable flow network, the recursive procedure ***Branch&Bound_prune***(), with parameters 'depth' and 'next', organises the tree of unique combination of edges to be removed. The function ***system_reliability***() determines the reliability of the network. If the calculated maximum reliability is smaller than the current record 'cur_max_reliability', the statement '**continue**' forces the loop to continue with the next iteration and skips the current combination of edges. As a result, no descending down the tree of removed edges is performed. This is because a further descending (pruning) will only result in smaller system reliability and the absolute maximum of the system reliability cannot possibly be found by further descending on the sub-tree of edges to be pruned. Descending down the tree of removed edges is initiated if the calculated system reliability from 's = ***system_reliability***()' is larger than the achieved-so-far global maximum (current record). Then, after replacing the current record with the calculated system reliability and saving the indices of the removed edges, a recursive call ***Branch&Bound_prune***(depth + 1,j) is performed which increases the depth on the tree of removed edges by one.

12 Repairable Networks with Merging Flows

12.1 The Need for Improving the Running Time of Discrete-Event Solvers for Repairable Flow Networks

The expected operating lives for many repairable flow networks, particularly oil and gas networks, can be many years; the number of sources of flow could reach hundreds and the number of components thousands. The value of the maximum throughput flow is required upon each failure of a component. With increasing the number of components, the time for calculating the maximum throughput flow upon failure of a component increases. Revealing the variation of the output flow over a large time interval requires generating thousands of failure histories, each of which could contain hundreds of failures. As a result, the maximum throughput flow algorithm will have to be executed not just once or several times, but hundreds of thousands of times.

Suppose that for a large network containing thousands of components, a single run of a maximum flow algorithm with polynomial complexity requires at least 0.5 s. Simulating 10,000 failure histories with approximately 100 failures in each failure history, will require more than 5.5 days. Such a long computational time is totally unacceptable for fast-track projects, where a large number of design alternatives must be assessed and compared. The expected computational time *is in the range of seconds and minutes*, not hours and days.

As a result, an important requirement from algorithms computing the throughput flow in production networks is their *computational speed*, if the discrete-event simulators of the network performance are to be efficient. In order to optimise production performance, a large number of alternative design layouts must be analysed in a short period of time and bottlenecks associated with large losses from failures must be identified quickly.

For very large production networks with merging flows, which may include thousands of components, the computational speed of existing algorithms for maximising the throughput flow, with polynomial time complexity, is not sufficient, particularly in cases where tens of thousands of failure histories need to be generated over large time intervals.

In the next section, it is shown that extremely efficient algorithms, orders of magnitudes faster than any of the existing algorithms for maximising the throughput flow, can be designed by exploiting the specific network topology. Because

Flow Networks. DOI: http://dx.doi.org/10.1016/B978-0-12-398396-1.00012-X

many production networks and manufacturing networks are networks with merging flows, the described algorithm will be based on this important network topology. Networks with merging flows are very common, for example, in subsea oil and gas production systems. A typical subsea oil and gas production system, for example, is composed of a number of production wells (sources of flow), from which the produced streams of hydrocarbons are carried through pipelines to manifolds, from which the hydrocarbons are in turn carried through pipelines to a riser. In what follows, a very efficient method is proposed for determining the throughput flow in networks with merging flows.

12.2 An Algorithm with Linear Running Time, for Maximising the Flow in a Network with Merging Flows

The building blocks of repairable networks with merging flows considered in this chapter are components with multiple input flows and a single output flow (Figure 12.1). A characteristic feature of a network with merging flows is that the flows *do not branch out after the components*. In other words, the entire flow feeding a particular component, passes through the next component.

Each component/edge is characterised by a flow capacity that decreases to zero upon failure of the component. The sink t is fed by several components, each of which is fed in turn by other components, etc. As a result, the merging flow network has a tree topology and can be modelled by a directed tree. In Figure 12.1A, the components are labelled by $C1, C2, \ldots, C11$. They are represented by edges in the

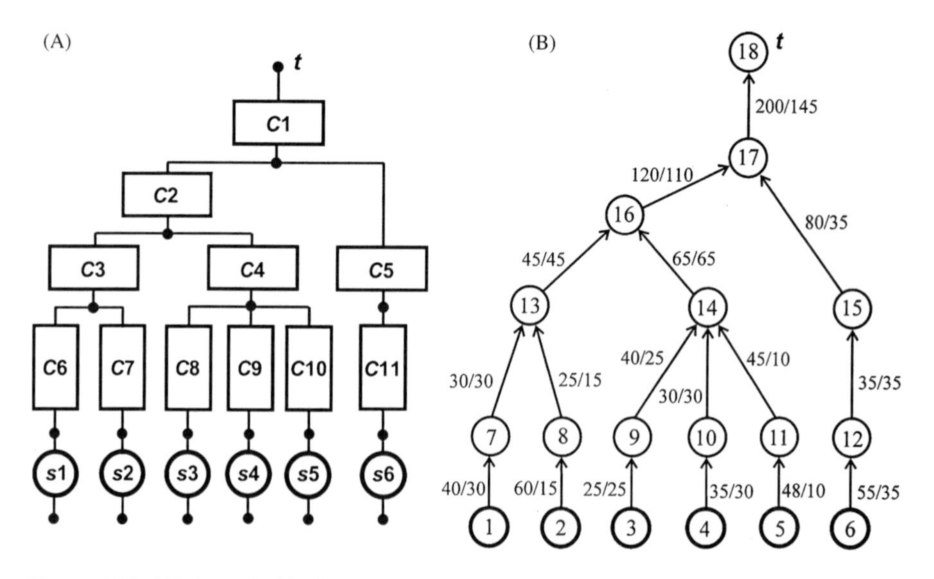

Figure 12.1 (A) A repairable flow network with merging flows, based on six production sources ($s1, s2, \ldots, s6$). (B) Modelling the network from figure (A) by a directed tree.

graph representation of the network shown in Figure 12.1B. Component $C6$, for example, has been represented by edge (7,13); component $C4$ has been represented by edge (14,16), etc. The sources are also treated as components and are also represented by edges. Thus, edge (1,7) represents the source $s1$; edge (2,8) represents the source $s2$, etc. The sink t for the network shown in Figure 12.1A has been represented by node 18 shown in Figure 12.1B.

The repairable networks with merging flows are *connected networks*. In other words, from any given component/edge, there is always a directed path to the sink.

The next lemma will be used to prove a theorem related to the maximum throughput flow in networks with merging flows. The lemma is a fundamental property of trees and can be proved by induction (Gross and Yellen, 2006; Johnsonbaugh, 1997).

Lemma 12.1 *The number of edges m in a connected network with tree topology is equal to the number of nodes n less than one* ($m = n - 1$).

Proof The property is certainly true for the simplest network with tree topology, based on two nodes only ($k = 2$). This network is represented by two nodes connected by an edge. For the number of edges ($m = 1$) and the number of nodes ($n = 2$), $m = n - 1$ holds.

Suppose that this property holds for any network with tree topology, based on k nodes. It can be shown that the property also holds for a network with tree topology based on $k + 1$ nodes.

Indeed, select the node with index q with the largest distance from the sink t (in terms of number of separating edges). This node certainly does not have a predecessor. Otherwise, if the node q has a predecessor with index p, the node p and not the node q will be the node characterised by the largest distance from the sink t, which is a contradiction with the choice of node q. Now, if node q with the connecting edge is removed, exactly one edge and one node will disappear from the original network. In addition, the remaining network will be a *connected network with tree topology* based on exactly k nodes. In the remaining connected tree network, by the induction assumption, the number of edges must be equal to $k - 1$. Because exactly one edge has been removed from the original network, the number of edges in the original network is equal to $(k - 1) + 1 = k$. Therefore, for a tree network with $k + 1$ nodes, the number of edges is k. This, together with the trivial case $k = 2$, proves the statement $m = n - 1$ for networks with tree topology.

The maximum throughput flow in a network with merging flows can be determined in worst-case running time that depends linearly on the number of components/edges in the network. To demonstrate this, consider the following procedure related to a network with merging flows, originating from N_p sources. All sources are connected to a super-source s through edges with capacity equal to the flow production capacity of the sources ($N_p = 6$ in Figure 12.2). In this way, the sources of flow disappear and ordinary nodes appear instead. The nodes corresponding to the sources are then marked with non-negative numbers, showing the flow production

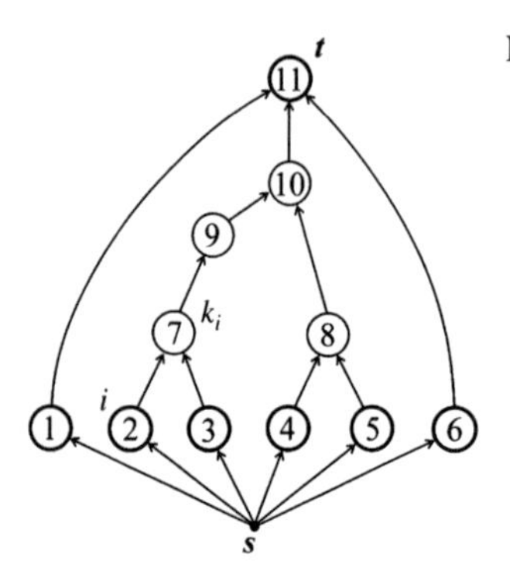

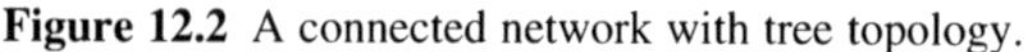
Figure 12.2 A connected network with tree topology.

capacity of the sources. The rest of the nodes are marked with the negative number '-1'. All marked nodes are then placed on a queue.

One by one, nodes are taken from the head of the queue. Because the network has a connected tree topology, each node has exactly one outgoing edge (exactly one descendent/successor node). For each node i, taken from the head of the queue, the descendent/successor node k_i is considered and if all of its predecessor nodes have non-negative labels, the sum of the labels of the predecessors of node k_i is compared with the capacity of the edge coming out of node k_i (Figure 12.2). The smaller of the two numbers is then assigned as a label of node k_i and node k_i is placed at the end of the queue. Thus, the label of node k_i is numerically equal to the maximum amount of flow that can pass through node k_i, without violating the flow conservation law at the node. If node k_i is the sink t, there are no outgoing edges and the label of the sink is equal to the sum of the labels of all of its predecessor nodes. If there is at least one predecessor of node k_i, with a negative label, node k_i *is not* placed in the queue. After processing the original node i in this way, node i is excluded from the queue. The next node in the queue is then extracted and the procedure is repeated. It is clear from this algorithm that a non-negative label at each node is equal to the flow through the outgoing edge from this node.

The size of the input in the described algorithm is equal to the total number of considered nodes-predecessors. The number of considered nodes-predecessors in the network, however, is equal to the number of components/edges. Each predecessor is considered exactly once because, for a directed tree network, there are no two nodes with the same predecessor. Consequently, the size of the input is equal to the number of edges m in the network.

Theorem 12.1 *Given that the source nodes have been labelled, the described procedure guarantees that the sink will be labelled in worst-case running time $O(m)$, where m is the number of edges in the network. The label of the sink contains the maximum throughput flow in the network.*

Proof The theorem can be proved by induction. For the simplest network with merging flows containing only two nodes ($k = 2$) (one of which is a source and the other is a sink) and a single edge $m = 1$, the theorem is obviously true. The maximum throughput flow in this trivial network will be reached after a single basic step consisting of considering the only predecessor of the sink.

It can also be easily verified that the theorem is also true for a network with merging flows containing three nodes, one of which is a sink and the other two are either two sources or one is a source and the other is an internal node. In this case, the maximum throughput flow in the network is reached after $m = 2$ basic steps, equal to the number of all existing predecessors in the network. Therefore, the theorem is true for networks with any number of nodes up to 3.

Suppose that the theorem is true for any number of nodes up to k nodes, one of which is a sink, and at least one of which is a source. It can be shown that the theorem is also true for any number of nodes up to $k + 1$ nodes.

Indeed, consider the sink node t in the network with $k + 1$ nodes. Consider all immediate predecessor nodes of the sink t. Suppose that there are p predecessors $(1,2, \ldots ,p)$. Remove the p edges connecting the predecessors with the sink t. Because of the connected tree topology, each of the predecessor nodes of the sink t will be the root of a tree, and the trees corresponding to the separate predecessors $(1,2, \ldots ,p)$ will not have common edges or sources. Each source can feed exactly one predecessor of the sink t and no source can feed two different predecessors of the sink t. Now, if the predecessors of the sink are in turn considered as 'sinks', together with the sources feeding them and the connecting edges, they will form separate sub-trees. These sub-trees are, in fact, connected networks with merging flows. Let the number of nodes in these sub-trees be $k_1,k_2, \ldots ,k_p$, correspondingly ($k_1 + k_2 + \cdots + k_p = k$ and $k_i \leq k$, $i = 1,2, \ldots ,p$).

For networks with up to k nodes, however, the theorem is true, which means that the predecessor nodes of the sink t will be reached in at most $k_1 - 1$, $k_2 - 1$, $\ldots$, $k_p - 1$ steps and the labels at the predecessor nodes will be the maximum throughput flows in the sub-tree networks. Consequently, all predecessor nodes will be labelled after at most $k_1 + k_2 + \cdots + k_p - p = k - p$ steps. Labeling the sink t involves going through all p predecessor nodes and requires p additional steps. Consequently, labeling the sink will be done in at most $k - p + p = k$ steps, equal to the number of edges in a connected network with $k + 1$ nodes and tree topology. Consequently, the worst-case running time of labelling the sink t will be $O(k)$ which, together with the trivial cases $k = 2$ and $k = 3$, proves the theorem. □

Theorem 12.1 and the associated procedure define an algorithm for determining the maximum throughput flow in a network with merging flows. Here is the algorithm in pseudo-code:

Algorithm 12.1

// Marks the nodes corresponding to the sources with labels showing the flow production capacity of the sources, which is a non-negative number.
for i = 1 **to** N_p **do** {label[i] = production from source *i*; q[i]=i; }

// Marks the rest of the nodes with the negative number '− 1'.
for i = $N_p + 1$ **to** *n* **do** label[i] = −1;

```
qhead = 0; qend = N_p
while (qhead < = qend) do
{
        qhead = qhead + 1;
        node = q[qhead]; // takes the current node from the head of the queue;
        succ_node = arr_s[node]; // take the only successor of node 'node';
        np = pred[succ_node][0]; // gives the number of predecessors of
                                       'succ_node'
        sum = 0; flag = 1;
        for i = 1 to np do
            {
                pred_node = pred[succ_node][i]; // extracts a predecessor of 'succ_node'
                if (label[pred_node]) > 0 then sum = sum + label[pred_node];
                            else {flag = 0; break;}
            }
            if (flag = 1) then {
                    if (succ_node = n) then {label[n] = sum; break;}
                    cap_outg_edge = cap[succ_node];
                    if (cap_outg_edge > sum) then label[succ_node] = sum;
                        else label[succ_node] = cap_outg_edge;
                    qend = qend + 1;
                  q[qend] = succ_node;
            }
  }//end while loop
```

The array pred[] contains the predecessors of a given node; pred[i][0] gives the number of predecessors of node i. The array entry arr_s[i] contains the index of the successor of node i. The array cap[i] contains the flow capacity of the outgoing edge from node i; the variable n contains the total number of nodes in the network and is also the index of the sink t. All nodes i in a merging flow network have a single successor, whose index is kept in arr_s[i]. The loop,

```
  sum = 0; flag = 1;
  for i = 1 to np do
    {
          pred_node = pred[succ_node][i];
          if (label[pred_node]) > 0 then sum = sum + label[pred_node];
          else {flag = 0; break;}
    }
```

extracts all predecessors of node succ_node and if they all have positive labels, their values are accumulated in the variable 'sum'. The variable 'sum' contains the amount of flow with which node 'succ_node' is fed. If the capacity of the outgoing edge from 'succ_node' is greater than 'sum', the node 'succ_node' is labelled with the sum of the flows going into it. If the converse is true, the flow through node 'succ_node' is constrained by the capacity of the outgoing edge, and node 'succ_node' is labelled with the capacity 'cap_outg_edge' of the outgoing edge.

If the 'succ_node' is labelled with a positive flow, this node is placed in the queue and the process continues.

In the network from Figure 12.1B, the first number c from the labels c/f is the edge capacity and the second number is the actual edge flow, obtained by the described algorithm. The maximum throughput flow obtained by the described algorithm is 145 units (Figure 12.1B).

12.3 Optimising the Topology of a Repairable Flow Network with Merging Flows to Minimise the Losses from Failures

A very important issue related to the design of repairable flow networks is the link between the topology of the network and its dynamic properties. An important question here is what features of the network topology make the network least sensitive to edge failures and provide a maximum transmitted flow in the presence of edge failures. It can be shown that, for a given number of independently working edges in a network with merging flows, the maximum throughput flow is associated with a topology where each edge failure is associated with choking a single source and the number of sources of flow is maximal.

Indeed, let us consider a network based on m components and N_p sources, with a life cycle of a years. Suppose that each of the sources is characterised by the same flow rate P_{max}/N_p units/per day, where P_{max} is the total maximum possible flow per day that can be produced.

For the sake of simplicity, let us assume that all edges work independently from one another and are characterised by the same hazard rate λ and the same repair time d. Suppose that, upon edge failure, the capacity of the edge is reduced to zero. Because of the topology of a network with merging flows, any failure of an edge is associated with at least one fully choked source of flow. Let k_1 be the number of edges in the network whose failure results in a full choking of exactly one source; k_2 be the number of edges in the network whose failure results in a full choking of exactly two sources; ... , and k_{Np} be the number of edges in the network whose failure results in a full choking of all sources. Repeated failures of an edge chokes the same number of sources. Consequently, with respect to the number of sources they fully choke, the edges can be separated in N_p mutually exclusive classes (1,2, ... , N_p) and $k_1 + k_2 + \cdots + k_{Np} = m$ holds.

For a repair time d, significantly smaller than the life cycle a of the network ($d << a$), overlapping failures can be neglected and, as a result, the losses from failures can be approximated by the sum of the losses generated by the separate edge failures.

For a network life cycle of a years, each edge will be associated with λa expected number of failures. Choking of a single source during the repair time of d days is associated with $(P_{max}/N_p)d$ amount of lost flow. Consequently, edges whose failure chokes a single source only will be associated with $k_1 \times (\lambda a \times (P_{max}/N_p)d)$ expected amount of lost flow. Edges whose failure results in choking exactly two sources will be associated with $k_2 \times (\lambda a \times (2P_{max}/N_p)d)$ expected amount of lost flow and so on. The total expected lost flow L can then be determined from

$$L = \lambda a k_1 \frac{P_{max}}{N_p} d + \lambda a k_2 \frac{2P_{max}}{N_p} d + \cdots + \lambda a k_{Np} \frac{N_p P_{max}}{N_p} d \tag{12.1}$$

which can also be presented as

$$L = \lambda a d \frac{P_{max}}{N_p} (k_1 + 2k_2 + 3k_3 + \cdots + N_p k_{Np}) \tag{12.2}$$

Since $k_1 + k_2 + \cdots + k_{Np} = m$, where m is the total number of edges in the network, the minimum of expression (12.2) is attained for the maximum possible value $N_p = N_p^{max}$ of the number of sources and when failures of each edge in the network results in choking of only a single source. In other words, when $k_1 = m$, $k_2 = k_3 = \ldots = k_{Np} = 0$ and $N = N_p^{max}$. Substituting in Eq. (12.2) yields

$$L_{min} = \lambda a d \frac{P_{max}}{N_p^{max}} m \tag{12.3}$$

In other words, the minimum lost flow is attained when failure of each edge results in choking no more than a single source and the number of sources through which the total possible flow P_{max} is produced is the maximal possible number.

Now, suppose that the tree-like topology with hierarchy, where failure of a component leads to choking of more than a single source, is already specified and cannot be altered, but the hazard rates of the edges can be altered in order to reduce the losses. Let $c_1, c_2, \ldots, c_m$ denote the number of choked sources associated with the separate edges ($1 \le c_i \le N_p$). Initially, suppose that all edges are characterised by the same hazard rate λ. The expression that yields the lost flow then becomes

$$L = ad \frac{P_{max}}{N_p} (\lambda c_1 + \lambda c_2 + \lambda c_3 + \cdots + \lambda c_m) \tag{12.4}$$

Despite the same hazard rate, upon failure, the different edges generate *different* losses. The more sources are fully choked upon failure of an edge, the higher the losses. The edge going to the sink, whose failures entail the largest number of choked sources, will be associated with the largest losses. To make the expected losses from all edges comparable (e.g. equal), edges higher up in the tree hierarchy *must be designed to higher reliability levels* (smaller hazard rates). To ensure the same amount of expected losses from each edge failure, the hazard rate of edges whose failure results in two choked sources must be half of the common hazard rate ($\lambda/2$). The hazard rate of edges, whose failure results in three choked sources must be one-third of the common hazard rate ($\lambda/3$) and so on. Substituting in Eq. (12.4) results in

$$L = ad\frac{P_{\max}}{N_{\mathrm{p}}}(c_1(\lambda/c_1) + c_2(\lambda/c_2) + \cdots + c_m \times (\lambda/c_m)) = \lambda ad\frac{P_{\max}}{N_{\mathrm{p}}}m \qquad (12.5)$$

In general, if all edges are characterised by the same expected number of failures U during the time interval (0,a), the total expected losses from failures in this interval are approximately

$$L = d\frac{P_{\max}}{N_{\mathrm{p}}}(Uc_1 + Uc_2 + Uc_3 + \cdots + Uc_m) \qquad (12.6)$$

To ensure the same amount of expected losses from each edge, the expected number of failures characterising the edges must be inversely proportional to the number of choked sources. As a result, the total lost flow becomes

$$L = d\frac{P_{\max}}{N_{\mathrm{p}}}\left[c_1(U/c_1) + (U/c_2)c_2 + \cdots + (U/c_m)c_m\right] = Ud\frac{P_{\max}}{N_{\mathrm{p}}}m \qquad (12.7)$$

Now, a principle for a risk-based design of repairable networks with merging flows can be formulated. The expected number of failures characterising the edges in a network with merging flows should be inversely proportional to the lost flow upon edge failure (the lost flow upon edge failure is proportional to the number of choked sources upon failure). In short, edges associated with a larger amount of lost flow upon failure should be designed to higher reliability levels (smaller hazard rates).

13 Flow Optimisation in Non-Reconfigurable Repairable Flow Networks

13.1 Lost Flow Caused by Edge Failures

Often, component failures cause the flow through the failed components to be lost until the flow is restored. Such is the case in *non-reconfigurable flow networks*, where on a component failure, the flow through the failed component cannot be redirected through alternative working components. A typical example of a non-reconfigurable network is the network with merging flows, which is characterised by a tree-like topology (Figure 12.1A), in which the edges going out of the sources are fully saturated with flow.

Failure of any edge in such networks causes the flow through the edge to be lost during the duration of its repair.

Treating the repairable flow network as a static flow network, where no edge failures are present, and maximising the throughput flow in it, *in general, does not maximise the throughput flow in the non-reconfigurable network*. Consider the manufacturing system shown in Figure 13.1, where two sources of workpieces $s1$ and $s2$, with maximum production capacities 2000 work-pieces per day, supply two parallel production lines $L1$ and $L2$. The production lines carry the workpieces to a machine centre MC, with production capacity 2400 work-pieces per day, after which, the work-pieces are collected at the sink t. In Figure 13.1, λ_{L1} and λ_{L2} denote the failure frequencies of the parallel production lines and λ_{MC} denotes the failure frequency of the machine centre. Suppose that the downtime for repair for each section of the production lines and for the machine centre is the same, 45 min. Flow of workpieces can be generated from each of the two sources $s1$ and $s2$, whose flow production, once set up, cannot be altered. The flow in the static flow network, where no component failures occur, can be maximised by saturating both flow lines with flows of magnitude 1200 workpieces per day. The maximum throughput flow rate of the manufacturing system is therefore 2400 workpieces per day. The lost flow due to the failures of sections $L1$, $L2$ and the machine centre means lost production.

The lost production can be reduced significantly by directing most of the flow through the more reliable production line (the second flow path $s2$, $L2$, MC, t). In this flow path, production line $L2$ fails less frequently than production line $L1$ from

Flow Networks. DOI: http://dx.doi.org/10.1016/B978-0-12-398396-1.00013-1

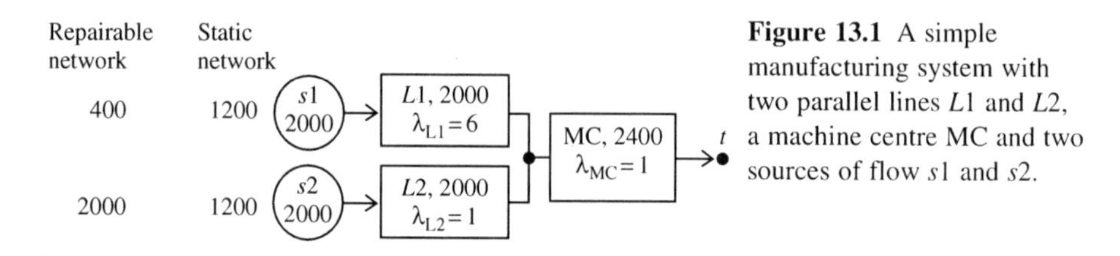

Figure 13.1 A simple manufacturing system with two parallel lines $L1$ and $L2$, a machine centre MC and two sources of flow $s1$ and $s2$.

the flow path ($s1$, $L1$, MC, t) and the lost flow will be smaller. The cost of intervention for repair will also be smaller because of the fewer failures and repairs.

The flow of workpieces can also be distributed as follows: 400 work-pieces per day through the flow line $L1$ and 2000 workpieces per day through the flow line $L2$. This allocation of the production flows sets the same throughput flow rate of 2400 units of flow per day in the manufacturing system, but it minimises significantly the lost production caused by failures of production line $L1$. The loss of flow from a flow path due to failures is proportional to the failure rates of the components along the flow path, their downtimes for repair, and the magnitude of the flow through the flow path.

Networks which do not have a tree topology may also be non-reconfigurable. In some networks with complex topology, once the flow has been set along some particular paths, it cannot be redirected. As a result, failures of components along these preset paths are associated with lost flow. Such are, for example, the computer networks with complex topology, in which a *virtual circuit* type of connection-oriented service is used (Tanenbaum, 2003). In this case, a path from the source router to the destination router must be established before any data can be transmitted. If failure of a router or a transmission line occurs during the time of transmission, the data flow through the preset flow path will be lost during the time needed for restoring the service. Transportation networks also exhibit a similar feature. Accidents in motorways far from exit roads, cause the traffic flow through the affected section to be fully or partially lost until the section is cleared. Some failure modes in supply networks and production networks are even associated with physical loss of flow during the repair time.

To reduce congestion in telecommunication networks, an optimal routing algorithm based on an exponential cost function has been proposed by Aspens et al. (1993). The edges are assigned costs that are exponential in the current edge loads. If a particular request occurs, the algorithm chooses the shortest path from source to destination in terms of the sum of exponential costs. The drawback of this algorithm is that it is suitable only in the case where the whole amount of requested traffic can fit in the bandwidth of the virtual circuit. If the required traffic exceeds the bandwidth of the virtual circuit, this approach does not utilise the existing alternative paths and the available residual capacity in the network.

Consider a repairable flow network with a single source s and a single sink t. The amount of flow per unit time, that needs to be transferred through the network from the source s to the sink t, is f^* $(0 < f^* \leq f_0)$, where f_0 is the maximum possible flow per unit time which could be transferred through the network, in the absence of component failures. The central problem is *how to set up the edge flows in the network, so that the throughput flow rate in the network, in the absence of failures, is f^* and the expected lost flow due to component failures is minimised.*

13.2 Resistance of a Path

In Chapter 2, a *path* has been defined as unique sequence of edges between two nodes. However, the absolute magnitude of the lost flow along a path does not provide a good measure of the quality of the path. Consider a path containing a single component with flow capacity 10 units of flow per day, a failure rate $\lambda_1 = 3$ year^{-1} and downtime for repair $d = 5$ days. This path is characterised by a smaller amount of lost flow due to failures compared to a more reliable path containing a single component with hazard rate $\lambda_2 = 1$ year^{-1}, downtime for repair $d = 5$ days and flow capacity 80 units of flow per day.

Suppose that the flow through a particular path can be increased by $\Delta_{\min}$, without violating the capacity constraints on the edges and the flow conservation law at the nodes along the path. Suppose that the average fraction of time during which edge i is unavailable is β_i. The time interval during which the network performance is studied is t_{op} time units. Commonly, for each edge, β_i is considerably smaller than unity ($\beta_i < 1$). Consider a forward edge i along the path. Suppose that $\Delta_{\min}$ is the bottleneck residual capacity characterising the path. Because, during the path augmentation, the flow along forward edges is increased, the lost flow due to failures of forward edges will also increase. For the ith forward edge, the expected lost flow L_i is equal to

$$L_i = \Delta_{\min} \times t_{\mathrm{op}} \times \beta_i \tag{13.1}$$

where $0 \leq \beta_i < 1$ is the average unavailability of edge i (the average fraction of time during which edge i undergoes repair). The availability α of a directed path composed of M_f forward edges is a product of the availabilities of the separate edges: $\alpha = \alpha_1 \times \cdots \times \alpha_{Mf}$. The unavailability β of the path is given by $\beta = 1 - \alpha$. Because the availability of each edge is $\alpha_i = 1 - \beta_i$, for the unavailability of the path, the expression

$$\beta = 1 - (1 - \beta_1) \times (1 - \beta_2) \times \cdots \times (1 - \beta_{Mf}) \tag{13.2}$$

holds. During the expansion of expression (13.2), products of order greater than one can be ignored because β_i are small numbers ($\beta_i << 1$). As a result, the unavailability β of the directed path can be approximated by

$$\beta \approx \sum_{i=1}^{Mf} \beta_i \tag{13.3}$$

The expected lost flow from a directed path composed of M_f forward edges, during a time of operation t_{op}, can be approximated by

$$L_f \approx \Delta_{\min} \times t_{op} \times \sum_{i=1}^{M_f} \beta_i \tag{13.4}$$

Suppose that a path containing both forward and backward edges is augmented with $\Delta_{\min}$ bottleneck flow.

On backward edges, the flow $\Delta_{\min}$ is not lost but is prevented from being lost because the flow along backward edges has been *decreased* by $\Delta_{\min}$. Therefore, the expected 'lost flow' L_j, on a backward edge with index 'j', has a negative sign:

$$L_j = -\Delta_{\min} \times t_{op} \times \beta_j \tag{13.5}$$

For M_b backward edges, the expected 'lost flow' can be approximated by

$$L_b \approx -\Delta_{\min} \times t_{op} \times \sum_{j=1}^{M_b} \beta_j \tag{13.6}$$

Consequently, the total lost flow L along a path including both forward and backward edges can be approximated by

$$L = L_f + L_b \approx \Delta_{\min} \times t_{op} \left(\sum_{i=1}^{M_f} \beta_i - \sum_{j=1}^{M_b} \beta_j \right) \tag{13.7}$$

The quantity $\rho = \sum_{i=1}^{M_f} \beta_i - \sum_{j=1}^{M_b} \beta_j$ $(-1 \le \rho \le 1)$ will be referred to as a *resistance of a path*. This new concept is of fundamental importance to understanding the properties of non-reconfigurable repairable flow networks.

A positive sign of the path resistance ρ means that along the flow path, flow is lost due to component/edge failures. A negative sign of non-reconfigurable ρ means that flow is prevented from being lost. The larger the magnitude of ρ, the larger is the amount of expected lost flow due to edge failures or the amount of flow which is prevented from being lost. Note that the resistance $\rho = \sum_{i=1}^{M_f} \beta_i - \sum_{j=1}^{M_b} \beta_j$ of a path traversed in one direction equals the resistance ρ' of the path traversed in the opposite direction taken with a negative sign. Clearly, changing the traversing direction of a path makes the forward edges backward edges and vice versa. Therefore,

$$\rho' = -\sum_{i=1}^{M_f} \beta_i + \sum_{j=1}^{M_b} \beta_j = -\rho \tag{13.8}$$

The as-defined path resistance ρ approximates adequately the lost flow from paths where the edges experience multiple failures during the operational time interval t_{op}. In the case where the edges along the path experience few failures, the path resistance will also approximate adequately the lost flow, provided that the path is sufficiently long.

13.3 Cyclic Paths: Necessary and Sufficient Conditions for Minimising the Lost Flow in Non-Reconfigurable Repairable Flow Networks

In a single-source, single-sink repairable flow network, a cyclic path is a path which (i) may include separately the source or the sink or both; (ii) the start node coincides with the end node and (iii) there are no other repeating nodes. It is important to point out that augmenting the flow along a cyclic path leads to a *feasible flow*, for which all capacity constraints of the edges and the flow conservation at the nodes are honoured.

The set of all flows along the edges of a network ($i,j \in E$) will be denoted by $f(i,j)$. The throughput flow in the network will be denoted by f^*. Augmenting the flow by Δ_{min}, along a cyclic flow path, does not alter the magnitude f^* of the throughput flow.

A throughput flow of a specified magnitude f^* can be attained through different feasible edge flows $f(i,j)$. There are *particular edge flows*, however, which minimise the lost flow due to edge failures in the network. By using the new concept '*resistance of a path*', an algorithm can be devised for minimising the lost flow, based on sequentially identifying and augmenting the s−t path with the smallest resistance, until no more augmentable paths can be found. The next two theorems serve as a justification of the proposed algorithm. According to Theorem 13.1, if there are no augmentable cyclic paths with negative resistance, the lost flow due to edge failures is minimal. According to Theorem 13.2, after a process of preferential saturation of $s-t$ paths characterised by the smallest resistance, there will be no augmentable cyclic paths with negative resistance. To prove Theorem 13.1, it is necessary to use a theorem, stated by Ahuja et al. (1993). Here, this theorem will be stated as Lemma 13.1, to reflect the circumstance that it will be used to prove Theorem 13.1.

Lemma 13.1 *If there are two different feasible edge flows $f_1(i,j)$ and $f_2(i,j)$, the flow $f_2(i,j)$ can be presented as a sum of the flow $f_1(i,j)$ and the augmented flows along a set of augmentable cyclic paths.*

The cyclic paths may include forward as well as backward edges. The cyclic flow paths may also include separately or simultaneously the source s and the sink t.

Theorem 13.1 *In a non-reconfigurable flow network, the necessary and sufficient condition for the smallest expected lost flow due to edge failures is the absence of augmentable cyclic paths with negative resistance* (Todinov, 2012c).

Proof Proving that the non-existence of augmentable cyclic paths with negative resistance is a necessary condition for a minimum lost flow due to edge failures is straightforward. Suppose that for a given throughput flow f^* in the network, the lost flow L due to edge failures is the smallest. If there exists an augmentable cyclic path with negative resistance, this path could be augmented, which will result in different feasible edge flows, with the same throughput flow f^*, but with a smaller expected lost flow L' due to component failures. However, this is impossible because, by assumption, the expected lost flow L is the smallest possible. We arrive at a contradiction.

Suppose now, that in a network characterised by edge flows $f_1(i,j)$ and throughput flow f^*, there is no augmentable cyclic path with negative resistance. Then, the lost flow L_1, due to component failures, associated with edge flows $f_1(i,j)$ is the smallest.

Indeed, suppose that edge flows $f_2(i,j)$ exist, associated with smaller losses $L_2 < L_1$. According to Lemma 13.1, the set of feasible edge flows $f_2(i,j)$ resulting in a throughput flow f^* can be obtained from the set of feasible edge flows $f_1(i,j)$, resulting in the same throughput flow f^*, by a series of augmentations along cyclic paths only. Without loss of generality, suppose that the edge flows $f_2(i,j)$ have been obtained from edge flows $f_1(i,j)$, after augmenting k cyclic paths. The expected lost flow L_2 associated with an augmentation along k cyclic paths is therefore given by

$$L_2 = L_1 + \rho_1 \Delta_1 t_{\text{op}} + \rho_2 \Delta_2 t_{\text{op}} + \cdots + \rho_k \Delta_k t_{\text{op}}$$

where ρ_i $(i = 1,2, \ldots ,k)$ are the resistances of the cyclic paths and $\Delta_i > 0$, $(i = 1,2, \ldots ,k)$ are the flows with which each cyclic path has been augmented. Because of the assumption $L_2 < L_1$, that edge flows $f_2(i,j)$ are associated with smaller losses than edge flows $f_1(i, j)$, the relationship

$$\rho_1 \Delta_1 t_{\text{op}} + \rho_2 \Delta_2 t_{\text{op}} + \cdots + \rho_k \Delta_k t_{\text{op}} < 0 \tag{13.9}$$

must necessarily hold. This relationship however is impossible because, according to our assumption, there is no augmentable cyclic path i with negative resistance $\rho_i < 0$. This completes the proof. Consequently, the lost flow L_1 associated with edge flows $f_1(i,j)$ is indeed the smallest possible. □

The property expressed by Theorem 13.1 is an analogue of the minimal cost property encountered in minimal cost problems (Ahuja et al., 1993).

Suppose that the $s-t$ path in Figure 13.2A is characterised by the smallest resistance. The section (b,d,e) is a *forward side section* spanning the section (b,p,e) which belongs to the $s-t$ path. Section (b,d,e) is referred to as a *forward side section,* because its starting node b is closer to the starting node s of the $s-t$ path than its end node e (Figure 13.2A). If the end node e is closer to the starting node s of the $s-t$ path, the side section is referred to as a *backward side section* (Figure 13.2B). The section (b,p,e) is the respective section on the $s-t$ path, spanned by the side section.

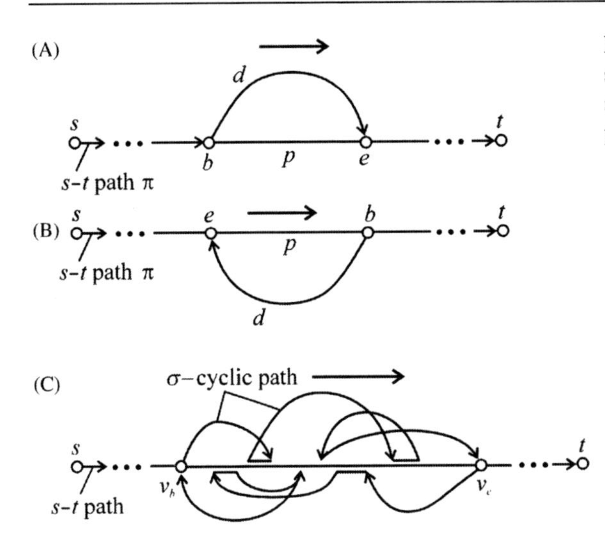

Figure 13.2 (A) A forward side section and (B) a backward side section, spanning part of the $s-t$ path; (C) a cyclic path σ.

Both the $s-t$ paths and the side sections (b,d,e) can include forward as well as backward edges. The following two lemmas are true.

Lemma 13.2 *Given an $s-t$ path π with the smallest resistance, the resistance of any augmentable forward side section is not smaller than the resistance of the respective spanned section.*

Proof Suppose that the resistance of a forward side section (b,d,e) (Figure 13.2A) is smaller than the resistance of the spanned section (b,p,e) on the π-path. By concatenating the side section (b,d,e) with the rest of the π-path, outside the spanned section (b,p,e), an alternative $s-t$ path (s,b,d,e,t) can be created. The resistance of the alternative $s-t$ path (s,b,d,e,t) is now smaller than the resistance of the π-path (s,b,p,e,t) which, by assumption, is the $s-t$ path with the smallest resistance. We arrive at a contradiction. Hence, the resistance of a forward side section cannot be smaller than the resistance of the respective spanned section on the $s-t$ path. □

This has been summarised in the inequality (13.10)

$$\rho_{\text{fd}} \geq \rho_{\text{fp}} \tag{13.10}$$

where ρ_{fd} is the resistance of the forward side section and ρ_{fp} is the resistance of the respective spanned section on the π-path. The index 'fd' stands for a *forward side section* and the index 'fp' stands for a *forward spanned section*.

Lemma 13.3 *Given that no cyclic paths with negative resistance exist before augmenting a particular $s-t$ path, the resistance of any augmentable backward side* *section connected to the augmented s−t path, is greater than the resistance of the respective spanned section on the s−t path, traversed in a backward direction.*

Proof According to the condition of the lemma, no cyclic path with negative resistance exists before augmenting the $s-t$ path. If the $s-t$ path can be augmented, so can the spanned section (e,p,b) in a direction from node e to node b (Figure 13.2B). The cyclic path (b,d,e,p,b) can also be augmented before augmenting the $s-t$ path, because it consists of the augmentable, by definition, side section (b,d,e) and the augmentable spanned section (e,p,b). Therefore, the resistance of the cyclic path (b,d,e,p,b) can be presented as

$$\rho_{\mathrm{bd}} + \rho_{\mathrm{f}p} \geq 0 \tag{13.11}$$

In inequality (13.11), ρ_{bd} is the resistance of the backward side section (b,d,e) and ρ_{fp} is the resistance of the forward spanned section (e,p,b). The index 'bd' stands for a *backward side section* and the index 'fp' stands for a *forward spanned section*.

From inequality (13.11), it follows that $\rho_{\mathrm{bd}} > -\rho_{\mathrm{fp}}$. Considering that the resistance ρ_{fp} associated with traversing the spanned forward section (e,p,b) in $s-t$ direction is equal to $-\rho_{\mathrm{bp}}$, where ρ_{bp} is the resistance of section (b,p,e), traversed in the opposite, $t-s$ direction ($\rho_{\mathrm{fp}} = -\rho_{\mathrm{bp}}$), the inequality

$$\rho_{\mathrm{bd}} \geq \rho_{\mathrm{bp}} \tag{13.12}$$

holds for each backward side section, which proves the lemma. □

The next fundamental property essentially establishes a sufficient condition for the absence of augmentable cyclic paths with negative resistance (Todinov, 2012c).

Theorem 13.2 *Augmenting sequentially the s−t paths with the smallest resistance guarantees that no augmentable cyclic path with negative resistance will exist in the network.*

Proof This theorem will be proved by induction. First, it will be shown that the theorem holds for a single augmented $s-t$ path π_1 ($k = 1$), characterised by the smallest resistance.

Indeed, a single augmented path from the source s to the sink t, in an empty network, must necessarily be a path consisting of forward edges only, because no flow can be subtracted from any backward edge (all backward edges are inadmissible for augmentation because they are empty). Further, if a cyclic path with negative resistance existed, it must necessarily have common edges with the single $s-t$ path π_1, characterised by the smallest resistance. This is because in the network outside the augmented $s-t$ path π_1, the edge flows are zero ($f(i,j) = 0$, for all edges). Consequently, in the network outside the augmented $s-t$ path π_1, only cyclic paths with forward edges can be augmented. All cyclic paths with forward edges however are characterised by a non-negative resistance. Therefore, no augmentable cyclic path with a negative resistance can possibly exist, if it has no common edges with the augmented $s-t$ path π_1.

Now suppose that an augmentable cyclic path σ with negative resistance exists, which has common edges with the augmented $s-t$ path π_1. Suppose that node v_b is the closest node to the source s where the cyclic path σ joins the augmented $s-t$ path π_1 and node v_e is the closest node to the sink t (Figure 13.2C). Note that no matter how complex the cyclic path σ is, it can only include (i) *side sections* joining the $s-t$ path π_1 at two nodes and not having common edges with the $s-t$ path π_1 and (ii) *common sections* whose edges are common to both the $s-t$ path π_1 and the cyclic path σ. The resistances of the ith side forward and backward sections are denoted by $\rho_{\text{fd,i}}$, $\rho_{\text{bd,i}}$, where the indices 'fd' and 'bd' stand for a *forward side section* and a *backward side section*, respectively. The resistances of the ith forward and backward common sections are denoted by $\rho_{\text{fc,i}}$ and $\rho_{\text{bc,i}}$, where the indices 'fc' and 'bc' stand for a *forward common section* and a *backward common section*, respectively. The resistances of ith forward and backward spanned sections are denoted by $\rho_{\text{fp,i}}$ and $\rho_{\text{bp,i}}$, respectively, where the indices 'fp' and 'bp' stand for a *forward spanned section* and a *backward spanned section*, respectively.

If the resistances of all forward side sections, backward side sections, forward common sections and backward common sections of the cyclic path σ are added together, the resistance ρ_σ of the whole cyclic path σ will be obtained:

$$\rho_\sigma = \sum_i \rho_{\text{fd},i} + \sum_i \rho_{\text{bd},i} + \sum_i \rho_{\text{fc},i} + \sum_i \rho_{\text{bc},i} \tag{13.13}$$

Considering inequalities (13.10) and (13.12), the inequality

$$\rho_\sigma \geq \sum_i \rho_{\text{fp},i} + \sum_i \rho_{\text{bp},i} + \sum_i \rho_{\text{fc},i} + \sum_i \rho_{\text{bc},i} \tag{13.14}$$

is obtained. The right-hand side of inequality (13.14) contains only terms related to sections belonging to the π_1-path. The right-hand side of inequality (13.14) however is equal to zero:

$$\sum_i \rho_{\text{fp},i} + \sum_i \rho_{\text{bp},i} + \sum_i \rho_{\text{fc},i} + \sum_i \rho_{\text{bc},i} = 0 \tag{13.15}$$

because the number of times each edge from the π_1-path has been counted as a forward edge is exactly equal to the number of times the same edge has been counted as a backward edge.

Consequently, from Eqs. (13.14) and (13.15), it follows

$$\rho_\sigma \geq 0 \tag{13.16}$$

We arrive at a contradiction. Consequently, no augmentable cyclic path σ with negative resistance exists.

Suppose now that the theorem is valid for k augmented $s{-}t$ paths. In other words, there are no augmentable cyclic paths with negative resistance anywhere in the network, before the next $(k+1)st$ $s{-}t$ path augmentation. The inductive step can be proved by showing that no augmentable cyclic paths with negative resistance exist after augmenting the $(k+1)st$ $s{-}t$ path. Indeed, suppose that the $(k+1)$ st $s{-}t$ path π (s, v_b, v_e, t) (Figure 13.2C) is an augmentable $s{-}t$ path characterised by the smallest resistance.

Now, suppose that there exists a cyclic path σ with negative resistance, along which the flow can be augmented (Figure 13.2C). Note that the cyclic path σ must necessarily have common edges with the $s{-}t$ path π. Indeed, if the paths σ and π were edge-disjoint, a cyclic path σ with negative resistance will exist before augmenting the π-path, which contradicts the inductive assumption that for k augmented $s{-}t$ paths, there are no augmentable cyclic paths with negative resistance.

Consequently, the augmentable cyclic path σ, with negative resistance, must necessarily have common edges with the $s{-}t$ path π. Again, the cyclic path σ can only include (i) *side sections* joining the $s{-}t$ path π at two nodes and not having common edges with the $s{-}t$ path and (ii) *common sections*, whose edges are common to the $s{-}t$ path π and the cyclic path σ. Similar to the proof of the case involving a single augmented $s{-}t$ path π_1 $(k=1)$, according to Lemma 13.2, for each forward side section of the σ-path, the resistance ρ_{fd} along the forward side section is never smaller than the resistance ρ_{fp} along the spanned forward section from the π-path. According to Lemma 13.3, the resistance of each backward side section from the σ-path is never smaller than the resistance of the respective spanned section traversed in backward direction. Following the same reasoning as that used for $k=1$ augmented $s{-}t$ paths, the inequality $\rho_\sigma \geq 0$ is proved for the resistance of the cyclic σ-path, which contradicts the assumption that the cyclic path σ has a negative resistance. Consequently, the resistance of the cyclic path σ is non-negative. This, together with the trivial case $k=1$, completes the inductive proof. □

13.4 Guaranteeing Throughput Flow Associated with the Smallest Lost Flow Due to Edge Failures

13.4.1 Non-Reconfigurable Networks with Complex Topology

To guarantee the specified magnitude f^* of the throughput flow, an extra 'edge' with flow capacity equal to f^* and zero failure rate (the edge never fails) is introduced between the source (which has unlimited capacity) and the rest of the network.

The unavailabilities of the edges can be interpreted as 'costs'. This transforms the problem of determining the network flow associated with the smallest lost flow due to edge failures into the problem of determining the minimum-cost flow in a network. The successive shortest-path algorithm, discussed in detail in Chapter 3, can be applied for determining the edge flows guaranteeing a specified throughput flow associated with the smallest lost flow due to edge failures. An important part of the successive shortest-path algorithm is determining the smallest path resistances from the source node '1' to each of the nodes of the network.

All forward edges in the network have a positive resistance. However, the algorithm searching for augmentable paths also scans paths containing backward edges. The same edge of the network can appear as both an edge with a positive and a negative resistance, depending on the direction of traversing the edge.

The backward edges have a negative resistance. They appear only during the search for an augmentable $s-t$ flow path. The Dijkstra algorithm for determining the shortest distances from the source to every other node in a network is capable of handling edges with positive weights only and cannot be applied directly. This predicament, however, can be overcome by introducing modified weights $\beta_{ij}^{d} = \beta_{ij} + \rho(i) - \rho(j)$, where β_{ij} is the unavailability of edge (i,j); $\rho(i)$ and $\rho(j)$ are the smallest possible resistances from the source s to nodes i and j, respectively. The quantities $\rho(i)$ and $\rho(j)$ correspond to the distances $d(i)$ and $d(j)$ from Algorithm 3.5. The problem of determining the throughput flow associated with the smallest lost flow due to edge failures has essentially been reduced to the minimum-cost flow problem. Algorithm 3.5 can then be applied without modification.

The application of this algorithm will be illustrated by an example featuring the communication network shown in Figure 13.3A, where a throughput data transfer rate of magnitude $f^* = 160$ GB/h is required from the source s to the sink t. The network consists of communication lines, each characterised by a data transfer rate (gigabytes per hour), failure rate and downtime for repair (in hours).

The communication lines connect a source of data and communication hubs which fail and get repaired. It is assumed that each communication hub has a sufficient capacity to accommodate the data transfer rates of all communication lines entering the hub. Each communication line is characterised by a data transfer rate of 100 GB/h.

It is required to minimise the lost data due to edge failures during a period of operation of the network of 360 h. The source coincides with hub '1'; the sink coincides with node '8'.

The unavailabilities of the hubs (except the source s and the sink t) are 1.4%; the unavailabilities of the communication lines are given by the first number on the labels. The second number gives the data throughput capacity. In order to model the communication network as a graph with perfectly reliable (notional) nodes and unreliable edges, each hub of the communication network has been presented by a pair of perfectly reliable nodes, connected with an unreliable edge. The connecting unreliable edge has the failure frequency, the downtime for repair and the flow capacity of the real hub. For each hub, the first node belonging to the hub collects the flow from all communication lines entering the hub. The second node belonging to the hub is incident to all communication lines leaving the hub.

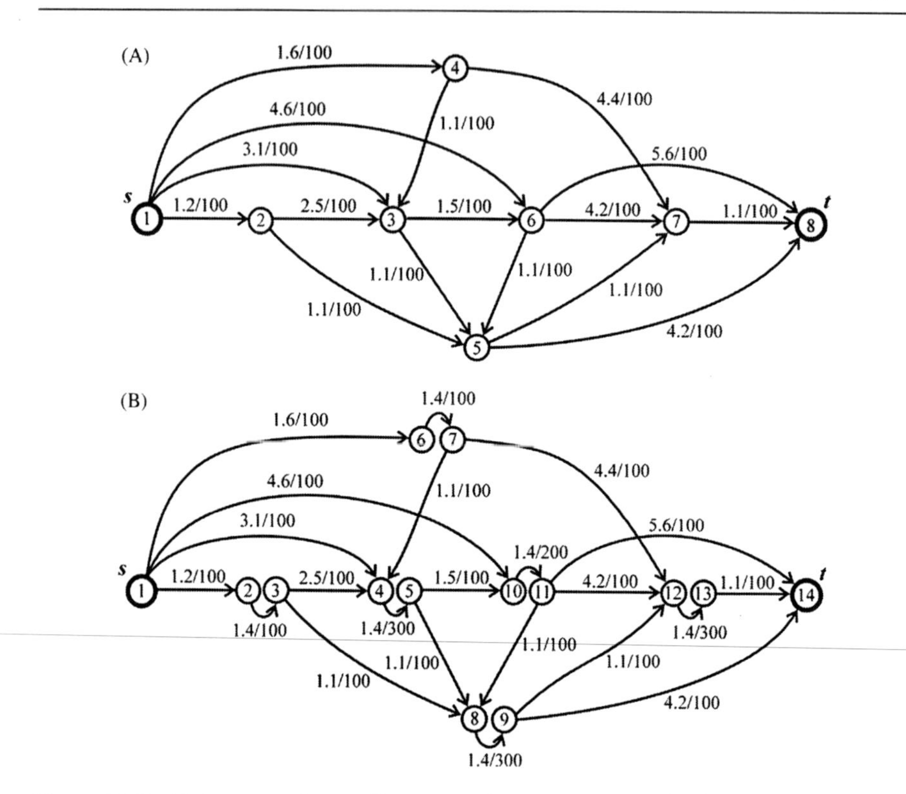

Figure 13.3 (A) A communication flow network with communication hubs and lines undergoing failure and repair. (B) A transformed communication network (each hub has been replaced by a pair of perfectly reliable nodes and an unreliable connecting edge).

According to the successive shortest-path augmentation algorithm, the first path to be saturated is the path (1,2,3,8,9,12,13,14) (Figure 13.3B), characterised by the smallest resistance

$$\beta_1 = 1.2 + 1.4 + 1.1 + 1.4 + 1.1 + 1.4 + 1.1 = 8.7\%$$

The path is saturated with 100 GB/h data flow. The next path to be saturated is the path (1,6,7,12,9,14), which is characterised by the next smallest resistance:

$$\beta_2 = 1.6 + 1.4 + 4.4 - 1.1 + 4.2 = 10.5\%$$

This path is saturated with 60 GB/h data flow. Note that edge (12,9) is a backward edge and its resistance of 1.1% is taken with a negative sign. After the augmentation of this edge with 60 GB/h data flow, the resultant flow along the edge is 40 GB/h.

The required throughput flow of 160 GB/h has been attained. As can be verified, there is no cyclic path (which may also include the source s, the sink t or both)

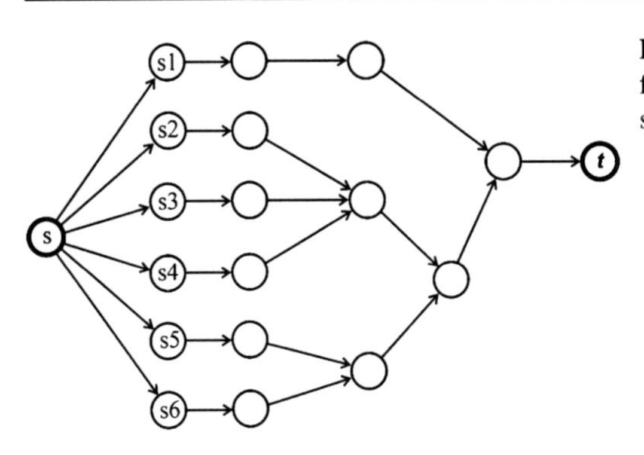

Figure 13.4 A network with merging flows from six sources converted to a single-source single-sink network.

with negative resistance. According to Theorem 13.1, the obtained edge flows are associated with the smallest expected lost flow from edge failures.

13.4.2 Non-Reconfigurable Networks with Merging Flows

The networks with merging flows are an important special case of non-reconfigurable repairable flow networks. Typical non-reconfigurable flow networks are flow networks with merging flows, where several sources of flow give rise to separate flow paths (e.g. subsea oil production wells connected to production trees, manifolds and pipelines). The flow paths (streams) join into bigger streams, etc. until the streams from all sources join into a single stream. The edges coming out of the sources are fully saturated with flow. For these networks, failure of any edge along one of the flow paths causes all the flow in the flow path to be lost during the repair because there is no possibility for redirecting the flows blocked by the failed edges. The next theorem is related only to networks with merging flows originating from multiple sources.

Theorem 13.3 *The lost flow due to edge failures is minimised by determining the directed unsaturated flow path with the largest average availability, fully saturating it with flow, and repeating this procedure until no more directed flow paths can be saturated* (Todinov, 2011a).

Proof This theorem is a corollary from Theorems 13.1 and 13.2. To see this, let us connect a super-source s to the sources of flow $s1, s2, \ldots, s6$ (Figure 13.4), with edges whose capacities are equal to the production capacities of the sources. The sources will disappear and the network with merging flows will be converted to a single-source, single-sink network. Because no s–t paths other than directed s–t paths may exist in a network with merging flows, any augmented s–t path will be a directed s–t path. A directed s–t path with the largest average availability means also a directed path with the smallest resistance. According to Theorem 13.2, augmenting sequentially the s–t paths with the smallest resistance guarantees that no augmentable cyclic path with negative resistance will exist in the network.

According to Theorem 13.1, the obtained throughput flow will be associated with the smallest losses from edge failures. □

An alternative proof of Theorem 13.3 has been given by Todinov (2011a).

13.5 Determining the Edge Flows which Minimise the Probability of Flow Disruption Caused by Edge Failures

The analysis presented in the previous sections applies to a case where the time t_{op} during which the network is operated is large and the separate edges fail and get repaired many times. Often, an alternative problem is present. It is required that at a given time of demand, a flow path is available for uninterrupted flow transmission during a required minimum period t_u. This problem is particularly relevant for telecommunication networks and computer networks, oriented towards media applications, where the flow path (virtual circuit) for the media file transfer is selected in advance (Tanenbaum, 2003). The probability of having uninterrupted connection of duration at least t_u determines the quality of the selected transmission path and ultimately the quality of service received from the network, which is a key performance characteristic.

Let us determine the probability $\Pr(T > t_u)$ that at a given time of demand, a particular flow path, consisting of M forward, independently working edges, will deliver flow undisturbed by edge failures during a required minimum time interval t_u. It is assumed that the edges are empty and the time to failure of the ith edge follows a negative exponential distribution with hazard rate λ_i.

A selected directed flow path will deliver undisturbed connection of duration at least t_u, if and only if each of its edges delivers undisturbed connection for this duration. In other words, no edge failure must occur during the required transmission time t_u.

Because the flow path consists of M independently working and empty edges, the probability of undisturbed operation for a minimum period t_u is equal to the product of the probabilities of undisturbed operation, characterising the edges:

$$\Pr(T \geq t_u) = \prod_{i=1}^{M} \exp(-\lambda_i t_u) = \exp[-(\lambda_1 + \lambda_2 + \cdots + \lambda_M)t_u] \tag{13.17}$$

The probability $\Pr(T \geq t_u)$ is an important reliability measure for directed flow paths composed of empty edges. The larger the minimum period of undisturbed operation t_u and the larger the probability $\Pr(T \geq t_u)$ with which it is guaranteed, the higher the quality of service on demand.

The susceptibility of the flow path to disruption caused by edge failures can be measured with the probability $\Pr(T \geq t_u)$ from Eq. (13.17). Alternative directed flow paths can be compared in terms of their resistance to flow disruption. Taking a logarithm from both sides of Eq. (13.17) yields the expression

$$\ln[1/\Pr(T \geq t_u)] = \sum_{i=1}^{M} \lambda_i t_u \tag{13.18}$$

Because $\ln(x)$ is a monotonically increasing function, a larger value $\ln[1/\Pr(T \geq t_u)]$ corresponds to a larger susceptibility to flow disruption caused by edge failures ($\ln[1/\Pr(T \geq t_u)] > 0$ because $1/\Pr(T \geq t_u) \geq 1$). The product $\lambda_i t_u \geq 0$ is the expected number of failures during the required time interval t_u of uninterrupted transmission. Each empty edge i is characterised by a non-negative weight

$$w_i = \lambda_i t_u \tag{13.19}$$

The weight w_i reflects the decrease of the probability of uninterrupted flow on demand, caused by failures of the ith empty edge, after it has been saturated with flow. Indeed, the larger the expected number of failures $\lambda_i t_u$ of the edge, the larger is the probability of flow disruption.

The value $\sum_{i=1}^{M} \lambda_i t_u$ from the right-hand part of Eq. (13.18) is a measure of the susceptibility of the directed flow path to flow disruption caused by edge failures, and will be referred to as *flow disruption number*. The larger the flow disruption number $\sum_{i=1}^{M} \lambda_i t_u$ of the flow path, the larger is the susceptibility of the flow path to flow disruption.

For a flow path in a network with empty edges, all edges transmitting flow are forward edges. Consider now the case where some of the edges in the network are partially saturated with flow.

Only failures of empty forward edges, saturated with flow, increase the flow disruption number. The expected number of failures of forward edges which are *already* partially saturated with flow is not added if the residual capacity of the edges is used for carrying additional flow. This is because the flow disruption number of these edges has already been added upon changing the state of the edges from empty to partially saturated. A flow disruption from a failed edge will exist as long as there exists some flow through the edge. The probability of flow disruption caused by edge failure does not depend on the magnitude of the flow through the edge. Once the decrease in the probability of flow disruption has been accounted for by adding the flow disruption number of the edge, the probability of flow disruption does not change by increasing the flow through the edge.

Similarly, the expected number of failures of a backward edge, *already* partially saturated with flow, is not subtracted if, as a result of a flow path augmentation, the flow through the edge has been decreased to a value greater than zero. Only when the flow through the backward edge has been decreased to zero, its flow disruption number is subtracted, because by preventing flow from entering the edge, the probability of disturbing the required throughput flow f^* has been decreased.

Consequently, the *flow disruption number* q for a new augmented $s-t$ path including both forward and backward edges is defined as

$$q = \sum_{i=1}^{M_{f0}} \lambda_i t_u - \sum_{j=1}^{M_{b0}} \lambda_j t_u$$

where M_{f0} is the number of empty forward edges to be saturated with flow during the path augmentation and M_{b0} is the number of backward edges to be fully emptied during the $s-t$ path augmentation.

A positive sign of the flow disruption number q indicates that after augmenting the $s-t$ path with flow, the probability of flow disruption will increase. A large positive magnitude of q indicates a large increase in the probability of flow disruption. A negative sign of q means that the probability of flow disruption will decrease.

In the case where the magnitude of the flow f^* to be transmitted is smaller than the bottleneck flow Δ_{min} of a single augmented $s-t$ path, the optimal edge flows minimising the probability of flow disruption can be obtained by augmenting the path with the smallest flow disruption number q. This path can be determined by running the Dijkstra shortest-path algorithm.

Now consider the more common case where the magnitude of the required transmitted flow f^* requires augmenting more than a single $s-t$ path. This is an important problem which can be stated as follows. For a specified throughput flow f^* and a required time interval t_u, find the edge flows which correspond to the largest probability of undisturbed throughput flow. Here we must point out that the method developed earlier, related to determining the edge flows guaranteeing a specified throughput flow f^* and characterised by the smallest lost flow from failures, cannot be applied to this problem. The reason has been illustrated in Figure 13.5.

The edge labels stand for hazard rate/edge capacity. For the sake of simplicity, $t_u = 1$ and the repair times of all edges are the same. The required throughput flow is $f^* = 200$ flow units.

According to the algorithm for minimising the lost flow from edge failures, the path (1,2,3,5,6), characterised by the smallest sum of edge unavailabilities, is to be saturated first, with 100 units of flow. Next the path (1,2,4,5,6) should be saturated

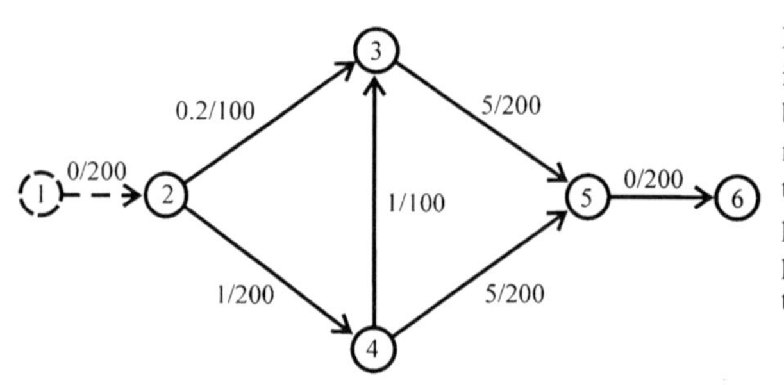

Figure 13.5 A network illustrating the difference between the problem of minimising the lost flow due to edge failures and the problem of maximising the probability of undisturbed throughput flow.

with 100 units of flow, because this is now the augmentable path with the smallest sum of edge unavailabilities. The obtained edge flows minimise the lost flow from edge failures.

However, if the probability of undisturbed throughput flow for $t_u = 1$ needs to be maximised, the optimal strategy will be very different. After augmenting with 100 units of flow the path (1,2,3,5,6), characterised by the smallest sum of expected edge failures during the time interval $t_u = 1$, the next path to be augmented with 100 units *is no longer* path (1,2,4,5,6). The next path to be augmented is (1,2,4,3,5,6). This is because, in non-reconfigurable networks, once the flow path connections have been preset, they cannot be altered. Hence, a disruption in the required throughput flow of $f^* = 200$ flow units *can be caused by a failure of any edge carrying nonzero flow*. The optimal combination of paths (1,2,3,5,6) and (1,2,4,3,5,6) is characterised by 7.2 total expected number of edge failures during the time interval $t_u = 1$. In contrast, the alternative combination of paths (1,2,3,5,6) and (1,2,4,5,6) is characterised by 11.2 total expected number of failures during the time interval $t_u = 1$, which means a higher probability of a throughput flow disruption.

Edge (3,5), which was partially saturated with flow as part of augmenting the path (1,2,3,5,6), has been used for transmitting flow again, as part of the path (1,2,4,3,5,6). However, the expected number of failures of edge (3,5) are not added again. They have already been added during the first saturation of edge (3,5) with flow. This constitutes the principal difference between the discussed procedure and the procedure for minimising the lost flow during edge failures. In the procedure for minimising the lost flow, the unavailability of edge (3,5) should be added twice – as part of path (1,2,3,5,6) and as part of path (1,2,4,3,5,6). This double addition is necessary, because the unavailability $\beta_{(3,5)}$ of edge (3,5) causes flow of $\beta_{(3,5)} \times 100$ units to be lost from path (1,2,3,5,6) and also flow of magnitude $\beta_{(3,5)} \times 100$ units to be lost from path (1,2,4,3,5,6).

However, the case related to disrupted throughput flow is different because once any part of the throughput flow f^* has been disrupted by an edge failure, there is no longer undisturbed flow of magnitude f^* during the time interval t_u. Failures of edge (4,5) have no influence on the probability of undisturbed throughput flow because the edge is empty.

Here is a necessary and sufficient condition for a maximum probability of undisturbed flow due to edge failures.

Theorem 13.4 *The maximum probability of undisturbed throughput flow from edge failures, during a specified time interval, is attained if and only if the network contains no augmentable cyclic paths with a negative flow disruption number.*

Proof The proof of this theorem is very similar to the proof of Theorem 13.1. Suppose that for a required throughput flow f^*, the edge flows $f(i,j)$ are associated with the largest probability of undisturbed flow. If there is an augmentable cyclic path with a negative flow disruption number, the cyclic path could be augmented, which will result in a different set of feasible edge flows, yielding the same throughput flow f^* and associated with a larger probability of undisturbed flow. However, this contradicts the assumption that the edge flows $f(i,j)$ are associated with the largest probability of undisturbed flow.

Now suppose that there is no augmentable cyclic path with a negative flow disruption number in a network characterised by edge flows $f(i,j)$. In this case, the probability of undisturbed throughput flow, associated with the edge flows $f(i,j)$, is the largest possible.

Indeed, suppose that edge flows $f'(i,j)$ exists, associated with a larger probability of undisturbed flow $P_{f'}(T \geq t_u) > P_f(T \geq t_u)]$, which essentially means that the inequality

$$x' = \ln[1/P_{f'}(T \geq t_u)] < x = \ln[1/P(T \geq t_u)]$$

holds.

According to Lemma 13.1, any feasible set of edge flows $f'(i,j)$ resulting in a throughput flow f^* can be obtained from another set of feasible edge flows $f(i,j)$ resulting in the same throughput flow f^*, by adding augmented flows along cyclic paths only. Without loss of generality, suppose that the edge flows $f'(i,j)$ have been obtained from the edge flows $f(i,j)$, after augmenting k cyclic paths. The quantity x' associated with augmenting k cyclic paths is given by $x' = x + q_1 + q_2 + \cdots + q_k$, where $q_m = \sum_{i=1}^{M_{f0}^{(m)}} \lambda_i t_u - \sum_{j=1}^{M_{b0}^{(m)}} \lambda_j t_u$, $m = 1,2, \ldots ,k$; λ_i is the hazard rate of the ith forward edge and $M_{f0}^{(m)}$ is the number of forward edges of the mth cyclic path; λ_j is the hazard rate of the jth backward edge and $M_{b0}^{(m)}$ is the number of backward edges of the mth cyclic flow path.

Because of the assumption $x' < x$, the relationship

$$q_1 + q_2 + \cdots + q_k < 0 \tag{13.29}$$

must necessarily hold. This relationship however is impossible because, according to our assumption, there is no augmentable cyclic path m with a negative flow disruption number $q_m < 0$. This completes the proof. Consequently, the probability $P_f(T \geq t_u)$ of undisturbed flow associated with the edge flows $f(i,j)$ is indeed the largest possible. □

It is very tempting, following Theorem 13.2, to state a sufficient condition for the absence of cyclic flow paths with a negative flow disruption number. For example, to state the 'sufficient condition': *'Augmenting sequentially the s−t paths with the smallest flow disruption numbers guarantees that no augmentable cyclic path with a negative flow disruption number will exist in the network'*. The counterexample in Figure 13.6, however, shows that this statement is false.

On the labels of each edge, the first number denotes the edge capacity, the second number denotes the actual flow through the edge and the last number (in parentheses) denotes the hazard rate of the edge. For the sake of simplicity, suppose that $t_u = 1$. Then the flow disruption number of a path is numerically equal to the sum of the hazard rates of the edges, taken with plus or minus sign, depending on whether the edges are forward edges or backward edges.

The first augmentable shortest path in Figure 13.6A is the path (s,6,4,12) because it has the smallest sum of edge hazard rates (6), and therefore it is characterised by the smallest flow disruption number. This path can be augmented with 13 units of flow. The next augmentable shortest path shown in Figure 13.6A is the path (s,3,2,5,9,10,11,12), with a sum of edge hazard rates equal to 10. It can be augmented with five units of flow. Finally, the next augmentable path is path (s,2,5,9,10,11,12), characterised by a sum of edge hazard rates equal to 6. It can be augmented with eight units of flow. The result is the network shown in Figure 13.6B. In this network, however, the cyclic path (2,4,6,7,8,9,5,2) can be augmented with 13 units of flow. The sum of the edge hazard rates along this path is -2 which means that the flow disruption number is negative.

As a result, the successive shortest-path algorithm is no longer applicable for minimising the probability of disrupted throughput flow. Instead, the successive shortest-path algorithm can be used to produce an initial approximation, which can be subsequently improved by an algorithm for identifying, augmenting and cancelling cycles with negative flow disruption number. The algorithm for cancelling directed loops of flow, presented in Chapter 4, can also be used for this purpose after modifying it to consider backward saturated edges. Note that any possible loop with a negative flow disruption number must necessarily contain at least a single non-empty backward edge. At the same time, the bottleneck flow with which the cyclic path can be augmented, must be equal to the smallest amount of flow along a backward edge.

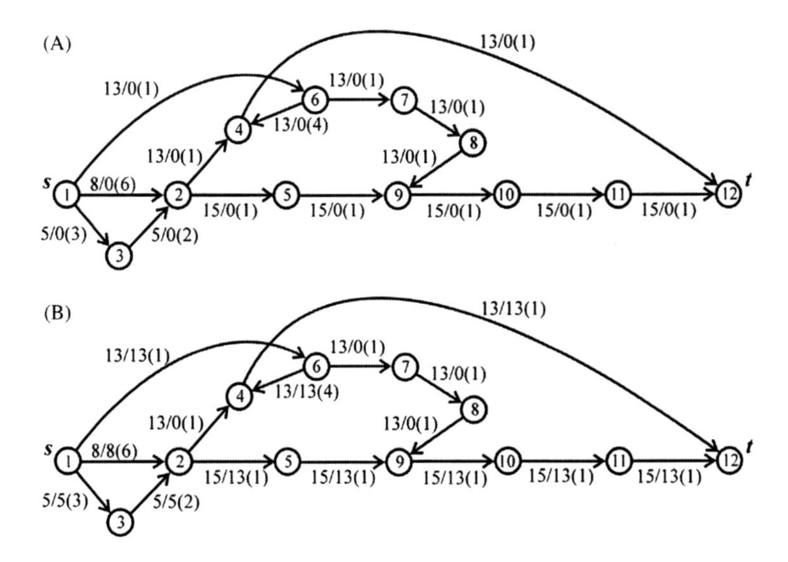

Figure 13.6 A counterexample showing that the application of the successive augmented path algorithm does not guarantee the absence of negative cycles (Todinov, 2012c).

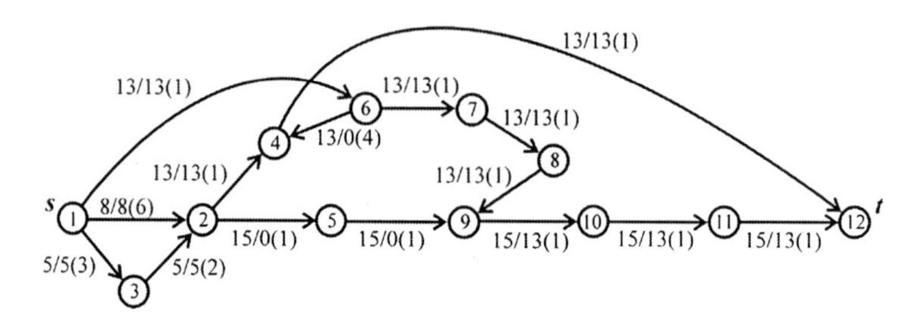

Figure 13.7 The network from Figure 13.6 has been optimised by augmenting the cyclic path (2,4,6,7,8,9,5,2), characterised by a negative flow disruption number.

After identifying the cyclic path (2,4,6,7,8,9,5,2) with a negative flow disruption number, the cyclic path is augmented with the bottleneck flow of 13 units. The result is the network shown in Figure 13.7. The throughput flow remains unchanged (26 flow units). There are no cyclic paths with a negative flow disruption number, which means that the edge flows shown in Figure 13.7 are associated with the largest probability of undisturbed throughput flow.

14 Virtual Accelerated Life Testing of Repairable Flow Networks

14.1 Acceleration Stresses and Acceleration Life Models

Acceleration stress is anything that leads to accumulation of damage and wear. Examples of acceleration stresses are the temperature, humidity, cycling, vibration, speed, pressure, voltage, current, concentration of particular ions, etc. This list is only a sample of possible acceleration stresses and can be extended significantly. Because acceleration stresses lead to a faster wearout, they entail a higher propensity to failure for groups of components. Components affected by an acceleration stress, acting as a *common cause*, are more likely to fail, which reduces the overall system reliability.

A typical example of a *common cause* is the high temperature which simultaneously increases the susceptibility to deterioration of several electronic components. By simultaneously increasing the hazard rates of the affected components, the deterioration due to a high temperature increases the probability of system failure. Humidity, corrosion or vibrations affect all exposed components. A common cause failure is usually due to a single cause, with multiple failure effects, which are not consequences of one another (Billinton and Allan, 1992). Acceleration stresses acting as common causes increase the joint probability of failure for groups of components or for all components in a complex system. Even in blocks with a high level of built-in redundancy, if a common cause is present, all redundant components in the block may fail within a short period of time and the advantage from the built-in redundancy will be lost. Such is the case of a flow network with built-in redundant seals isolating toxic production fluids from the environment. Increasing temperature acts as a common cause which compromises simultaneously the sealing properties of all redundant seals. Failure to account for the acceleration stresses acting as common causes usually leads to optimistic reliability predictions – the actual reliability is smaller than predicted.

For a number of common engineering components, accelerated life models already exist. They have been built by using a well-documented methodology (Kececioglu and Jacks, 1984; Nelson, 2004; Porter, 2004).

Building an accelerated life model for a component starts with the time-to-failure model of the component (Nelson, 2004; Porter, 2004). The most common time-to-failure model is the *Weibull distribution*:

$$F(t) = 1 - \exp[-(t/\eta)^{\beta}] \tag{14.1}$$

Flow Networks. DOI: http://dx.doi.org/10.1016/B978-0-12-398396-1.00014-3

where $F(t)$ is the cumulative distribution of the time to failure, β (*shape parameter*) and η (*characteristic life/scale parameter*) are constants determined from experimental data. This model is commonly used in the case where the hazard rate depends on the age of the component. Another common time-to-failure model is the *negative exponential distribution*:

$$F(t) = 1 - \exp[-(t/\text{MTTF})] \tag{14.2}$$

where $F(t)$ is the cumulative distribution of the time to failure and MTTF is the mean time to failure. The negative exponential distribution can be obtained as a special case of the Weibull distribution for $\beta = 1$ and is used in cases where the hazard rate characterising the component does not practically depend on its age.

The scale parameter η in the Weibull distribution and the MTTF in the negative exponential distribution depend on the acceleration stresses, through the *stress–life relationships* (Kececioglu and Jacks, 1984; Nelson, 2004; Porter, 2004; ReliaSoft, 2007). When the stress–life dependence is substituted in Eqs. (14.1) and (14.2), the acceleration time to failure model for the component is obtained. The acceleration time-to-failure model *is the time-to-failure model at particular levels of the acceleration stresses.*

14.1.1 Arrhenius Stress–Life Relationship and Arrhenius-Type Acceleration Life Models

For this type of accelerated life model, the relationship between the life and the level V of the acceleration stress is

$$L(V) = C \times \exp(B/V) \tag{14.3}$$

where $L(V)$ is a quantifiable life measure and C and B are constants obtained from experimental measurements. The Arrhenius stress–life relationship is appropriate in cases where the acceleration stress is temperature. The temperature values must be in absolute units (K).

In the case of a Weibull time-to-failure model, $L(V) \equiv \eta = C \times \exp(B/V)$, where η is the characteristic life (scale parameter). Substituting this in the Weibull time-to-failure model (14.1), yields the *Arrhenius–Weibull* time-to-failure accelerated life model:

$$F(t, V) = 1 - \exp(-[t/(C\exp(B/V))]^{\beta}) \tag{14.4}$$

14.1.2 Inverse Power Law Relationship and IPL-Type Acceleration Life Models

The relationship between the life of the component and the level V of the acceleration stress is

$$L(V) = 1/(KV^{n}) \tag{14.5}$$

where $L(V)$ is a quantifiable life measure; K and n are constants obtained from experimental measurements. The inverse power law (IPL) stress–life relationship is appropriate for non-thermal acceleration stresses like 'load', 'pressure', 'contact stress'. It can also be applied in cases where V is a stress range or even in cases where V is a temperature range (e.g. in the case of fatigue caused by thermal cycling).

For a Weibull time-to-failure model, the life measure is assumed to be the characteristic life $L(V) \equiv \eta = 1/(KV^n)$, where η is the characteristic life (scale parameter). Substituting this in the Weibull time-to-failure model (14.1) yields the *IPL–Weibull* accelerated life model:

$$F(t, V) = 1 - \exp[-(tKV^n)^\beta] \tag{14.6}$$

14.1.3 Eyring Stress–Life Relationship and Eyring-Type Acceleration Life Models

The relationship between the life and the acceleration stress level V is

$$L(V) = \frac{1}{V}\exp\left[-(A - B/V)\right] \tag{14.7}$$

where $L(V)$ is a quantifiable life measure; A and B are constants obtained from experimental measurements. Similar to the Arrhenius stress–life relationship, the Eyring stress–life relationship is appropriate in the case of thermal acceleration stresses. It can also be used, however, for non-thermal acceleration stresses, such as humidity. In the case of a Weibull time-to-failure model, $L(V) \equiv \eta = (1/V)\exp[-(A - B/V)]$, where η is the characteristic life (scale parameter). Substituting this in the Weibull time-to-failure model (14.1), yields the *Eyring–Weibull* accelerated life model:

$$F(t, V) = 1 - \exp[-(tV\exp(A - B/V))^\beta] \tag{14.8}$$

There also exist stress–life models involving simultaneously two acceleration stresses, for example 'temperature' and 'humidity' (ReliaSoft, 2007). Such are the *Temperature–Humidity (TH) relationship and TH-type acceleration life models* and *Temperature-Non-thermal relationship (T-NT) and T-NT-type acceleration life models*.

14.1.4 A Motivation for the Proposed Method

Despite the increasing research in both the area of accelerated life testing (Meeker and Escobar, 1993) and the area of common cause failure modelling (Kvam and Miller, 2002; Parry, 1991; Prentice, 1978), *no models and software tools are currently available* for building the accelerated life model of a complex system, from the accelerated life models of its components. Building an accelerated life model for a complex system has another significant advantage. Quantifying the system's availability under normal operating conditions requires tests involving a large amount of time and resources. This could be a very complex and expensive task which does not have to be addressed, if a method is developed for building an accelerated life model of a complex system from the accelerated life models of its components. Deducing the availability of a repairable system under normal operating conditions from the accelerated life models of its components will be referred to as '*virtual accelerated life testing*' of repairable systems. The significant advantages of the virtual accelerated life testing can be summarised as follows (Todinov, 2011e):

- Virtual accelerated life testing reveals the effect of environmental stresses on the performance of repairable systems.
- Virtual accelerated life testing permits extrapolating the availability of the repairable system under normal operating conditions, from the accelerated life models of its components.
- Virtual accelerated life testing reveals the interdependencies among the building components, caused by environmental stresses acting as common causes.

The virtual accelerated life testing offers significant flexibility in specifying various levels for the acceleration stresses. Consequently, the primary objective of this chapter is to describe a method of extrapolating the network availability under normal operating conditions from the accelerated life models of its components.

14.2 Determining the Availability of a Repairable System

Consider for example the system in Figure 14.1A, whose logic of operation and failure has been represented by the reliability network in Figure 14.1B (see Chapter 8 for more details about modelling the system logic by reliability networks). Edge e1 models the power supply PS; edges e2 and e3 model the power umbilicals PU and PU′; edges e4–e11 model the switches S1–S4 and S1′–S4′ and finally, edges e12–e15 model the electro-mechanical devices EMD1–EMD4. The system is operational only if power is supplied to each electro-mechanical device and if each electro-mechanical device is operational.

The reliability of edges (e1–e11) depends on temperature, which accelerates the degradation of the components in the power supply unit, and the corrosion processes and material deterioration in the switches and power umbilicals. The reliability of edges (e12–e15) depends on the average contact pressure, which accelerates the wearout of the electro-mechanical devices.

The network is in operation if a path exists from the source s to each of the terminal nodes (sinks) $t1$, $t2$, $t3$ and $t4$. Availability is understood as the average percentage of time during which paths through working components exist between the start node s and each of the end nodes 9, 10, 11 and 12. The edges in the reliability network are undirected.

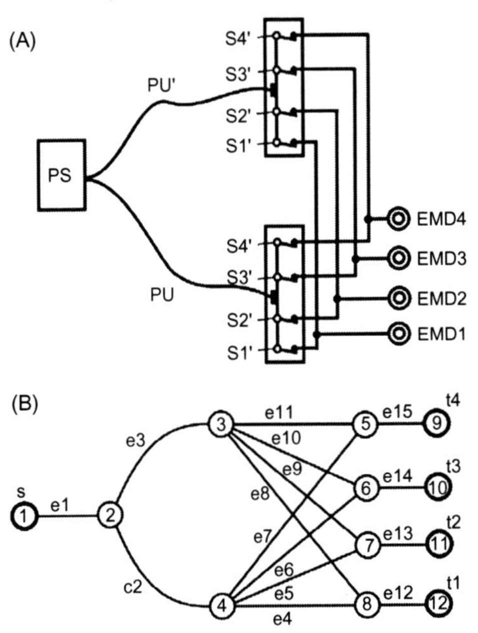

Figure 14.1 (A) A system with four electro-mechanical devices with dual power supply. (B) The reliability network modelling the system in A.

Revealing the system's availability is done on the basis of a discrete-event solver, whose algorithm is similar to the discrete-event solver described in Chapter 9 for revealing the production availability of a repairable flow network.

Again, times to failure are generated for all components in the system, and placed in a queue of events, in ascending order. A second loop is subsequently entered, which is terminated when the end of the operating period of the system is reached. The smallest time to event is obtained at the head of the queue of events. If the current event is a 'component failure', the components are essentially excluded from the network by reducing their flow capacity to zero. After a delay determined by the downtime for repair of the component, a new life is generated for the component, and put in the ordered list of events.

The function '**paths_to_all_sinks**()' from Chapter 8 is used to test whether paths to all sinks exist through working edges (edges with nonzero flow capacities). The function '**paths_to_all_sinks**()' is executed after each component failure and component repair. This is necessary to keep track of the time during which the system is functioning (the fraction of time during which paths to all four terminal nodes exist).

The availability characterising the current simulation trial is determined by taking the ratio of the total time during which paths to all sinks exist, to the specified time interval of operation. The availabilities characterising each simulation trial are recorded in an array, which is sorted in ascending order at the end of the simulation. This array essentially gives the distribution of the system availability.

14.2.1 A Solved Test Example

For the network in Figure 14.1, suppose that the power supply unit *PS* (edge *e1*) is characterised by Arrhenius stress–life model (14.3) where the acceleration stress, the temperature, is set at a level $V = 333$ K and the constants in the equation are $B = 461.64$ and $C = 2$. Suppose that the power umbilicals (edges e2 and e3) are also characterised by Arrhenius stress–life models, with constants $B = 118.77$ and $C = 1.4$, and the acceleration stress is again temperature, set also at a level $V = 333$ K. Suppose that all switches (edges e4–e11) are characterised by Eyring stress–life relationship (Eq. (14.7)), with constants $A = 1.8$ and $B = 3684.8$, where the acceleration stress is also the temperature, set at a level $V = 413$K.

Finally, suppose that the electro-mechanical devices (edges e12–e15) are characterised by IPL stress–life relationship (Eq. (14.5)), with constants $K = 9e - 5$ and $n = 1.7$, where the acceleration stress is 'radial contact pressure', set at a level $V = 60$ MPa.

The text file '*acceleration_stresses.txt*' specifies the indices and the corresponding levels of the acceleration stresses. The structure of the file is presented in Table 14.1. For the acceleration stress with index 1 (e.g. temperature), the value 333K has been specified. The acceleration stress with index 2 is also temperature, and its level has been set to 413K. The third acceleration stress is radial contact pressure, and its level has been set to 60 MPa.

Because the acceleration stresses are accessed by their indices, this is a simple and efficient solution of the problem related to representing various types and sets of acceleration stresses in a large repairable system.

All edges are characterised by a negative exponential time-to-failure distribution.

The duration of the time interval for which the availability was calculated is $a = 10$ years.

For the repairable system in Figure 14.1B, edge $e1$ is characterised by a normal distribution of the times to repair with mean 0.06 years and standard deviation $\sigma = 0.001$. Edges e2 and e3 are characterised by a negative exponential distribution of the times to repair with MTTR = 0.2 years. Edges e4–e11 are characterised by a constant time to repair with mean 0.1 years. Finally, edges e12–e15 are characterised by a Weibull time-to-repair distribution with characteristic life $\eta = 0.07$ years and $\beta = 1.5$ (Eq. (14.1)).

For the system availability, the execution of the programme yielded: 0.897.

Table 14.1 Structure of the Input Text File *Acceleration_Stresses.txt*

Acceleration Stress Index	Acceleration Stress Level
1	333
2	413
3	60

Ten thousand simulations have been performed within 0.953 s, on a laptop with processor *Intel(R) T7200 @ 2.00 GHz*. The distribution of the availability of the system in Figure 14.1B is shown in Figure 14.2 (the curve with label 'elevated levels of the acceleration stresses').

Finally, an extrapolation of the system availability under normal levels of the acceleration stresses (normal operating conditions) has been made. This constitutes the main advantage of the developed method: estimating the availability of a complex repairable system working at normal conditions, without allocating time and resources for real testing. The normal conditions correspond to the specified (in Table 14.2) levels of the acceleration stresses in the input file (room temperature in Kelvin, and a very small radial contact pressure).

For the system availability, the programme yielded 0.967.

Ten thousand simulations have been performed within 0.39 s, on a laptop with processor *Intel(R) T7200 @ 2.00 GHz*.

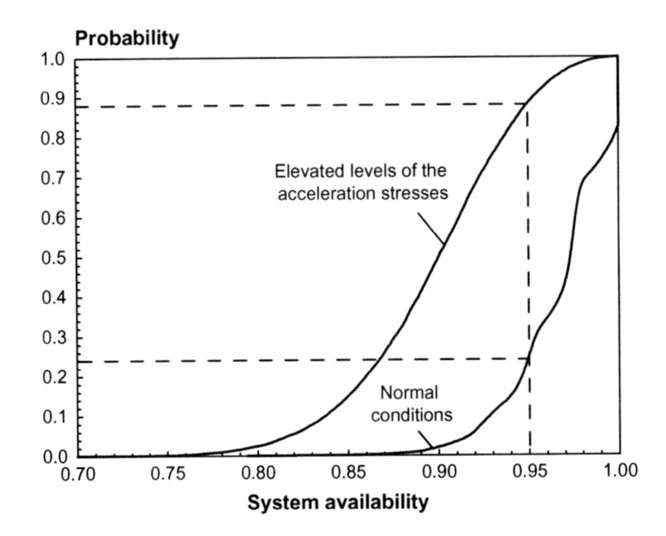

Figure 14.2 Distribution of the availability for the system in Figure 14.1A.

Table 14.2 Structure of the Input Text File *Acceleration_Stresses.txt*, Under Normal Operating Conditions

Acceleration Stress Index	Acceleration Stress Level
1	293
2	293
3	1

In Figure 14.2, the system availability distribution which corresponds to the levels of the acceleration stresses from Table 14.2, has been represented by the curve with label 'Normal conditions'.

The correctness of the developed software tool has been verified on numerous test cases involving series and series–parallel networks with well-known theoretical solutions (Ebeling, 1997; Sherwin and Bossche, 1993). The predicted values confirmed the analytical results.

14.2.2 Interpretation of the Results

The network availability is the average percentage of time during which the network is performing its intended function. The developed software produces two characteristics of the system availability – the *average availability* and the *distribution of the system availability*. The system availability distribution gives the probability that the network availability will be smaller than a specified value, at the specified levels of the acceleration stresses.

The most important function of the developed software is the capability of extrapolating the average network availability and the availability distribution under normal operating conditions, without conducting tests which consume time and resources. Thus, for a specified time of 10 years, the availability distribution graph in Figure 14.2, which corresponds to normal levels of the acceleration stresses (normal operating conditions), yields approximately 24% probability that the system availability will be smaller than 95%. For elevated levels of the acceleration stress, the probability that the system availability will be smaller than 95% is already 88%. These trends are also mirrored by the average (expected) availability. While the average network availability which corresponds to normal levels of the acceleration stresses is 0.967, at elevated levels of the acceleration stresses, the average network availability decreases to 0.897. The graph in Figure 14.2 clearly illustrates the impact of the increased levels of acceleration stresses on the system availability distribution.

References

Ahuja RK, Orlin JB: Distance-directed augmenting path algorithms for maximum flow and parametric maximum flow problems, *Nav Res Logistics*. 38:413–430, 1991.

Ahuja RK, Goldberg AV, Orlin JB, Tarjan RE: Finding minimum-cost flows by double scaling, *Math Program*. 53:243–266, 1992.

Ahuja RK, Magnanti TL, Orlin JB: *Network flows: theory, algorithms and applications*, New Jersey, 1993, Prentice Hall.

Amrbruster A, Gosnell M, McMillin B, Crow ML: Power transmission control using distributed max-flow. In *Proceedings of 29th annual international computer software and applications conference*, Edinburgh, 2005.

Andrews JD, Moss TR: *Reliability and risk assessment*, London, 2002, Professional Engineering Publishing.

Asano T, Asano Y: Recent developments in maximum flow algorithms, *J Oper Res Soc Jpn*. 43(1):2–31, 2000.

Ascher H, Feingold H: *Repairable systems reliability*, New York, 1984, Marcel Dekker.

Aspens J, Azar Y, Fiat A, Plotkin SA, Waarts O: Online load balancing with applications to machine scheduling and virtual circuit routing. In *Proceedings of the 25th ACM symposium on theory of computing (STOC)*, San Diego, 1993, 623–631.

Assad A: Multicommodity network flows – a survey, *Networks*. 8(1):37–91, 1978.

Barlow RE, Proschan F: *Mathematical theory of reliability*, New York, 1965, John Wiley & Sons.

Barlow RE, Proschan F: *Statistical theory of reliability and life testing*, New York, 1975, Rinehart and Winston.

Bazovsky I: *Reliability theory and practice*, Englewood Cliffs, New Jersey, 1961, Prentice Hall.

Bellman R: *Dynamic programming*, Princeton, NJ, 1957, Princeton University Press.

Bennington GE: An efficient minimal cost flow algorithm, *Manage Sci*. 19(9):1042–1051, 1973.

Billinton R, Allan RN: *Reliability evaluation of engineering systems*, ed 3, New York, 1992, Plenum Press.

Blake IF: *An introduction to applied probability*, New York, 1979, John Wiley & Sons.

Blischke WR, Murthy DN: *Reliability: modelling, prediction, and optimisation*, New York, 2000, John Wiley & Sons.

Chartrand G, Lesniak L, Zhang P: *Graphs & digraphs*, ed 5, Boca Raton, 2011, Chapman & Hall/CRC.

Cherkaski BV: Algorithm of construction of maximum flow in networks with complexity $O(|V^2|\sqrt{|E|})$ operations, *Math Methods Solut Econ Probl*. 7:117–125, 1977.(in Russian)

Colbourn CJ: *The combinatorics of network reliability*, New York, NY, 1987, Oxford University Press.

Cook WJ, Cunningham WH, Pulleyblank WR, Schrijver A: *Combinatorial optimisation*, New York, 1998, John Wiley & Sons.

Cormen TH, Leiserson TCE, Rivest RL, Stein C: *Introduction to algorithms*, ed 2, Boston, 2001, MIT Press and McGraw-Hill.

Dantzig GB: *Application of the simplex method to a transportation problem*, In: *Activity analysis and production and allocation*, New York, NY, 1951, Wiley, pp 359–373.

Dasgupta S, Papadimitriou C, Vazirani U: *Algorithms*, Boston, 2008, McGraw-Hill.

DeGroot M: *Probability and statistics*, Reading, Massachusetts, 1989, Addison-Wesley.

Dijkstra EW: A note on two problems in connexion with graphs, *Numer Math.* 1: 269–271, 1959.

Dinic EA: Algorithm for solution of a problem of maximum flow in a network with power estimation, *Sov Math Dokl.* 11(8):1277–1280, 1970.

Divan D, Johal H: A smarter grid for improving system reliability and asset utilization. In *CES/IEEE fifth international power electronics and motion control conference*, Shanghai, 2006, pp 1–6.

Dong J, Li W, Cai C, Chen Z: Draining algorithm for the maximum flow problem. In *International conference on communications and mobile computing*, Yunnan, 2009, pp 197–200.

Ebeling CE: *An introduction to reliability and maintainability engineering*, New York, 1997, McGraw-Hill.

Edmonds J, Karp RM: Theoretical improvements in algorithmic efficiency for network flow problems, *J ACM.* 19(2):248–264, 1972.

Elias P, Feinstein A, Shannon CE: Note on maximum flow through a network, *IRE Trans Inf Theory.* IT2:117–119, 1956.

Elsayed AE: *Reliability engineering*, Reading MA, 1996, Addison Wesley Longman.

Evans JR: Maximal flow in probabilistic graphs the discrete case, *Networks.* 6:161–183, 1976a.

Evans JR: A combinatorial equivalence between a class of multi-commodity flow problems and the capacitated transportation problem, *Math Program.* 10:401–404, 1976b.

Fishman GS: The distribution of maximum flow with applications to multistate reliability systems, *Oper Res.* 35(4):607–618, 1987.

Ford LR, Fulkerson DR: Maximal flow through a network, *Can J Math.* 8(5):399–404, 1956.

Ford LR, Fulkerson DR: A suggested computation for maximal multi-commodity network flows, *Manage Sci.* 5(1):97–101, 1958.

Ford LR, Fulkerson DR: *Flows in networks*, Princeton, NJ, 1962, Princeton University press.

Friesdorf H, Hamacher H: Weighted min cost flows, *Eur J Oper Res.* 11:181–192, 1982.

Gibbons A: *Algorithmic graph theory*, Cambridge, 1985, Cambridge University Press.

Glasserman P: *Monte Carlo methods in financial engineering*, New York, 2003, Springer.

Goldberg AV, Rao S: Beyond the flow decomposition barrier, *J ACM.* 45(5):783–797, 1998.

Goldberg AV, Tarjan RE: Solving minimum-cost flow problems by successive approximation, *Proceedings of the 19th annual ACM symposium on theory of computing*, New York, NY, 1987, ACM.

Goldberg AV, Tarjan RE: A new approach to the maximum flow problem, *J ACM.* 35:921–940, 1988.

Goldberg AV, Tarjan RE: Finding minimum-cost circulations by cancelling negative cycles, *J ACM.* 36(4):883–886, 1989.

Goldberg AV, Tardos E, Tarjan RE: *Network flow algorithms, Paths, rows and VLSI-layout (algorithms and combinatorics)*, vol 9, Berlin Heidelberg, 1990, Springer-Verlag.

Goodrich MT, Tamassia R: *Algorithm design*, New York, 2002, John Wiley & Sons.

Gross JL, Yellen J: *Graph theory and its applications*, 2 ed, Boca Raton, 2006, CRC Press.

Hochbaum DS: The pseudoflow algorithm: a new algorithm for the maximum-flow problem, *Oper Res*. 56(4):992–1009, 2008.

Holzner S: *C++*, Black book, Scottsdale, Arizona, 2001, Coriolis.

Hoyland A, Rausand M: *System reliability theory*, New York, 1994, John Wiley & Sons.

Hu TC: Multicommodity network flows, *Oper Res*. 11:344–360, 1963.

Hu TC: *Integer programming and network flows*, Reading, MA, 1969, Addison-Wesley.

Itai A, Shiloach Y: Maximum flows in planar networks, *SIAM J Comput*. 8:135–150, 1979.

Jane C, Lin J, Yuan J: Reliability evaluation of a limited-flow network in terms of minimal cutsets, *IEEE Trans Reliab*. 42(3):354–368, 1993.

Jewell WS: Optimal flow through networks with gains, *Oper Res*. 10(4):476–499, 1962.

Johnsonbaugh R: *Discrete mathematics*, ed 4, NJ, Upper Saddle River, 1997, Prentice Hall.

Karzanov A: Determining the maximal flow in a network by the method of preflows, *Sov Math Dokl*. 15:434–437, 1974.

Kececioglu D, Jacks JA: The Arrhenius, Eyring, inverse power law and combination models in accelerated life testing, *Reliab Eng*. 8:1–6, 1984.

Kelly F: Charging and rate control for elastic traffic, *Eur Trans Telecomm*. 8:33–37, 1997.

Kim BH: Comparison of distributed optimal power flow algorithms, *IEEE Trans Power Syst*. 15(2):599–604, 2000.

Kirschen D, Strbac G: Why investments do not prevent blackouts, *Electr J* :29–36, 2004.

Klein M: A primal method for minimal cost flows with applications to the assignment and transportation problems, *Manage Sci*. 14(3):205–220, 1967.

Kleinberg J, Tardos E: *Algorithm design*, Boston, 2006, Addison-Wesley.

Kuo W, Prasad VR, Tillman FA, Hwang CL: *Optimal reliability design*, Cambridge, 2001, Cambridge University Press.

Kvam PH, Miller JG: Common cause failure prediction using data mapping, *Reliab Eng Syst Saf*. 76:273–278, 2002.

Lawler E: *Combinatorial optimisation*, New York, NY, 1976, Dover Publications.

Lee SH: Reliability evaluation of a flow network, *IEEE Trans Reliab*. R-29(1):24–26, 1980.

Lin YK: A simple algorithm of reliability evaluation of a stochastic-flow network with node failure, *Comput Oper Res*. 28(13):1277–1285, 2001a.

Lin YK: Study on the multicommodity reliability of a capacitated flow network, *Comput Math Appl*. 42(1–2):255–264, 2001b.

Lin YK: Using minimal cuts to evaluate the system reliability of a stochastic-flow network with failures at nodes and arcs, *Reliab Eng Syst Saf*. 75(1):41–46, 2002.

Lin YK: Evaluate the performance of a stochastic-flow network with cost attribute in terms of minimal cuts, *Reliab Eng Syst Saf*. 91:539–545, 2006.

Lin YK, Yuan J: A new algorithm to generate d-minimal paths in a multistate flow network with non-integer arc capacities, *Int J Reliab Qual Saf Eng*. 5(3):269–285, 1998.

Lin JC, Jane C, Yuan J: On reliability evaluation of a capacitated flow network in terms of minimal pathsets, *Networks*. 25:131–138, 1995.

Meeker WQ, Escobar LA: A review of recent research and current issues in accelerated testing, *Int Stat Rev*. 61(1):147–168, 1993.

Mutale J, Strbac G: Transmission network reinforcement versus FACTS: an economic assessment, *IEEE Trans Power Syst*. 15(3):961–967, 2000.

Nelson W: *Accelerated testing, statistical models, test plans and data analysis*, New York, 2004, Wiley.

Nguyeu PH, Kling WL, Myrzik JMA, 2009. Power flow management in active networks. In *Proceedings of the 2009 Bucharest power tech conference*, Bucharest.

Okamura H: Multicommodity flows in graphs, *Discrete Appl Math.* 6:55–62, 1983.

Orlin JB: A faster strongly polynomial minimum cost flow algorithm, *Oper Res.* 41(2): 338–350, 1993.

Overbeeke, F: Active networks: distribution networks facilitating integration of distributed generation. In *Proceedings of second international symposium on distributed generation: power system and market aspects*, Stockholm, 2002.

Papadimitriou CH, Steiglitz K: *Combinatorial optimisation: algorithms and complexity*, New York, NY, 1998, Dover Publications.

Park SK, Miller KW: Random number generators: good ones are hard to find, *Commun ACM.* 31(10):1192–1201, 1988.

Parry G: Common cause failure analysis: a critique and some suggestions, *Reliab Eng Syst Saf.* 34:309–326, 1991.

Peixin Z, Xin Z: A survey on reliability evaluation of stochastic-flow networks in terms of minimal paths. In *International conference on information engineering and computer science*, 19–20 December, 2009, Wuhan.

Porter A: *Accelerated testing and validation*, Oxford, 2004, Newnes.

Prentice RL: The analysis of failure times in the presence of competing risks, *Biometrics.* 34:541–554, 1978.

Ramakumar R: *Engineering reliability: fundamentals and applications*, Upper Saddle River, New Jersey, 1993, Prentice Hall.

ReliaSoft, *Accelerated life testing on-line reference.* ReliaSoft's eTextbook for accelerated life testing data analysis, 2007.

Ross SM: *Simulation*, 2 ed, San Diego, 1997, Harcourt Academic Press.

Ross SM: *A first course in probability*, 6 ed, NJ, Upper Saddle River, 2002, Prentice Hall.

Rubinstein RY: *Simulation and the Monte-Carlo method*, New York, NY, 1981, John Wiley & Sons.

Sakarovitch M: Two commodity network flows and linear programming, *Math Program.* 4:1–20, 1973.

Sedgewick R: *Algorithms in C++*, Reading, MA, 1992, Addison-Wesley.

Sherwin DJ, Bossche A: *The reliability, availability and productiveness of systems*, London, 1993, Chapman & Hall.

Shigeno M: A survey of combinatorial maximum flow algorithms on a network with gains, *J* Oper Res Soc Jpn vol. 47(4):244–264, 2004.

Shiloach Y, Vishkin U: An O(n^2log n) parallel max-flow algorithm, *J Algorithms.* 3:128–146, 1982.

Sleator DD, Tarjan RE: *An O(*nm *log* n*) algorithm for maximum network flow, Technical report STAN-CS-80-831, Department of computer science*, Stanford, CA, 1980, *Stanford University.*

Tanenbaum AS: *Computer networks*, 4 ed., Prentice Hall, Upper Saddle River, NJ, 2003, Pearson Education International.

Tardos E: A strongly polynomial minimum cost circulation algorithm, *Combinatorica.* 5(3): 247–255, 1985.

Tarjan RE: *Data structures and network algorithms*, Philadelphia, PA, 1983, SIAM.

Tillman FA, Hwang FA, Kuo W: *Optimisation of systems reliability*, New York, 1985, Marcel Dekker.

Todinov MT: Reliability and risk models: setting reliability requirements, Chichester, 2005, Wiley.

Todinov MT: *Risk-based reliability analysis and generic principles for risk reduction*, Amsterdam, 2007, Elsevier.

Todinov MT: Analysis and optimization of repairable networks with merging flows. In *Proceedings of the ESREL conference*, Prague, 2009.

Todinov MT: Analysis and optimization of repairable flow networks with complex topology, *IEEE Trans Reliab*. 60(1):111–124, 2011a.

Todinov MT: Fast augmentation algorithms for maximising the flow in repairable flow networks after a component failure. In *Proceedings of the 11th IEEE international conference on computer and information technology*, Paphos, 2011b, pp. 505–512.

Todinov MT: A fast augmentation algorithm for optimizing the performance of repairable flow networks in real time. In *Proceedings of ESREL conference*, Troy, 2011c, pp 1951–1958.

Todinov MT: Topology optimization of repairable flow networks and reliability networks, *Int J Simul Syst Sci Technol*. 11(3):75–84, 2011d.

Todinov MT: Virtual accelerated life testing of complex systems. In: Bouvry, P., González-Vélez, H., Kolodziej, J. (Eds.), Intelligent decision systems in large-scale distributed environment. Springer, 2011e Berlin, pp 293–314.

Todinov MT: Fast augmentation algorithms for maximising the output flow in repairable flow networks after edge failures, *Int J Syst Sci* doi: 101080/00207721.2012.670294, 2012a.

Todinov MT: Topology optimisation of repairable flow networks for a maximum average availability, *Comput Math Appl*. 64:3729–3746, 2012b.

Todinov MT: Algorithms for minimising the lost flow due to failed components in repairable flow networks with complex topology, *Int J Reliab Saf*. 6(4):283–310, 2012c.

Todinov MT: The dual network theorem for static flow networks and its application for maximising the throughput flow, *Artif Intell Res*, 2(1):81–106, 2013, published online 2012; doi: 10.5430/air.v2n1p81.

Trivedi KS: *Probability and statistics with reliability, queuing and computer science applications*, New York, 2002, John Wiley & Sons.

Truemper K: On max flows with gains and pure min-cost flows, *SIAM J Appl Math*. 32(2): 450–456, 1977.

Van Hertem D, Verboomen J, Belmans R, Kling WL: Power flow controlling devices: an overview of their working principles and their application range. In *International conference on future power systems*, Amsterdam, 2005, pp. 1–6.

Vose D: *Risk analysis: a quantitative guide*, 2 ed, Chichester, 2000, John Wiley & Sons.

Wollmer RD: Stochastic sensitivity analysis of maximum flow and shortest route networks, *Manage Sci*. 19(9):551–564, 1968.

Yarlagadda R, Hershey J: Fast algorithm for computing the reliability of a communication network, *Int J Electron*. 70(3):549–564, 1991.

Yeh WC: A simple approach to search for all d-MCs of a limited-flow network, *Reliab Eng Syst Saf*. 71(1):15–19, 2001a.

Yeh WC: A simple algorithm to search for all d-MPs with unreliable nodes, *Reliab Eng Syst Saf*. 73(1):49–54, 2001b.

Yeh WC, Ho HC, Chen YC, Yeh YM: A new algorithm for finding all minimal cuts in modified networks, *Int J Innovative Comput Inf Control*. 8(1(A)):419–430, 2012.

Yi Y, Chiang M: Stochastic network utility maximization, *Eur Trans Telecommun*:1–22. doi:10.1002/ett.0000, 2008.

Zhou H, Chen H-L, Bruck J: On the synthesis of stochastic flow networks. In *IEEE international symposium information theory proceedings (ISIT)*, Austin, 2010, pp 1330–1334.

CPSIA information can be obtained at www.ICGtesting.com
Printed in the USA
BVOW020318060213

312487BV00006B/202/P